Lecture Notes in Physics

Volume 994

The series Lecture Notes in Physics (LNP), founded in 1969, reports new developments in physics research and teaching - quickly and informally, but with a high quality and the explicit aim to summarize and communicate current knowledge in an accessible way. Books published in this series are conceived as bridging material between advanced graduate textbooks and the forefront of research and to serve three purposes:

- to be a compact and modern up-to-date source of reference on a well-defined topic;
- to serve as an accessible introduction to the field to postgraduate students and non-specialist researchers from related areas;
- to be a source of advanced teaching material for specialized seminars, courses and schools.

Both monographs and multi-author volumes will be considered for publication. Edited volumes should however consist of a very limited number of contributions only. Proceedings will not be considered for LNP.

Volumes published in LNP are disseminated both in print and in electronic formats, the electronic archive being available at springerlink.com. The series content is indexed, abstracted and referenced by many abstracting and information services, bibliographic networks, subscription agencies, library networks, and consortia.

Proposals should be sent to a member of the Editorial Board, or directly to the responsible editor at Springer:

Dr Lisa Scalone
Springer Nature
Physics
Tiergartenstrasse 17
69121 Heidelberg, Germany
lisa.scalone@springernature.com

More information about this series at https://link.springer.com/bookseries/5304

Daniel Beysens

The Physics of Dew,
Breath Figures
and Dropwise
Condensation

Daniel Beysens
ESPCI-PMMH and OPUR
Paris, France

ISSN 0075-8450   ISSN 1616-6361   (electronic)
Lecture Notes in Physics
ISBN 978-3-030-90441-8   ISBN 978-3-030-90442-5   (eBook)
https://doi.org/10.1007/978-3-030-90442-5

This Springer imprint is published by the registered company Springer Nature Switzerland AG
The registered company address is: Gewerbestrasse 11, 6330 Cham, Switzerland

Foreword

露とくとく

A monk meets a gyrovague and asks him to teach something that could open his heart. The gyrovague shows him a hedge with beautiful flowers covered by the morning dew and answers: *Can you see how vain is the flowering of these convolvuli? However, the morning dew will disappear before them...* This story, attributed to the great poet Saigyô, reveals how dew, as early as during the twelfth century, was considered in Japan as a pure and melancholic symbol of ephemerality—vanishing even faster than the fragile convolvuli. Among others, Matsuo Bashô five centuries later developed (assuming that a haiku can constitute a development) this line of thought in his *Record of a Weather-exposed Skeleton:*

Wishing dewdrops

could somehow wash

this perishing world...

Dew thus plays in Japan the role of soap bubbles in Europe—humble objects, on the one hand, yet worth of contemplation and metaphysical thoughts. And like soap bubbles, the scientists who looked at it realized that dew has indeed a lot to say and to teach: the way it grows, as we blow on a cool surface, achieving a so-called breath figure; the way it diffuses light, turning opaque a transparent glass, from which we can deduce the micrometric scale of its individual droplets and the way it evaporates, as it should indeed for the fleeting bidimensional cloud that it was.

Daniel Beysens is one of the most eminent scientists in the field of dew and water condensation, a pioneer in the study of breath figures in the 80s, who devoted the second half of his career to these fragile and elusive objects, with a successful balance of fundamental and applied approaches. Speaking of applications, controlling dew is a challenging program in itself, be it Peruvian villagers whose unique source of water relies on harvesting the morning dew (or fog) with tall nets, or sophisticated texture that can encourage the spontaneous jump and evacuation of condensing droplets as they form.

We can be grateful to Daniel Beysens to have written a thick book on this light subject—not only for the philosophical satisfaction to see something (even a lot) coming out of (nearly) nothing, but also because such a book did not exist; a book that spans the entire field, from basic principles that tell us under which conditions water condenses, to classical laws of dew growth, and further to recent developments—in a domain characteristic of the world of soft matter, with its ability to hybridize different subject areas, here physics of interfaces, fluid dynamics, statistical physics, surface chemistry and optics; a book which is also timely, since all the progresses in our understanding of wetting phenomena have benefited to dew, with the added charm of the reduction in scale: a drop of ten micrometers does not always behave as its big sister of one millimeter.

The comprehensive and concrete character of Daniel Beysens' book together with its amazing collection of experimental illustrations makes it suitable for all kinds of readers. Students will learn the physics of dew without any prerequisite; researchers will be stimulated by the latest evolution of the field and engineers will benefit from a complete state of the art and from the light shed on practical applications. Dew may quickly fade out, but this book will certainly remain!

David Quéré
ESPCI and École Polytechnique
Paris, France

Preface

My interest in dew and Breath Figures (BFs, the pattern that forms on a cold surface when breathing) started long ago when I was intrigued to see a ring around a streetlight from the vantage point of a dewy window on an early morning bus. The ring, which shrank after breathing on the window, was the signature of a well-defined order in the growing drop pattern on the window. Several years of study were then carried out in my laboratory at the French Atomic Energy Commission (CEA) and at the University of California, Los Angeles (UCLA). While at UCLA, I collaborated with my friend Charles Knobler and we published the first paper (Beysens and Knobler, 1986) concerning this simple but very rich phenomenon whereby droplet growth, once constrained to grow on a substrate with lower dimensionality, exhibits quite unusual features and, in particular, self-organization. During these pioneering times, I had the great opportunity to discuss and work with brilliant scientists (P. G. de Gennes, Y. Pomeau, J.-L. Viovy, D. Roux, J.-P. Bouchaud) and met Roger Mérigoux, who was the first scientist to become interested in BFs on liquids back in the 1930s.

Dew and breath figures are well known and ubiquitous phenomena although they form under different cooling process: outdoor radiative deficit for dew and indoor conduction cooling for breath figures. They exhibit, however, the same dropwise condensation characteristics and the fact that droplet growth proceeds from humid air or, to be more precise, from a gradient of water vapor pressure in non-condensable gases. The same dropwise condensation pattern is also observed when pure vapor condenses, however the growth characteristics can be markedly different, with no vapor pressure gradient but pronounced thermal effects.

Dropwise condensation is a type of condensation that engineers have known for long to be more efficient for heat transfer than filmwise condensation, simply because it leaves dry a fraction of the condenser surface. In addition to their application in the properties of heat condensers, many technological applications used the self-organization of BFs as, e.g. thin film growth (Beysens et al., 1990) or producing nanofiltering membranes (Rodríguez-Hernández and Bormashenko, 2020). However, curiously, nobody had previously investigated the statistical properties of drop pattern evolution in terms of scaling.

After a period of investigating the self-similar properties of droplet patterns, another approach followed where different substrates, with their specificities, were considered: liquid surface (e.g. paraffinic oil), phase-change materials (e.g. cyclohexane) and micropatterned surfaces. The possibility to considerably change the wetting properties by micro- or nanomanufacturing a solid substrate, together with large improvements in micro- and nanolithography, opened the way to a great variety of substrate, superhydrophobic or superhydrophilic patterned surfaces, with micro-grooves, micro-pillars or other shapes. In addition, the ability of trapping oil in micropatterned substrates produced liquid-like surfaces. At the time of writing, a large number of studies have dealt with the extremely rich varieties of such patterned surfaces, giving rise to enhanced condensation and water collection on inclined substrates and improved heat transfer, to cite only two examples.

Another aspect is water harvesting from dew, a complementary facet of the study of their physics. Natural dew is an omnipresent phenomenon on earth. Water collected this way is enhanced by using the last developments of dropwise condensation (see my recent book *Dew water* (Beysens, 2018)) which deals with different aspects of this alternative source of water.

This book aims to give to students, researchers and engineers the detailed processes involved in dew and breath figures and, more generally, in dropwise formation and collection from humid air or pure water vapor. Heat and mass transfer, nucleation and growth on smooth substrates are considered (solid, liquid, plastic, undergoing phase-change or micropatterned substrates). The particular role of thermal or geometrical discontinuities where growth can be enhanced or reduced, dynamical aspects of self-diffusion, problems related to drop collection by gravity and the optics of dropwise condensation are all discussed. Although the content mainly deals with condensation from humid air, it can be readily generalized to condensation of any substance. The specificities of pure vapor condensation (e.g. steam) are also examined.

This foreword would not be complete without mentioning the many friends with whom I collaborated on this subject: Anand S., Andrieu C., Bardat A., Berkowicz S., Bintein P.-B., Bouchaud J.-P., Bourouina T., Broggini F., Chatterjee R., Chelle M., Cui T., Doppelt E., Flura D., Fritter D., Godreche C., González-Viñas W., Guadarrama-Cetina J., Guenoun P., Huber L., Jacquemoud S., Khandkar M. D., Knobler C. M., Lhuissier H., Limaye A. V., Liu X., Marcos-Martin M., Marty F., Medici M.-G., Milimouk-Melnytchouk I., Mongruel A., Narhe R., Nikolayev V. S., Pomeau Y., Roux D., Royon L., Rukmava C., Rykaczewski K., Saint-Jean S., Schmitthaeusler R., Shelke P. B., Sibille P., Steyer A., Subramanyam S. B, Tixier N., Trosseille J., Varanasi K. K, Viovy J.-L., Yekutieli I. and Zhao H. I am very grateful to them for their friendly support.

I thank David Quéré for his thoughtful Forewords. I am much indebted to the CEA and the City of Paris Industrial Physics and Chemistry Higher Educational Institution (ESPCI) for having provided me with their everlasting support. I thank S. Berkowicz for English corrections and L. Limat and A. Eddi for helpful discussions.

I must express my deep gratitude to my beloved wife, Iryna, who supported me throughout the writing of this book.

Paris, France   Daniel Beysens

Contents

About the Author

Daniel Beysens holds Ph.D. in Physics and in Engineering and is a specialist of phase transition with emphasis on dropwise condensation. He is the co-founder and a president of OPUR—the International Organization for Dew Utilization. He teaches at the University Paris Diderot in Paris and is also an honorary director of Research at Ecole Supérieure de Physique et Chimie, Paris. With his team, he carries out experimental and theoretical study on dew condensation and phase transition, having started the field at the Alternative Energies and Atomic Energy Commission (CEA) when he was the head of an institute. He serves as an associate editor for several scientific journals and has authored many books and scientific publications. He was awarded various prices and honors in Physics and Environmental Sciences.

Humid Air

The first documented study of humid air is the report to the *Académie Royale des Sciences* of Leroy (1751), a medical doctor in Montpellier (France). He reported that water could be dissolved in air according to air temperature, with higher temperatures corresponding to larger dissolution. In support of his claim, he described several experiments. One demonstration was made with a bottle of air closed at daytime temperature. After cooling at night, the air was unable to hold all water previously dissolved at higher daytime temperatures and the excess water led to well-visible condensed droplets *inside* the bottle.

Air is indeed never completely dry; it always holds invisibly some water vapor in different concentrations depending on its temperature. In addition to vapor, humid air can also contain water in visible condensed states: liquid (fog droplets) and solid (frosty fog). In the latter cases where vapor and condensed phases coexist, humid air is said to be supersaturated.

Humid air can thus be considered of being formed of (1) dry air unlikely to condense in the conditions of temperature and pressure considered here, and (2) water vapor likely to condense in liquid or ice. Dry air (see Table 1.1) is mainly composed of nitrogen ($\approx$78%) and oxygen ($\approx$21%). For regular conditions of temperature and pressure found at the Earth's surface, both gases are far from their critical point coordinates (N_2: 126 K, 33.5 bar; O_2: 155 K, 50 bar) and both fluids can be accepted as ideal gases. Air is thus considered as a single ideal gas. Water is also far from its critical point coordinates (647 K, 218 bar) and can be considered as an ideal gas. Useful data concerning air and water are provided in Table 1.2.

1.1 Humid Air Characteristics

A volume V of a mixture of dry air and water vapor at temperature T is considered. The pressure of the mixture, p_m, is considered constant (atmospheric pressure).

© Springer Nature Switzerland AG 2022

D. Beysens, *The Physics of Dew, Breath Figures and Dropwise Condensation*,
Lecture Notes in Physics 994, https://doi.org/10.1007/978-3-030-90442-5_1

Table 1.1 Major
constituents of dry air, by
volume

Name	Formula	Percentage
Nitrogen	N_2	78.084
Oxygen	O_2	20.946
Argon	Ar	0.9340
Carbon dioxide	CO_2	0.0397
Neon	Ne	0.001818
Helium	He	0.000524
Methane	CH_4	0.000179

1.1.1 Dalton Law

The partial pressure of a gas is the pressure that would have if the gas were alone
in volume V. As both dry air and water are ideal gases, the total pressure is equal
to the sum of the partial pressures, i.e. the Dalton law. With p_a (resp. p_v) the
partial pressure of air (resp., water), one obtains

$$p_m = p_a + p_v \tag{1.1}$$

This additivity rule is also valid for the partial volumic mass and entropy. It
neglects intermolecular forces among the gases molecules. As the pressure is due
to impacts of moving gas molecules, the total pressure is simply the addition of
impacts of each type of molecules.

1.1.2 Humid Air Equation of State

The equation of state for dry air and water vapor can be written as

$$p_i V = n_i RT \tag{1.2}$$

Here i stands for air ($i = a$) or water vapor ($i = v$); $n_i = m_i/M_i$ is the number
of moles (i) in V, with mass m_i and molar mass M_i ($= 29$ g for dry air and 18 g
for water). $R = 8.314$ J mole^{-1} K^{-1} is the molar gas constant.

Equation 1.2 can also be rewritten as

$$p_i V = \frac{m_i}{M_i} RT = m_i r_i T \tag{1.3}$$

with specific (mass) constant $r_i = R/M_i$ ($= 287$ J kg^{-1} K^{-1} for air and
462 J kg^{-1} K^{-1} for water).

From the Dalton law, Eq. 1.1 and the equations of state for dry air and humid
air, Eqs. 1.2 and 1.3, one obtains the same relationship for humid air, using mass

Table 1.2 Useful data (temperature ≈ 5 °C)

	Latent heat (kJ kg⁻¹)	Liquid–air surface tension (mN m⁻¹)	Liquid–air surface tension thermal derivative (mN m⁻¹ K⁻¹)	Density (kg m⁻³)	Molar gas constant R (J mole⁻¹ K⁻¹)	Molar mass (10⁻³ kg)	Specific mass constant (J kg⁻¹ K⁻¹)	Kinematic viscosity (10⁻⁶ m² s⁻¹)	Thermal conductivity (W m⁻¹ K⁻¹)	Specific heat (kJ kg⁻¹ K⁻¹)	Thermal diffusivity (10⁻⁶ m² s⁻¹)	Water–air diffusion coefficient (10⁻⁶ m² s⁻¹)	Volumetric thermal expansion coefficient (10⁻³ K⁻¹)
Water (liquid)	Evaporation 2.5×10^3	72.2	−0.15	1000	8.314	18	–	15	0.6	4.18	0.143	–	0.2
Water (vapor)	condensation 2.5×10^3	–	–	0.0086*	8.314	18	462	20	0.019	1.86	13	24	–
Water (ice)	solidification 335	–	–	918	8.314	18	–	–	2.25	2.03	12	–	–
Air	–	–	–	12	8.314	29	287	14	0.026	1.006	23	–	3.4

*Calculated from Eq. (1.2) for RH = 90% and $T = 283$ K

conservation $m = m_a + m_v$:

$$(p_a + p_v)V = (m_a r_a + m_v r_v)T$$
$$p_m V = mrT \tag{1.4}$$

One can thus define a specific "constant" of humid air:

$$r = \frac{m_a}{m}r_a + \frac{m_v}{m}r_v = Y_a r_a + Y_v r_v \tag{1.5}$$

The quantity $Y_i = m_i/m$ is the mass fraction of gas i in humid air.

The humid air contains $n = n_a + n_v$ moles. One can thus define M, the molar mass of humid air, from $nM = n_a M_a + n_v M_v$ or:

$$M = \frac{n_a}{n}M_a + \frac{n_v}{n}M_v = X_a r_a + X_v r_v \tag{1.6}$$

The quantity $X_i = n_i/n = p_i/p_m$ is the mole fraction of gas i in humid air.

1.1.3 Humid Air Density

The density or mass per unit volume of humid air, ρ_h, is given by

$$\rho_h = \frac{m}{V} = \frac{p_m}{rT} \tag{1.7}$$

After some algebraic manipulations, it becomes

$$\rho_h = \frac{p_m}{T}\left(\frac{X_a}{r_a} + \frac{X_v}{r_v}\right) = \frac{p_m}{T}\left[\frac{1}{r_a} - X_v\left(\frac{1}{r_a} - \frac{1}{r_v}\right)\right] = \frac{p_m}{T}\left[\frac{1}{r_a} - \frac{p_v}{p_m}\left(\frac{1}{r_a} - \frac{1}{r_v}\right)\right] \tag{1.8}$$

After applying the relation $X_a = 1 - X_v$. As $r_v > r_a$, the second term in the square bracket is negative and shows that humid air density is lower than dry air density. This seemingly paradoxical effect simply results from the fact that air molecules (nitrogen: molar mass 28 g and oxygen: molar mass 32 g) are replaced by lighter water molecules (molar mass: 18 g).

1.1.4 Saturated Vapor Pressure

Let us consider the cooling process at constant pressure p_m of a mass m of humid air that contains a mass m_v of water. Mass conservation requires that both m_v and m remain constant during the process. This is therefore also the case for the number of moles of vapor, n_v, and humid air, n, and the corresponding molar fraction $n_v/n = p_v/p_m$. It results that the water vapor pressure remains constant during the cooling process. In the atmosphere, humid air cooling thus occurs *at constant water vapor pressure.*

During cooling, condensation into liquid can occur (see the Clapeyron phase diagram in Fig. 1.1). Let us consider a mass of humid air initially at point A on isotherm T_1. When temperature decreases at constant pressure p_v, its volume also decreases. The liquid–vapor coexistence curve (the saturation curve) is reached (point B) at some temperature T_2 and liquid drops can appear (provided that metastability is weak, see Chap. 3). Point B is called the dew point and T_2 is the dew point temperature T_d. When air is cooled further, condensation proceeds at constant pressure p_v and temperature T_d. Cooling energy only compensates the release of the condensation latent heat. In C, all water contained in the humid air has condensed. Zone BC is the fog zone where liquid droplets coexist with vapor. Further cooling (until point D) is only concerned with liquid.

The liquid–vapor saturation curve represents in the plane p_v–T the liquid–vapor equilibrium (Fig. 1.2). At a given temperature, the maximum pressure above which water vapor changes into liquid water is the saturated vapor pressure p_s. Therefore,

Fig. 1.1 Cooling at constant pressure p_v in the clapeyron phase diagram. CP: critical point; B: dew point; T_d: dew point temperature where $p_v = p_s(T_d)$, the saturated vapor pressure at temperature T_d

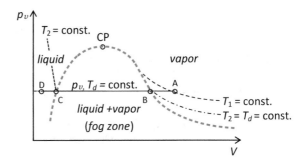

Fig. 1.2 Phase diagram of water highlighting the saturation line $p_s(T_d,)$. CP: critical point; TP: triple point. B: dew point

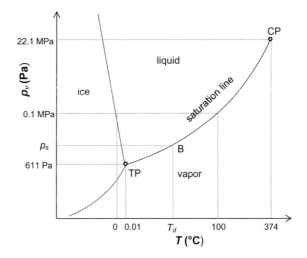

in a given mass of humid air, vapor pressure can be such that (i) $p_v < p_s$: water in humid air is in the vapor state; (ii) $p_v = p_s$: water in humid air is in both vapor state and liquid state as the phase change is at constant pressure $p_s = p_v$ (T_d). Humid air can be saturated (point B in Fig. 1.1) or supersaturated, when liquid droplets (fog) are present (line BC in Fig. 1.1).

Saturated vapor pressure can then be reached in a given humid air in two ways. (i) Cooling a given mass of humid air. Vapor pressure remains constant at p_v but p_s decreases until the equality $p_v = p_s$ (T_d) is fulfilled. (ii) Adding a mass m_v of water to a given humid air volume at constant temperature. The vapor pressure $p_v = m_v r_v T/V$ increases until it reaches at same temperature the vapor pressure $p_v = p_s = (m_v)_{max} r_v T/V$. If more water at constant temperature is added, one obtains the coexistence of saturated vapor pressure and liquid. Humid air is then supersaturated.

The saturation curve can be described by integrating the Clausius-Clapeyron relation (see Appendix A):

$$\frac{dp_s}{dT} = \frac{L_v p_s}{r_v T^2}$$

$$p_s = Ce^{-\frac{L_v p_s}{r_v T}} \tag{1.9}$$

Here L_v is the latent heat of evaporation ($\approx 2.5 \times 10^6$ J kg^{-1} at 0 °C) and r_v (= 462 J kg^{-1} K^{-1}, see Table 1.2) is the specific constant for water. C depends on the reference temperature for which the value of L_v is chosen ($C = 2.53 \times 10^{11}$ Pa at 0 °C).

Another well-known equation is the Antoine equation (Antoine, 1888), which is also derived from the Clausius-Clapeyron relation:

$$\log(p_s/p_0) = A_1 - \frac{B_1}{C_1 + T} \tag{1.10}$$

with $p_0 = 10^5$ Pa and T in K. The Antoine equation alone cannot describe the full saturation curve between the triple and critical points, which means that the coefficients depend on the temperature range. In the range 0–30 °C, $A_1 = 5.40221$, $B_1 = 1838.675$ K, $C_1 = -31.737$ K. In a slightly different form, the more accurate August-Roche-Magnus formula is widely used, with temperature in °C noted as θ:

$$p_s = A \exp\left(\frac{B\theta}{C + \theta}\right) \tag{1.11}$$

For current ground atmospheric temperatures (-20 ± 50 °C) and p_s in Pa, $A = 610.94$ Pa, $B = 17.625$ °C^{-1} and $C = 243.04$ °C (Alduchov and Eskridge 1996).

1.2 Specific Quantities

In any humid air evolution, the dry air mass remains constant. It is thus judicious to scale all mass quantities to the dry air mass (and not humid air mass). Those scaled quantities are generally called "specifics".

1.2.1 Moisture Content, Humidity Ratio, Mass Mixing Ratio, Absolute and Specific Humidity

The moisture content (also called humidity ratio, mass mixing ratio, absolute or specific humidity) is the ratio of water mass, m_v, to dry air mass, m_v, contained in a given volume V:

$$w = \frac{m_v}{m_a} \tag{1.12}$$

In the atmosphere, the moisture content varies from a few g/kg in the middle latitudes to 20 g/kg in the tropics.

The moisture content can be related to the mass fraction Y_v (see Sect. 1.1.2) through

$$Y_v = \frac{m_v}{m_a + m_v} = \frac{w}{1 + w} \approx w \tag{1.13}$$

The approximation is valid since w never exceeds a few percent.

The moisture content can also be expressed as a function of vapor pressure. From Eq. 1.3, one obtains

$$w = \frac{r_a}{r_v} \frac{p_v}{p_m - p_v} \approx 0.6212 \frac{p_v}{p_m} \tag{1.14}$$

since $p_m \gg p_v$. The moisture content is thus nearly proportional to the water vapor pressure and increases when atmospheric pressure p_m decreases. At saturation, $w_s \approx 0.6212 \frac{p_s}{p_m}$. The saturation moisture content thus increases with temperature along the saturation curve. It also increases with decreasing p_m.

1.2.2 Relative Humidity

The relative humidity (RH) is the ratio of the actual water vapor pressure to the saturation water vapor pressure at the prevailing temperature. It is expressed in %:

$$RH = 100 \frac{p_v(T_a)}{p_s(T_a)} \tag{1.15}$$

The water content of the air is not defined unless temperature is known. Notice that the air characteristics are not involved in the definition of RH (an airless volume can thus have a RH).

1.2.3   Dew Point Temperature and Relative Humidity

In the usual isobaric cooling, a mass of humid air cooled from temperature T to saturation or dew point temperature T_d obeys the following relation, since the mass and then the partial water vapor pressure p_v remain constant (see Sect. 1.1.4):

$$p_s(T_d) = p_v(T) = \frac{p_v(T)}{p_s(T)} p_s(T) = \mathrm{RH} p_s(T) \tag{1.16}$$

Substituting Eq. 1.15 in Eq. 1.11 yields the dew point temperature as a function of the ambient vapor pressure and temperature. Using the definition of RH, Eq. 1.15, one obtains with θ_d the dew point temperature in °C:

$$\theta_d = \frac{C\left[\ln\left(\frac{RH}{100}\right) + \frac{B\theta}{C+\theta}\right]}{B - \ln\left(\frac{RH}{100}\right) - \frac{B\theta}{C+\theta}} \tag{1.17}$$

A simplified expression can be obtained from Eq. 1.9, with T and T_d in K:

$$T_d = T\left[1 - \frac{T\ln\left(\frac{RH}{100}\right)}{L_v/r_w}\right]^{-1} \tag{1.18}$$

1.2.4   Dew Point Depression Temperature and Relative Humidity

In natural dew formation, the dew point depression temperature with respect to ambient air temperature $(T_d - T)$ is a key parameter as it determines the heat losses during condensation (see Beysens 2016, 2018). The depression, which remains in the range [0–10 °C], appears to be the most limiting parameter of the process. In this small temperature range, the $(T_d - T)$ dependence on RH is logarithmic (Eqs. 1.17 and 1.18) and depends weakly on T for usual temperatures (Fig. 1.3). If one considers that the condenser provides a mean cooling effect of 5 °C, the threshold in RH is 67% for air at -10 °C and 76% for air at 40 °C. The conclusion is that this threshold does not vary much with air temperature and is of the order of 70% for a cooling effect $T_c - T = -5$ °C. Since it is in this range that natural dew forms, the condition for its formation is thus having RH > 70–80%.

Note that, at large RH (>50%) corresponding to small $(T - T_d)$ (<10 °C), it is possible to rearrange and approximate Eq. 1.18. It follows the simple linear

Fig. 1.3 Difference $T_d - T_a$ versus relative humidity RH (semi-log plot). The interrupted lines are for $T_a = -10, 0, 10, 20, 30, 40$ °C. The full curve is Eq. 1.19. A condensation threshold $T_d - T_a = -5$ °C corresponds to 67% RH for air at -10 °C and 76% RH for air at 40 °C. Warmer air needs only slightly more relative humidity than cold air for the same $T_d - T_a$ threshold

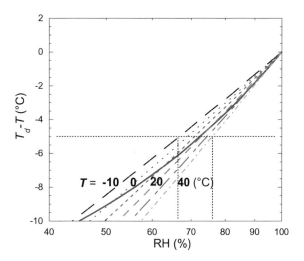

relationship (Lawrence 2005), where $A = \frac{L_v}{r_w T_a T_d} \frac{RH - 100}{\ln(RH/100)}$ varies only slightly around 5 K^{-1} in usual temperature and RH ranges:

$$RH \approx 100 - A(T - T_d) \tag{1.19}$$

1.2.5 Degree of Saturation

The degree of saturation is the ratio of the humidity ratio of moist air to the humidity ratio of saturated moist air at the same temperature and pressure:

$$\psi = \frac{w(T)}{w_s(T)} \tag{1.20}$$

Applying the ideal gas equation, Eq. 1.2, and the Dalton law, Eq. 1.1, one obtains the following relation between ψ and RH:

$$\psi = RH \frac{p_m - p_s}{p_m - RH p_s} \approx RH \tag{1.21}$$

since $p_m \gg p_s$.

1.2.6   Specific Volume

The specific volume, v', is defined as the ratio of humid air volume and dry air mass:

$$v' = \frac{V}{m_a} \tag{1.22}$$

From Eq. 1.8 describing the humid air density, ρ_h, and applying the ideal gas, Eq. 1.2, and making use of the definition of the moisture content, w, from Eq. 1.12, one can deduce the following relation between humid air density and the mixing ratio:

$$\frac{1}{\rho_h} = r_v \frac{\left(\frac{r_a}{r_v} + w\right)}{1 + w} \frac{T}{p_m} \tag{1.23}$$

The specific volume can then be easily deduced, making use of the relation $v' = (1+w)/\rho_h$:

$$v' = r_v \left(\frac{r_a}{r_v} + w\right) \frac{T}{p_m} \tag{1.24}$$

1.2.7   Specific Enthalpy

Enthalpy or heat content is an extensive quantity. Enthalpy H of a given mass of humid air is then the sum of dry air and water enthalpies. Noting $h_a(h_w)$ the mass enthalpy of dry air (water), one obtains

$$H = m_a h_a + m_v h_w \tag{1.25}$$

Water can be either in the vapor, liquid or solid state. In units of dry air mass, one obtains the specific enthalpy $h = H/m_a$ of humid air:

$$h = h_a + w h_w \tag{1.26}$$

Heating or cooling a given mass of humid air implies a sensible heat, characterized by air and water specific heats, and latent heats at constant temperature when water undergoes a phase change (liquid/vapor, liquid/solid, vapor/solid).

As enthalpy is defined within a constant, by definition dry air enthalpy is said to be zero at 273.15 K (0 °C). Water enthalpy will also be zero at 0 °C, in its liquid state and with its saturation pressure at 0 °C (610 Pa).

1.2.7.1 Dry Air Enthalpy

At temperature θ expressed in °C, it follows from the above definition the dry air enthalpy:

$$h_a = C_{pa}\theta \tag{1.27}$$

with $C_{pa} \approx 1.006$ kJ kg^{-1} K^{-1} the air specific heat at constant pressure.

1.2.7.2 Water Specific Enthalpy

Depending whether water is in a liquid, solid or vapor state, the water enthalpy takes different forms.

- liquid water h_l

Neglecting the effect of pressure to retain only that of temperature, it remains only the sensible heat term from the initial, liquid water state at 0 °C:

$$h_l \approx C_{pl}\theta \tag{1.28}$$

with $C_{pl} \approx 4.18$ kJ kg^{-1} and K^{-1} the liquid water specific heat at constant pressure.

- vapor water h_v

Assuming that vapor enthalpy is independent of pressure and the water vapor specific heat at constant pressure C_{pv} is constant (≈ 1.83 kJ kg^{-1} K^{-1}) with respect to temperature and pressure, one has to take into account, in addition to the sensible heat $C_{pv}\theta$, the latent heat of evaporation $L_v = 2.5 \times 10^3$ kJ kg^{-1} at 0 °C and 610 Pa:

$$h_v \approx L_v + C_{pv}\theta \tag{1.29}$$

- solid water h_s

Using the same kind of reasoning as above for water vapor, one obtains

$$h_s \approx -L_s + C_{ps}\theta \tag{1.30}$$

with $C_{ps} \approx 2.03$ kJ kg^{-1} K^{-1} the ice specific heat at constant pressure and $L_s = 335$ kJ kg^{-1} the latent heat of solidification.

1.2.7.3 Non-supersaturated Humid Air

Using Eq. 1.26 and replacing dry air and water vapor enthalpy by their expressions, Eqs. 1.27 and 1.29, it comes

$$h \approx C_{pa}\theta + w\left(L_v + C_{pv}\theta\right) \tag{1.31}$$

1.2.7.4 Supersaturated Humid Air

Humid air is formed by a vapor phase with a mixing ratio w_s and a condensed (liquid or solid depending on temperature) phase with mixing ratio w-w_s.

- vapor and liquid ($\theta > 0\ °C$)

The specific enthalpy is composed of contributions from dry air (Eq. 1.27), vapor (Eq. 1.29) with mixing ratio w_s and liquid (Eq. 1.28) with mixing ratio $w - w_s$:

$$h = C_{pa}\theta + w_s\left(L_v + C_{pv}\theta\right) + (w - w_s)C_{pl}\theta \tag{1.32}$$

- vapor and ice ($\theta < 0\ °C$)

The same reasoning leads to the following formulation, using Eq. 1.30:

$$h = C_{pa}\theta + w_s\left(L_v + C_{pv}\theta\right) + (w - w_s)\left(-L_s + C_{ps}\theta\right) \tag{1.33}$$

1.2.8 Wet Bulb Temperature and the Psychrometric Constant

The psychrometer, or wet and dry bulb thermometer, is made of a regular, "dry bulb", thermometer and another thermometer whose sensitive part (the bulb) is coated with a tissue imbibed of water, the "wet bulb" (see, e.g. Simões-Moreira 1999). The mass of water is small and its influence on the room wet air properties, supposedly unsaturated, can be neglected. Air enters the boundary layer region (see Chap. 2) at temperature θ_a and leaves it at a lower, "wet bulb" temperature $\theta_w < \theta_a$. Water indeed evaporates from the wet wick and the latent heat required for evaporation into the air flow around the wet bulb is taken from the wet surface, thus cooling the air and thermometer. Heat losses with ambient, hotter air eventually equilibrate the cooling process and a steady-state temperature is reached (θ_w). Water temperature and evaporated water flux are constant. The wet surface temperature, θ_w, lies between ambient temperature θ_a (case where no cooling or evaporation occurs, meaning that the room is at saturation temperature) and dew point temperature θ_d. The equation that describes this equilibrium can be written as

$$-aS_c(\theta_w - \theta_a) = -\left(\frac{dm_w}{dt}\right)L_v \tag{1.34}$$

The left-hand side corresponds to the heat losses (heating) between the wet bulb (temperature θ_w, surface S_c) and ambient air (temperature θ_a), with the coefficient of convective heat exchange a. The right-hand side is the cooling heat flux corresponding to the evaporation latent heat. For a given air, the condition defined by (θ_a, RH) corresponds to only one wet bulb temperature.

This process corresponds to saturation of an incoming moist airstream brought into contact with liquid (solid) water. The entire process is isenthalpic because in an ideal experiment there is no heat exchange with the environment. It thus follows that the wet bulb temperature remains constant on a line of constant enthalpy. On a moist air chart (Fig. 1.5), its value is the dew point temperature θ_s where the enthalpy constant line intersects the saturation line.

The property of moist air (vapor pressure, RH) can be related with the wet and dry bulb temperatures as follows. Dry and wet bulb temperatures are seen to follow the empirical equation:

$$p_v(\theta_a) = p_s(\theta_w) - \gamma(\theta_a - \theta_w) \tag{1.35}$$

where γ is the psychrometric constant and p_s is the saturation water pressure of vapor at θ_w. The psychrometric constant depends on atmospheric pressure (see discussion below). It is usually obtained in careful laboratory experiments where experimental conditions are well controlled: $\gamma = 65.5$ Pa K^{-1} at sea level atmospheric pressure.

Assuming as always in this chapter that humid air is an ideal gas, following Simões-Moreira (1999) it can be shown that the constancy of the psychrometric "constant" is a mere coincidence in the vicinity of $\theta_a = 20\,°C$. Let us consider the conservation of energy and mass in the wet bulb process, corresponding to an adiabatic process. Then the following equality holds:

$$h(\theta_a) + (w_w - w)L_v = h_w(\theta_w) \tag{1.36}$$

Here w is the moisture content of incoming air at temperature θ_a, w_w is the moisture content temperature of departing air at temperature θ_w, $h(\theta_a)$ is the specific enthalpy of humid air at θ_a and $h_w(\theta_w)$, the specific enthalpy of humid air at θ_w. Expressing h and h_w from Eq. 1.31, it becomes

$$w_w - w = \frac{C_{pm}}{L_v} + (\theta_a - \theta_w) \tag{1.37}$$

where C_{pm} is the humid air specific heat:

$$C_{pm} = C_{pa} + wC_{pv} \tag{1.38}$$

Substituting Eq. 1.38 into Eq. 1.36 and making use of moist air enthalpy (Eq. 1.26) and the relation water content-vapor pressure (Eq. 1.14), it becomes

$$\gamma = \frac{C_{pm}}{0.6212L_v} \frac{(p_m - p_v)(p_m - p_s)}{p_m} \approx \frac{C_{pm}p_m}{0.6212L_v} \tag{1.39}$$

Fig. 1.4 The thermodynamic psychrometric constant γ expressed in units of atmospheric pressure p_m as a function of the wet bulb temperature for several constant temperature curves (at normal pressure). Adapted from Simões-Moreira (1999)

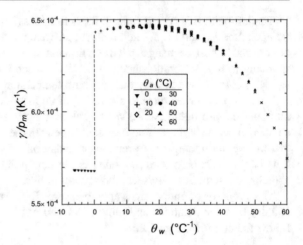

The expression $A = (\gamma/p_m)$ when plotted with respect to θ_w at different θ_a shows a maximum at 6.47×10^{-4} K^{-1} near $\theta_w = 20\,^{\circ}$C (Fig. 1.4). A good approximation of Eq. 1.39 for the psychrometric constant is then the constant:

$$\gamma \approx 65.5\,\text{Pa K}^{-1} \tag{1.40}$$

It follows from the definition of relative humidity, Eq. 1.15, and by using Eq. 1.35, the relation:

$$\text{RH} = 100\frac{p_s(\theta_w) - \gamma(\theta_a - \theta_w)}{p_s(\theta_a)} \tag{1.41}$$

Note that when the wet thermometer is frozen, the different values for solidification, evaporation and latent heat make the psychrometric constant change ($\gamma \approx$ 57 Pa K^{-1}, see Fig. 1.4).

It is interesting to note that the psychrometric constant allows a relation to be obtained between the heat transfer coefficient, a, from Eq. 1.34 and the mass transfer coefficient, a_w. The evaporation mass flux from surface with area S_e can indeed be written as

$$\left(\frac{dm}{dt}\right) = S_e a_w (p_v - p_s) \tag{1.42}$$

From 1.34, 1.41 to 1.35, expressing the vapor pressure difference in function of temperature difference, it becomes

$$\frac{a}{a_w} = \gamma L_v \tag{1.43}$$

This relation is also valid for condensation since the mass transfer coefficient must be the same for both evaporation and condensation processes. A precise calculation using thermal and mass boundary layers is given in Appendix B.

1.2.9 Mollier Diagram and the Psychometric Chart

The various characteristics of a humid air are related by somewhat complex relations. These relations are easily seen in the diagrams of humid air where a unique point defines a given moist air for a certain atmospheric pressure. On such a diagram, a representative point of a humid air is perfectly determined when there are only two characteristics. The most common diagrams are generated from specific enthalpy and absolute humidity (or equivalently water vapor pressure) as ordinates, with the atmospheric pressure as a parameter. These diagrams are called enthalpy diagrams of humid air. In order to improve readability, these diagrams are constructed in oblique coordinates. The absolute humidity axis is vertical. The specific enthalpy axis makes an angle (which varies according to authors) with respect to the preceding axis. These diagrams are constituted of families of isovalue curves (Fig. 1.5a–b).

The Mollier diagrams express the same psychrometric properties as the psychrometric charts, however the axes are not shown the same way. In order to transform a Mollier diagram into a psychrometric chart, the diagram must first be reflected in a vertical mirror and then rotated 90°.

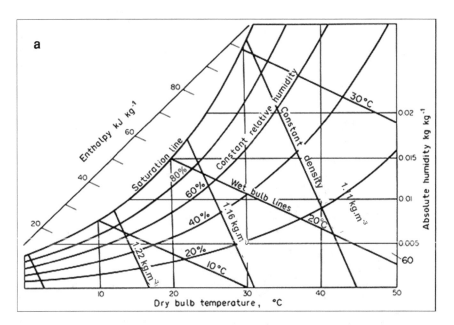

Fig. 1.5 Psychometric chart diagram. **a** Schematics. **b** Example for a humid air indicated by a full circle, whose characteristics are the following: dry bulb temperature 21 °C, absolute humidity 8 g kg^{-1}, vapor pressure 1.28 kPa, dew point temperature 10.8 °C, specific enthalpy 41.8 kJ kg^{-1}, wet bulb temperature 12.8 °C, relative humidity 52%, density 1.196 kg m^{-3}

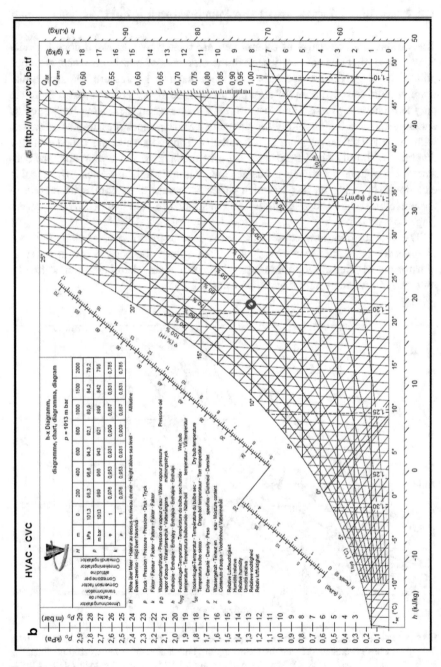

Fig. 1.5 (continued)

1.2.9.1 Relation Vapor Concentration-Vapor Pressure

Concentration c of water vapor is counted in mass per volume ($c = m_v/V$) and can be related to partial vapor pressure and atmospheric pressure. One first relates c to the moisture content (Eq. 1.12) w ($= m_v/m_a$). It follows $c = \rho_a w$. Then one applies Eq. 1.14 to relate w to the water partial vapor pressure p_v and total atmospheric pressure p_m:

$$c = w\rho_a \approx \rho_a \left(\frac{r_a}{r_v}\right)\left(\frac{p_v}{p_m}\right) = 0.745\left(\frac{p_v}{p_m}\right) \tag{1.45}$$

A more precise relationship, which includes temperature, can be deduced from Eq. 1.3:

$$c = \frac{p_v}{r_v T} \tag{1.46}$$

Boundary Layers

Boundary layers play a key role in the condensation process of humid air and, more generally, to vapor in the presence of non-condensable gases. Some characteristics of the different boundary layers are thus reviewed below. Gas flow can be forced or triggered by natural thermal convection. It results in a hydrodynamic boundary layer that defines, thanks to a diffuse or thermal Peclet number, a zone where mass or heat diffusion is more efficient for transport than convection. These zones are the diffuse and thermal boundary layers.

2.1 Hydrodynamic Boundary Layer

2.1.1 Forced Convection

Fluid velocity is, in general, zero at a surface (except some surfaces like micropatterned surfaces, see Chap. 10). Velocity decreases from the surface to the bulk due to the finite viscosity of the fluid. The region where the velocity varies from zero at the surface to a finite value U_0 associated with the flow is known as the hydrodynamic or velocity, boundary layer (Fig. 2.1). The thickness of the hydrodynamic boundary layer is usually defined as the distance from the solid surface to the point at which the viscous flow velocity is 99% of the free stream velocity.

There are two different types of boundary layer flow (Fig. 2.1): laminar and turbulent. The laminar boundary layer is a smooth flow, while the turbulent boundary layer contains eddies. The laminar flow creates less skin friction drag than the turbulent flow, but is less stable. As the flow continues from the leading edge, the laminar boundary layer increases in thickness.

The transition laminar-turbulent flow depends on the Reynolds number:

$$\text{Re} = \frac{U_0 x}{\nu} \qquad (2.1)$$

© Springer Nature Switzerland AG 2022
D. Beysens, *The Physics of Dew, Breath Figures and Dropwise Condensation*,
Lecture Notes in Physics 994, https://doi.org/10.1007/978-3-030-90442-5_2

Fig. 2.1 Development of a velocity boundary layer profile of thickness δ_H along a plate submitted to a flow with velocity U_0. Turbulence occurs after length x such as the Reynolds number $\mathrm{Re} > 5 \times 10^5$ (see text)

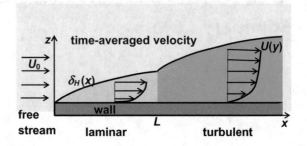

where x is length, ν is kinematic viscosity. With $\nu = 1.4 \times 10^{-5}$ m^2 s^{-1} for air (see Table 1.2), for an open planar structure, turbulence occurs for $\mathrm{Re} > 5 \times 10^5$ (Rohsenow et al. 1998). It corresponds to $x = 1$ m for $U_0 = 7$ m/s.

The characteristic dimension parallel to the flow is in the order of x whereas perpendicular to the flow it is in the order of the thickness of the boundary layer $\delta_H (x) \ll x$. These two very different length scales lead to several approximations in the Navier–Stokes equations (shown in Appendix C).

The flow being two dimensional and incompressible, the conservation equation of mass (Appendix C, Eq. C.2) can be written, with components U_x and U_z the components of velocity $\overrightarrow{U}$ in directions x and z:

$$\frac{\partial U_x}{\partial x} + \frac{\partial U_z}{\partial z} = 0 \tag{2.2}$$

This equation shows that the velocity component perpendicular to the wall, U_z, is much lower than the parallel component, U_x. Using the expression, Eq. 2.5, for the boundary layer thickness δ_H, one obtains

$$U_z \approx U_x \frac{\delta_H(x)}{x} \sim \frac{U_0}{\sqrt{\mathrm{Re}}} \ll U_x \tag{2.3}$$

Here the Reynolds number (Eq. 2.1) is apparent.

In the laminar regime (Blasius 1908), the velocity profile shows a sharp transition. In scaled coordinates $z/\sqrt{\frac{\nu x}{U_0}}$ and U/U_0, the profile is linear in a large domain with a sharp transition near $U/U_0 = 1$ (Fig. 2.2). The velocity profile in the linear part of the profile can be expressed as

$$U \approx \frac{1}{3} \frac{U_0^{3/2}}{(\nu x)^{1/2}} z \tag{2.4}$$

The thickness of the boundary layer corresponds to the thickness z where U/U_0 takes the arbitrary value 0.99. From the full velocity profile (see, e.g. Guyon et al. 2012; Schlichting 2017):

$$\delta_H = 5 \sqrt{\frac{\nu x}{U_0}} = 5 \frac{x}{\sqrt{Re}} \tag{2.5}$$

Fig. 2.2 Dimensionless
variation of the velocity in
the boundary layer

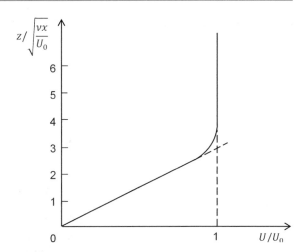

As an example, using current conditions for BFs formation.
$U_0 = 1$ m s^{-1} and Table 1.2 data for air, one finds $\delta_H = 5.9$ mm at $x = 0.1$ m
from the edge of a condensing plate.

2.1.2   Free Convection

Let us consider the case where air flow is produced by the temperature difference
between the plate and the surrounding air. This is precisely the case encountered in
the condensation process where the condensing plate is at temperature $T_c < T_\infty$,
the surrounding air temperature.

Following Gersten and Herwig (1992), one can obtain a rough estimation of
the boundary layer thickness by a scaling analysis of the natural convection above
a horizontal cooled plate. Due to buoyancy effects, air flow remains confined in a
hydrodynamic layer of thickness δ_F (Fig. 2.3) corresponding to non-zero values

Fig. 2.3 Free thermal
convection above a tilted
plane. Notations: see text

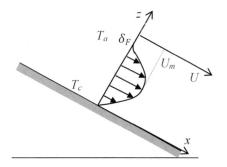

of air velocity. The boundary layer is a function of the Grashof number:

$$Gr = \frac{g\beta\Delta T L^3}{\nu^2} \tag{2.6}$$

such as

$$\delta_F = LGr^{-1/5} \tag{2.7}$$

Here L is the plate characteristic length, $\Delta T = T_a - T_c$, $\beta = 2/(T_a + T_c) = 2.4 \times 10^{-3}$ K^{-1} is the air volumetric thermal expansion coefficient (air considered as ideal gas) (Table 1.2). Taking typical values $\Delta T \approx 5$ K and $L = 0.1$ m gives $Gr \approx 8.5 \times 10^5$ and $\delta_F \approx 6.5$ mm. Note that, as $(\delta_F \sim \Delta T)^{1/5}$, its value is only weakly sensitive to the air-condenser temperature difference.

The maximal velocity in the hydrodynamic layer is given by

$$U_m \sim (\delta_F \beta g \Delta T)^{1/2} \tag{2.8}$$

Using the numerical values above, $U_m \approx 3$ cm s^{-1}.

2.2 Thermal and Mass Diffusion Boundary Layers

Thermal boundary layer (thickness δ_T), and mass diffusion boundary layers (thickness ζ) are related together as they are both associated to a diffusion process. Their values are dependent on the air flow configuration. Two types of air flows can be considered, forced flow, and free convection. The mass diffusion boundary layer will be thus noted below as $\zeta \equiv \delta_{FF}$ for forced flow and $\zeta \equiv \delta_{FC}$ for free convective flow.

2.2.1 Forced Flow

The condensation process involves three boundary layers corresponding to hydrodynamic, heat and mass diffusion. The hydrodynamic boundary layer in the laminar regime of forced convection can be considered as governed by a hydrodynamic diffusivity (viscosity), while the thermal boundary layer is governed by thermal diffusivity and mass boundary layer by mass diffusion. All of them can be viewed as balances between convection and diffusion perpendicular to the wall direction and can be treated the same way by considering diffusion coefficients such as (hydrodynamic) $D_h = \nu$; the kinematic viscosity, (thermal) D_T; the thermal diffusivity and D, the molecular diffusion coefficient (mutual diffusion coefficient for water molecules in air).

The ratio D_h/D_T represents the Prandtl number:

$$Pr = \frac{\nu}{D_T} \tag{2.9}$$

Thermal and diffuse boundary layers can be defined through thermal and diffuse Peclet numbers, which compare the convective time on mass diffusion or thermal diffusion times. When it is smaller than unity, diffusion is more efficient than convection for heat or mass transport. For heat transport, with U the flow velocity at the edge of the boundary layer, the thermal Peclet number is defined as

$$Pe_T = \frac{Ux}{D_T} = Pr.Re \tag{2.10}$$

With the definition of the Schmidt number, which corresponds to the Prandt number for mass diffusion:

$$Sc = \frac{\nu}{D} \tag{2.11}$$

one defines for mass transport the mass diffusion Peclet number:

$$Pe_M = \frac{Ux}{D} \tag{2.12}$$

The thermal boundary layers, δ_T, and mass diffusion boundary layer, δ_{FF}, correspond to, respectively, $Pe_T = 1$ and $Pe_M = 1$. For air temperatures corresponding to usual condensation conditions, Table 1.2 data shows that $Pr = 0.583$ and $Sc = 0.609$. It appears therefore that thermal and diffuse boundary layers exhibit nearly the same value and can be confounded. The hydrodynamic boundary layer is, however, slightly thinner, in the ratio $\sqrt{Sc} \approx 0.76$, giving

$$\delta_T \approx \delta_{FF} = 1.3\delta_H \tag{2.13}$$

When looking to typical numerical values (Sect. 2.1.1), the diffuse and thermal boundary layers generally extend over several mm. It means that for droplets of radius less than a few mm, growth remains limited by diffusion. When the drop diameter becomes larger, a thorough description of the convective flow in the exact experimental configuration where droplet size matters will be needed.

2.2.2 Free Convection

In the free convection regime, one assumes for simplicity that air velocity varies near the plate according to a parabolic-like flow (Fig. 5.3):

$$U(z) = 4U_m \frac{z(\delta_F - z)}{\delta_F^2} \tag{2.14}$$

The mass and thermal boundary layer thickness can be deduced from $\mathrm{Pe}_M (z = \delta_{FC}) = \frac{U(z)z}{D} = 1$, giving a third-degree polynomial equation:

$$4U_m z^2 (z - \delta_F) + \delta_F^2 D = 0 \tag{2.15}$$

This equation exhibits three solutions, one negative and two positives, with only one corresponding to a value close to the plate (see below). This value can be approximated, with $\delta_{FC} \ll \delta_F$, as

$$\delta_{FC} \approx \delta_T \approx \sqrt{\frac{D\delta_F}{4U_m}} = \frac{D^{1/2}}{2} \left(\frac{\delta_F}{\beta g \Delta T} \right)^{1/4} \tag{2.16}$$

Note that the $\delta_{FC} \approx \delta_T$ values are only weakly temperature dependent $((\sim \Delta T)^{1/4})$.

Using the numerical values of Sect. 2.1.2, the three solutions of Eq. 2.15 are $z = 1.3$ mm, -1.05 mm and 6.3 mm. The first solution is the only physical solution, to be compared to the approximated value 1.2 mm from Eq. 2.16. It means that for droplets of radius lower than 1.3 mm, growth remains limited by diffusion. As noted in Sect. 2.2.1, when the drop diameter becomes larger, a comprehensive description of the process would need a fine description of the convective flow in the exact experimental configuration where the influence of drops cannot be neglected.

Nucleation

3

Water contained in humid air condenses on a surface whose temperature is colder or equal to the dew point temperature T_d, the temperature at which water vapor contained in air is at saturation pressure p_s. It is currently observed that condensation initially proceeds only on a few particular sites of the substrate where the formation (nucleation) of liquid water is favored.

This chapter addresses the different aspects of the nucleation process of a single droplet in the bulk of the vapor (homogeneous nucleation) and on a substrate (heterogeneous nucleation).

3.1 Homogeneous Nucleation

Let us consider a supersaturated humid air at temperature T and pressure $p_v > p_s$, the saturation pressure. Saturation corresponds to the temperature $T = T_d$, the dew point temperature. Water vapor is in a metastable state and will eventually condense into a liquid phase, its state of minimal energy. In the framework of the classical Volmer theory (Volmer 1938, Landau and Lifshitz 1958), the first event is the formation or nucleation of the smallest cluster (liquid drop) which is thermodynamically stable, i.e. does not evaporate. This embryo develops from thermally activated local density fluctuations.

This model corresponds to having the size of the cluster fluctuating by addition or withdrawal of molecules. The growth is favored by the energy of formation of the liquid phase, while evaporation is supported by the cost in energy (surface tension) of the creation of the interface between the liquid cluster and its vapor environment. The two effects counterbalance for a particular size of cluster, known

The original version of the book was revised: Belated corrections have been updated. The correction to the book is available at https://doi.org/10.1007/978-3-030-90442-5_16

D. Beysens, *The Physics of Dew, Breath Figures and Dropwise Condensation*, Lecture Notes in Physics 994, https://doi.org/10.1007/978-3-030-90442-5_3

as "critical size". Consequently, the liquid clusters smaller than the critical size tend to re-evaporate under the effect of the thermal fluctuations. Conversely, the liquid clusters larger than the critical size tend to grow on average, until forming a macroscopic liquid phase.

Quantitatively, the gain in energy W^V for a volume V of vapor which transforms itself into liquid is

$$W^V = -\Delta e V \tag{3.1}$$

with Δe the gain in volumic free energy. When the gain is small $(d\Delta e)$, it corresponds to the work $d\Delta W$ of the Carnot cycle when expressing the Clausius–Clapeyron equation (see Appendix A, Eqs. A.1, A.2). With V_l the volume of condensed liquid, m_l its mass and ρ_l its density, one can write, with L_v the latent heat of vaporization:

$$d\Delta e \equiv \frac{d\Delta W^V}{V_l} = \frac{m_l L_v}{V_l} \frac{dT}{T} = \rho_l L_v \frac{dT}{T_d} \tag{3.2}$$

Making use of the Clausius–Clapeyron relation (Appendix A5, Eq. A.5) $dp/dT = pL_v/r_v T_d^2$ to make apparent the vapor pressure, Eq. 3.2 becomes, with r_v the water vapor specific constant (see Eq. 1.3),

$$d\Delta e = \rho_l r_v T \frac{dp}{p} \tag{3.3}$$

Let us now define the supersaturation ratio:

$$SR = \frac{c}{c_s} = \frac{p}{p_s} \tag{3.4}$$

where p and p_s are the vapor pressure and saturation vapor pressure, respectively, corresponding to the corresponding water molecules concentration in air, c and c_s (counted in mass per volume $c = m_v/V$, cf. Eq. 1.46). Equation 3.3 can be integrated to make apparent the supersaturation ratio corresponding to vapor temperature T and dew point or saturation temperature T_d:

$$\Delta e = \rho_l r_v T \int_p^{p_s} \frac{dp}{p} = \rho_l r_v T \ln(SR) \approx \rho_l L_v \frac{\Delta T}{T_d} \tag{3.5}$$

The approximation corresponds to Eq. 3.2 with small Δp and ΔT where, from the Clausius–Clapeyron, Eq. A5, one has

$$\ln(SR) \approx \frac{\Delta p}{p} \approx \frac{L_v \Delta T}{r_v T_d^2} \tag{3.6}$$

Equation 3.1 can be rewritten in terms of number n of molecules in the nucleating cluster and Boltzmann constant $k_B = R/N_A$, with R the ideal gas constant and N_A the Avogadro number with the help of Eqs. 1.2 and 3.5. After some algebra, one finds, with n the number of molecules,

$$W^V = -nk_BT\ln(SR) \tag{3.7}$$

In this process, however, an energy barrier must be crossed: the energy of formation of the liquid-gas interface. The corresponding surface energy, W^S, can be written as $S\sigma_{LG}$, where σ_{LG} is the liquid-gas interfacial tension and S the droplet surface area. The nucleating cluster being spherical, with radius ρ and volume V, one readily obtains

$$W^S = 4\pi\sigma_{LG}\rho^2 = 4\pi\sigma_{LG}\left(\frac{3V}{4\pi}\right)^{2/3} = 4\pi\sigma_{LG}\left(\frac{3nv_m}{4\pi}\right)^{2/3} \tag{3.8}$$

In this equation, v_m is the volume occupied by a molecule, with the number of molecules $n = V/v_m$. Considering the volume V with mass $m = \rho_l V$ of (liquid) condensate, with M the molar mass, one obtains

$$v_m = \frac{V}{N_a(m/M)} = \frac{M}{N_a\rho_l} \tag{3.9}$$

whose value for water is $v_m \approx 3.0 \times 10^{-29}$ m^3. Then the work of cluster formation W through nucleation is given as

$$W(n) = W^V + W^S = -\alpha n + \beta n^{2/3} \tag{3.10}$$

where

$$\alpha = k_BT\ln(SR); \quad \beta = 4\pi\sigma_{LG}\left(\frac{3v_m}{4\pi}\right)^{2/3} \tag{3.11}$$

The nucleation work exhibits a maximum before decreasing and becoming negative, corresponding to an energy barrier—the cost of nucleating an interface (Fig. 3.1). The maximum occurs for a critical cluster containing n^* molecules, corresponding to critical radius ρ^* and critical volume V^* below which the liquid drop is not stable since $dW/dn < 0$. The critical value is

$$n^* = \left(\frac{2\beta}{3\alpha}\right)^3 = \frac{32\pi}{3}\frac{\sigma_{LG}^3 v_m^2}{[k_BT\ln(SR)]^3} \tag{3.12}$$

Fig. 3.1 Homogeneous
nucleation, work of formation
W^* and critical cluster with
n^* molecules

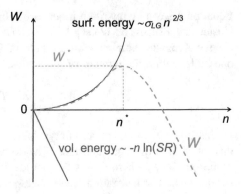

It corresponds to the critical droplet volume $V^* = nv_m$ or:

$$V^* = \frac{32\pi}{3}\left(\frac{\sigma_{LG}v_m}{k_BT\ln(SR)}\right)^3 \tag{3.13}$$

or critical droplet radius:

$$\rho^* = 2\frac{\sigma_{LG}v_m}{k_BT\ln(SR)} \approx 2\frac{\sigma_{LG}T}{\rho_l L_v \Delta T} \tag{3.14}$$

The work of formation, W^*, is

$$W^* = \frac{4\beta^3}{27\alpha^2} = \frac{16\pi\sigma_{LG}^3 v_m^2}{3[k_BT\ln(SR)]^2} \tag{3.15}$$

In terms of supersaturation, the work of formation Eq. 3.15 corresponds to a critical supersaturation SR^* such as (see Eq. 3.12) $\alpha = 2\beta/3n^{*1/3}$, or (Eq. 3.15):

$$\ln(SR^*) = \frac{2\beta}{3k_BTn^{*1/3}} = \frac{2W^S}{3k_BTn^*} \tag{3.16}$$

Then the critical supersaturation required to form a critical cluster of size n^* is only dependent upon the surface energy at a given temperature.

The associated nucleation rate (the number dN of droplets of critical size n^* per unit volume that has nucleated during the time dt) corresponds to the probability that a number density fluctuation gains energy W^*. The formation of liquid droplets is a thermally activated process with contribution of a Brownian walk over the thermodynamic barrier. One thus obtains the following exponential dependence (Arrhenius type) with prefactor A, which depends on the number of nucleation sites per unit time and unit volume and the probability that a nucleus of critical size will continue to grow and not dissolve:

$$\frac{dN}{dt} = Ae^{-\frac{W^*}{k_BT}} \tag{3.17}$$

The prefactor A is a very large number, in the order of 10^{25}–10^{40} m^{-3} s^{-1}, and can be evaluated by considering (Kashchiev 2000) as the following product:

$$A = c_0 Z f^* \tag{3.18}$$

Here c_0 is the volume concentration of available nucleation sites. In humid air, all molecules can receive a cluster. Making use of the ideal gas equation Eq. 1.2 c_0 can be calculated, with p_m the atmospheric pressure:

$$c_0 = \frac{p_m}{k_B T} \tag{3.19}$$

The Zeldovich factor, Z, in Eq. 3.18 is derived by assuming that the nucleus near the top of the barrier can grow diffusively into a larger nucleus that will grow into a new phase, or can lose molecules and vanishes. The Zeldovich factor basically takes into account the loss of nuclei during their Brownian motion. The runaway of nuclei from the nucleation region goes on to form other critical nuclei until a steady-state equilibrium is reached. The probability that a given nucleus goes forward is related to the flatness of the energy profile around the critical size, i.e. the curvature of the free energy, which the Zeldovich factor characterizes as

$$Z = \left(\frac{1}{2\pi k_B T} \frac{-d^2 W}{dn^2} \right)^{1/2} = \frac{3\alpha^2}{4\beta^2} \left(\frac{\beta}{\pi k_B T} \right)^{1/2} \tag{3.20}$$

The parameter f^* in Eq. 3.17 is the collision rate of monomers with the critical cluster, which is the attachment frequency. It depends on the saturation of condensing molecules with their environment. The transport of vapor molecules in air occurs through diffusion and obeys the Fick law (see Chap. 4, Eq. 4.15). According to Kashchiev (2020), f^* can be written as

$$f^* = \left(48\pi^2 v_m \right)^{1/3} \alpha_m D S R x_s n^{*1/3} \tag{3.21}$$

where D is the mutual diffusion coefficient of vapor molecules in air, SR is the saturation ratio (Eq. 3.4), α_m is the monomer sticking coefficient (accommodation coefficient ≤ 1) and x_s is the water molecule concentration in air at saturation, in units of molecules per unit volume. The latter is deduced from the ideal gas equation (Eq. 2.2):

$$x_s = \frac{p_s}{k_B T} \tag{3.22}$$

Using the numerical values of Table 1.2 for water and typical values $T = 293$ K; $p_s \approx 2.3 \times 10^3$ Pa, $p_m \approx 10^5$ Pa, one finds, assuming the accommodation coefficient $\alpha_m = 1$, $Z \approx 0.0685$, $f^* \approx 1.95 \times 10^{11} \frac{SR}{\ln SR}$ s^{-1}, $c_0 = 2.47 \times 10^{25}$

m^{-3}, $A \approx 3.39 \times 10^{35} \frac{SR}{\ln SR}$ $m^{-3}.s^{-1}$ and $W^*/k_B T = 88/(\ln SR)^2$. Equation 3.16 can thus take the following expression for typical water values (Table 1.2):

$$\frac{dN}{dt} = 3.39 \times 10^{35} \frac{SR}{\ln SR} e^{-\frac{88}{(\ln SR)^2}} \ (m^{-3}.s^{-1}) \tag{3.23}$$

In Fig. 3.2 is shown the nucleation rate with respect to SR. If one considers $dN/dt = \dot{N}_0$ as an experimental threshold and neglecting the weak SR dependence in A when compared to its influence in the exponential, one can deduce from Eq. 3.16 a critical supersaturation ratio SR_0^*, such that

$$SR_0^* \approx \exp\left[\left(\frac{16\pi\sigma_{LG}^3 v_m^2}{(k_B T)^3 \ln(A/\dot{N}_0)}\right)^{1/2}\right] \tag{3.24}$$

Considering $\dot{N}_0 \sim 10^6$ $m^{-3}.s^{-1}$ (1 $cm^{-3}.s^{-1}$) as an experimental threshold a supersaturation $SR_0^* \approx 3.1$ is needed to nucleate droplets. A very close value is observed in Fig. 3.2 by using Eq. 3.23 without approximations. The corresponding critical radius of nucleation is $\rho^* \approx 0.95$ nm from Eq. 3.14. In an air saturated at 20 °C, temperature must be lowered to about 2 °C for homogeneous nucleation to occur according to the approximated formula Eq. 3.5. These values are in agreement with the measurements of Heist and Reiss (1973).

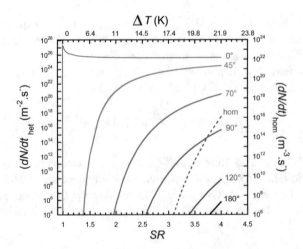

Fig. 3.2 Homogeneous (right ordinate, Eq. 3.23) and heterogeneous (left ordinate, Eq. 3.37) nucleation rate with respect to supersaturation $SR = p/p_s$ (lower abscissa) or (upper abscissa) approximate temperature difference ΔT using Eq. 3.5 (semi-log plot). For heterogeneous nucleation different drop contact angles are shown. The value $dN/dt = \dot{N}_0 = 10^6$ $m^{-3}.s^{-1}$ (1 $cm^{-3}.s^{-1}$) is considered here as an experimental nucleation threshold for homogeneous nucleation and $dN/dt = \dot{N}_0 = 10^4$ $m^{-2}.s^{-1}$ (1 $cm^{-2}.s^{-1}$) for heterogeneous nucleation

3.2 Heterogeneous Nucleation

Daily experience shows that dew forms for much lower supersaturation than calculated just above. This is because condensation occurs on substrates whose surface energy lowers or even suppresses the cost of forming the liquid-vapor interface. Such a nucleation is called "heterogeneous", in contrast to the "homogeneous" nucleation process of Sect. 3.1, which occurs in the bulk of the gas.

The energy barrier depends on the wetting properties of the substrates. Wetting is characterized by the balance of surface energy (or surface tension) between liquid and gas (σ_{LG}), liquid and solid (σ_{LS}) and solid-gas (σ_{SG}), determining the contact angle θ_c drop substrate (Fig. 3.3). The contact angle is zero for complete wetting (water forms a wetting film) and maximum for a liquid droplet that does not wet the substrate (complete drying). The work to obtain a stable nucleus is thus modified through a θ_c-dependent function, which accounts for the volume of the spherical cap and the surfaces of the cap and its base.

Going back to Eq. 3.8 to evaluate the new surface energy, one considers ρ, the drop radius of curvature and R, the drop cap radius. The drop cap volume (see Eqs. D.5, D.6 in Appendix D) can be written as

$$V = \pi \rho^3 F(\theta_c) \qquad (3.25)$$

with

$$F(\theta_c) = \left(\frac{2 - 3\cos\theta_c + \cos\theta_c{}^3}{3} \right) \qquad (3.26)$$

Fig. 3.3 Contact angle variation of $f(\theta_c)$, the cost of energy to form a nucleus on a substrate, relative to the bulk case (see text). Complete drying and bulk case: $\theta_c = 180°$; partial wetting: $180° > \theta_c > 0$; complete wetting: $\theta_c = 0$

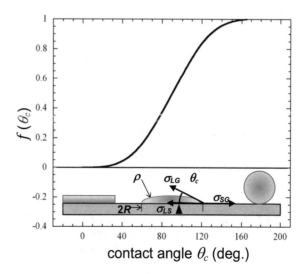

The volume contribution is (Eq. 3.1) $W_h^V = -V \Delta e$. The drop surface in contact with gas is written as

$$S_{LG} = 2\pi\rho^2(1 - \cos\theta_c) \tag{3.27}$$

The drop surface in contact with solid is

$$S_{LS} = \pi\rho^2 \sin\theta_c{}^2 = \pi\rho^2\left(1 - \cos\theta_c{}^2\right) \tag{3.28}$$

The surface contribution W_h^S corresponds to the cost in generating the GL and LS energies and the gain of losing the SG energy. With σ_{LS} and σ_{SG} the liquid-solid and solid-gas interfacial tensions, respectively:

$$W_h^S = S_{LG}\sigma_{LG} + S_{LS}\sigma_{LS} - S_{LS}\sigma_{SG} \tag{3.29}$$

Using the Young–Dupré equation $\sigma_{SG} = \sigma_{LS} + \sigma_{LG}\cos\theta_c$ (Eq. E.1 in Appendix E), it becomes

$$W_h^S = \sigma_{LG}(S_{LG} - S_{LS}\cos\theta_c) \tag{3.30}$$

Using Eqs. 3.27 and 3.28 and referring to W^s in Eq. 3.8, one obtains

$$W_h^S = 4\pi\rho^2\sigma_{LG}\left(\frac{2 - 3\cos\theta_c + \cos\theta_c{}^3}{4}\right) = W^s f(\theta_c) \tag{3.31}$$

The function $f(\theta_c)$ is related to the drop cap volume function $F(\theta_c)$ (Eq. 3.26):

$$f(\theta_c) = \frac{3}{4}F(\theta_c) = \frac{1}{4}(2 - 3\cos\theta_c + \cos\theta_c{}^3)$$

$$= \frac{1}{4}(2 + \cos\theta_c)(1 - \cos\theta_c)^2 \tag{3.32}$$

The surface energy can be written as a function of n, similar to Eq. 3.8,

$$W_h^S = 4\pi\sigma_{LG}\left(\frac{3V}{4\pi}\right)^{2/3} f(\theta_c)^{1/3} = 4\pi\sigma_{LG}\left(\frac{3nv_m}{4\pi}\right)^{2/3} f(\theta_c)^{1/3} \tag{3.33}$$

Similar to Eq. 3.10 for the homogeneous case, the total energy W_h for heterogeneous nucleation is the sum of surface and volume contributions, which can now be written as a function of θ_c and total energy W of the homogeneous nucleation case:

$$W_h = W_h^V + W_h^S = f(\theta_c)W \tag{3.34}$$

The critical energy barrier is derived from Eqs. 3.15 and 3.34 and appears as

$$W_h^* = f(\theta_c)W^* \tag{3.35}$$

The number of molecules in the critical droplet, n_h^*, the critical droplet volume V_h^*, its radius of curvature, ρ_h^* and drop cap radius, $R_h^* = \rho_h^* \sin \theta_c$ (see Appendix D, Eq. D1) follows from Eqs. 3.12–3.14, and 3.35

$$n_h^* = f(\theta_c)n^*; \quad V_h^* = f(\theta_c)V^*;$$

$$\rho_h^* = \rho^*; \quad R_h^* = \rho^* \sin \theta_c \tag{3.36}$$

The heterogeneous nucleation rate has to be expressed per unit surface area. According to Christian (1975), the prefactor in the heterogeneous nucleation rate can be deduced from the prefactor of the homogeneous rate by the ratio of molecular volume/molecular surface. The molecular radius, corresponding to the volume or surface occupied by a molecule, is indeed the natural lenghtscale of the process. With $R_0 \left(= \left(\frac{3v_m}{4\pi}\right)^{1/3} = 1.965 \times 10^{-10}$ m$\right)$ the molecular radius, Eq. 3.17, becomes, with A the prefactor for homogeneous nucleation (Eq. 3.18):

$$\frac{dN}{dt} = \frac{R_0}{3} A e^{-\frac{W_h^*}{k_B T}} = \frac{R_0}{3} A e^{-\frac{f(\theta_c)W^*}{k_B T}} \tag{3.37}$$

Figure 3.2 reports the nucleation rate data for the same typical conditions as for the homogeneous nucleation case in Sect. 3.1, with prefactor $R_0A/3 \approx 2.22 \times 10^{25} \mathrm{m}^{-2}\mathrm{s}^{-1}$. The values range for the homogeneous nucleation case when $\theta_c = 180°$ (purely hydrophobic case, $SR \approx 3.8$) to 1 (purely hydrophilic case, $\theta_c = 0°$). In the latter case, nucleation occurs exactly at the dew point temperature of humid air.

One cannot compare the nucleation rates for the heterogeneous and homogeneous situations because of the different units. However, considering as detection threshold $\dot{N}_0 = 10^6$ m^{-3}.s^{-1} (1 cm^{-3}.s^{-1}) for the homogeneous case and $\dot{N}_0 = 10^4$ m^{-2}.s^{-1} (1 cm^{-2}.s^{-1}) for the heterogeneous case, it is clear on Fig. 3.2 that for angles $\theta_c < 120°$ the supersaturation needed to observe nucleation becomes smaller than for homogeneous nucleation. Heterogeneous nucleation therefore permits condensation for temperature differences much less than required for homogeneous nucleation (Twomey 1959) (see also Sect. 6.5 and Fig. 6.18, where contact angle is continuously varied). In addition, geometrical defects of the substrate (e.g. scratches) and chemical heterogeneities (e.g. salt aerosols increasing the dew point temperature, nanoscale agglomerates from absorption of sulfuric acid and adsorption of volatile organic compounds, see Cha et al. 2017) favor nucleation. Experimentally speaking, it is then only on substrates completely wetted by water ($\theta_c = 0$) that condensation can be observed at the dew point.

Because of unavoidable contamination (e.g. by human manipulation), substrates that are not specially protected or designed are covered with fatty substances, which makes the contact angle water substrate to be around 40–70°. These angles often contrast with what is observed on clean substrates, e.g. glass where $\theta = 0°$ when ultra-clean. This is why dew and breath figures are most often the result of the condensation of water into tiny droplets, which scatter light and make condensation appear "white". This dropwise condensation contrasts with filmwise condensation, optically nearly invisible.

The fact that nucleation is favored by wetting conditions has important implications. For example, Sect. 10.3.2 describes how this property can be used in biphilic microstructures to favor nucleation while preventing drop pinning. Biological sterilization can be ensured by micro-condensation on wetting patches of biological materials (Marcos-Martin et al. 1996).

3.3 Growth Regimes Overview

Different stages of growth can be identified in dew formation on a flat surface (Fig. 3.4). They are detailed in Chaps. 4 and 6 The main characteristics are as follows:

(i) Droplets nucleate preferentially on substrate defects, with mean distance between nucleation sites $\langle d_0 \rangle$ ($\sim 1/\sqrt{4n_s}$, with n_s the surface density of nucleation sites and assuming the sites on a square lattice).

(ii) Droplets grow with no or rare coalescences corresponding to low surface coverage. The drop contact radius, R, of the droplet spherical cap varies with time, t, according to a power law, $R \sim t^\beta$. The exponent value depends on the relative values of droplet inter-distance $\langle d_0 \rangle$ compared to the water concentration boundary layer, ζ (as described in this chapter). If in this stage drops are far apart from each other (typically $\langle d_0 \rangle > 2\zeta$), then the drops grow independently in a hemispherical profile centered on each drop, resulting in $\beta = 1/2$. If the drops are more packed ($\langle d_0 \rangle \ll 2\zeta$), the individual concentration profiles overlap and a mean profile arises, directed perpendicularly to the substrate and the growth exponent is smaller, $\beta = 1/3$.

(iii) Then droplets touch each other and coalesce, leading to a constant surface coverage and a self-similar growth behavior. The concentration profiles overlap with a mean profile still directed perpendicularly to the substrate, corresponding again to $\beta = 1/3$ for each individual drop. The mean radius of the droplet pattern grows as $< R > \sim t^\gamma$ with $\gamma = 3\beta = 1$. The surface coverage, ε^2, being constant at this stage, implies that $< d >$ scales with $< R >$ (see Sect. 6.3.2, Eq. 6.41).

(iv) At some time, the mean distance between drops, $\langle d \rangle > 2\zeta$, and nucleation of new droplets can occur between neighboring drops. The new droplets that have nucleated then follow the same growth law behavior as described earlier.

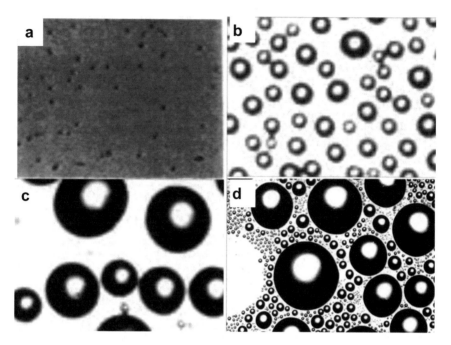

Fig. 3.4 Self-similar growth of a pattern of droplets condensing on a cooled hydrophobic glass substrate. The largest dimension of the photos **a, b, c** corresponds to 285 μm and **d** to 1.1 mm. **a** Nucleation on substrate defects of isolated droplets, stage (i), and further growth, stage (ii) without coalescences. **b** Pattern at $t = 1$ s after condensation started showing a dense droplet pattern with coalescences, stage (iii). **c** Same as **b** but later at $t = 6$ s, statistically equivalent to the pattern in **b** after rescaling, showing self-similarity of the growth. **d** Pattern at $t = 25$ s with new scale 0.25, stage (iv). Novel families of droplets have nucleated between the initial drops. When taken separately, these families present the self-similar properties of the first generation in **b, c**. (Photo **d** Briscoe and Galvin 1989)

(v) Gravity effects (drop shape deformation on a horizontal substrate, shedding of larger drops on an inclined substrate) generally occur during the late stages (iii)–(iv). Shedding determines the maximum drop size present on the surface (see Sect. 14.1.2).

Single Droplet Growth

4

Different aspects of the growth of a single droplet after its nucleation on a condensing substrate are discussed in this chapter. Growth in a pure vapor is first analyzed and then a discussion on condensation from vapor with non-condensable gases (e.g. air), whose growth is limited by diffusion. In this latter case, which corresponds to dew and BFs, thermal aspects will be neglected and the drop liquid–gas interface temperature assumed constant and equal to the substrate temperature. This is a simplification. The specific thermal aspects are treated in Chap. 13.

Single droplet growth corresponds to an actual isolated drop or, when diffusion-limited growth is considered, to a drop in a pattern whose diffusive water vapor profile does not overlap with the neighboring profiles. The latter occurs when the mean distance between the drops is much larger than the boundary layer thickness as defined in Chap. 3.

4.1 Growth in Pure Vapor

Let us consider a drop that has nucleated in its own vapor. Vapor is at saturation temperature $T_\infty = T_s$ and saturation pressure $p_v = p_s(T_\infty) = p_\infty$. Interface temperature is $T_i \approx T_s$ (due to the large interfacial heat transfer, see Eq. 4.10 and Chap. 13) and pressure is $p_i < p_\infty$. Both vapor temperature and pressure are considered uniform (Fig. 4.1). In the kinetic theory of gas, the interface can be viewed as dynamic, with molecules leaving it (evaporation) and joining it (condensation). At saturation ($p_i = p_v$), both processes balance each other. When $p_i < p_v$, net condensation takes place, the molecules are impinging on the surface outnumbering those that are absorbed by the liquid. This molecular exchange is accompanied

The original version of the book was revised: Belated corrections have been updated. The correction to the book is available at https://doi.org/10.1007/978-3-030-90442-5_16

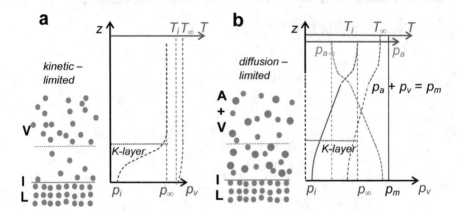

Fig. 4.1 Sketch of condensation in **a** pure vapor and **b** vapor and non-condensable gas (air). L: liquid phase; V: vapor phase; A: non-condensable gas. In (a) condensation is limited by the balance evaporation—aggregation of molecules in the Knudsen layer (K-layer). In (b) condensation is limited by molecule diffusion in gas A. Other notations: see text

with a thermal resistance at the interface, corresponding to the release of latent heat. The process is confined within the Knudsen layer, the layer near the interface where the medium outside can be considered as continuous and described by hydrodynamics. Its thickness in a gas is a few mean free paths, the mean distance l_{mfp} traveled by molecules before a collision. The effective cross-sectional area σ for spherical particles with diameter d is the area perpendicular to their relative motion within which they must meet in order to scatter from each other, that is, $\sigma = \pi d^2$. With s_v the number of target particles per unit volume, the mean free path corresponds to having one collision in the volume $l_{mfp}\sigma$, which corresponds to $s_v = 1/l_{mfp}\sigma$. However, this situation corresponds to an immobile target with a moving particle with velocity v; in reality both particles are moving, which leads to consider their mean relative velocity $v_r = \overline{v_1 - v_2}$ such as $\overline{(v_1 - v_2)^2} = \overline{v_1^2 + v_2^2 - 2\overline{v_1 v_2}} = \overline{v_1^2 + v_2^2}$ since v_1 and v_2 are uncorrelated. It thus becomes $v_r = \sqrt{2}v$, or a mean free path $\sqrt{2}$ times smaller, leading to

$$l_{mfp} = \frac{1}{\sqrt{2}\sigma s_v} \qquad (4.1)$$

From the equation of ideal gas (Eq. 1.2) where $s_v = p/k_B T$, Eq. 4.1 transforms into

$$l_{mfp} = \frac{k_B T}{\sqrt{2}\pi d^2 p} \qquad (4.2)$$

Here p is pressure, T is temperature and k_B is the Boltzmann constant. For saturated water pressure at 100 °C, $p = 10^5$ Pa and $l_{mfp} = 16$ nm. At 20 °C, $l_{mfp} = 5$ μm.

In the Knudsen region, the vapor pressure thus changes abruptly from p_v to p_i. Interface temperature is only slightly less than vapor temperature and should be uniform since any change in temperature will induce immediate evaporation or condensation.

The rate of condensation, expressed in mass per unit area $dm_s/dt = \dot{m}_s$ or volume per unit interface area $dh/dt = \dot{h}$, is classically expressed from the kinetic theory of gases (see, e.g. Tanasawa, 1991 and Refs. therein. The distribution of molecules velocity is assumed to obey a Maxwellian distribution. A fraction σ_c (the condensation coefficient) of molecules that hit the interface are incorporated and the fraction of molecules leaving it correspond to the evaporation coefficient σ_e. It follows the Hertz-Knudsen-Schrage or Kucherov-Rikenglaz equation:

$$\dot{m}_s = \rho_l \dot{h} = \sigma_c \Gamma \frac{p_v}{\sqrt{2\pi r_g T_\infty}} - \sigma_e \frac{p_i}{\sqrt{2\pi r_g T_i}} \tag{4.3}$$

ρ_l is liquid density, $r_g = R/M$ is the vapor specific mass constant with $R = 8.314$ J.mole^{-1}, K^{-1} the molar gas constant and M the gas molar mass (for water, $r_g = 462$ J.kg^{-1}.K^{-1}, see Table 1.2). The first term on the right-hand side of Eq. 4.3 (without Γ) represents the flux of molecules which incorporate the surface. The term

$$\Gamma = 1 + \rho_l \frac{\dot{h}}{\sqrt{2\pi r_g T_\infty}} \tag{4.4}$$

where ρ_l is the liquid density corresponds to the fact that the entire vapor progresses towards the surface during condensation; this "progress" velocity must be superimposed on the Maxwell velocity distribution. The second term corresponds to the emission of molecules at temperature T_i. Assuming that $\sigma_c = \sigma_e = \sigma$ (which is not strictly true, see Marek and Straub 2001) and $T_\infty - T_i \ll T_i$ and applying Eq. 4.2, the relation (Eq. 4.1) becomes

$$\dot{h} = \left(\frac{2\sigma}{2 - \sigma}\right) \frac{p_\infty - p_i}{\rho_l \sqrt{2\pi r_g T_\infty}} \tag{4.5}$$

When compared to experiments the coefficient σ is found in the range 1–0.003 depending on fluid and surface impurities (Tanasawa 1991; Anand and Son 2010; Chavan et al. 2016). It is expected that in pure vapor conditions σ is closer to 1, whereas at lower pressures its value is expected to be lesser than 1 (Marek and Straub (2001, Anand and Son 2010). For example, concerning water in air around room temperature, the value $\sigma \approx 0.04$ (Umur and Griffith 1965) is generally adopted.

The pressure difference $p_\infty - p_i$ when not too large can be expressed as a function of temperature difference $T_\infty - T_i$ from the Clausius-Clapeyron relation (Appendix A1, Eq. A1.5). With ρ_v the vapor density is

$$p_\infty - p_i \approx \frac{\rho_v L_v}{T_\infty}(T_\infty - T_i) \tag{4.6}$$

Equation 4.5 becomes, using Eq. 4.6,

$$\dot{h} = \left(\frac{2\sigma}{2 - \sigma}\right) \frac{\rho_v L_v}{T_\infty^{3/2} \rho_l \sqrt{2\pi r_g}} (T_\infty - T_i) \tag{4.7}$$

Drop growth is limited by the heat flux per unit area, q_i, corresponding to the release of latent heat L_v at the interface:

$$q_i = \dot{m}_s L_v = \rho_l L_v \dot{h} \tag{4.8}$$

When $\dot{h}$ is expressed by Eq. 4.7 in the equation above, one can define an interfacial heat transfer coefficient a_i (see Chap. 13) such as

$$q_i = a_i (T_\infty - T_i) \tag{4.9}$$

leading to

$$a_i = \left(\frac{2\sigma}{2 - \sigma}\right) \frac{\rho_v L_v^2}{T_\infty^{3/2} \sqrt{2\pi r_g}} \tag{4.10}$$

The case of filmwise condensation on a plane tilted horizontal with angle α (Nusselt film) is treated in Sect. 13.4. The growth rate is, with ξ the distance from the top of the plate, λ_l the liquid thermal conductivity, η the dynamic viscosity and g the earth's gravity acceleration constant (Eq. 13.27),

$$h = \left(\frac{4\eta\lambda_l(T_i - T_c)}{\varrho_l^2 L_v g \sin\alpha}\right)^{1/4} \xi^{1/4} \tag{4.11}$$

When dealing with dropwise condensation, the heat flux in the drop must be estimated. Both conduction and convection are present. Since temperature gradients cannot be present at the interface, Marangoni flows are absent. However, buoyancy and expansion of the drop with motion of the contact line can still induce convection. Simulations of convections and conduction (Xu et al. 2018; see Sect. 13.3.2) show that in small drops (below 200 μm) conduction is the dominant process for heat transfer.

Conduction can be treated analytically for a drop of curvature radius ρ and contact angle θ_c (Anand and Son 2010; Kim and Kim 2011). By expressing the heat flux $Q(\rho, \theta_c)$ through the drop to the substrate through a coating by accounting for all thermal resistances interface-drop-drop curvature-coating-substrate (Eq. 13.56) and using Eq. 4.8, one obtains the following differential equation (Rykaczewski 2012):

$$\frac{d\rho}{dt} = \frac{\left(1 - \frac{\rho^*}{\rho}\right)(T_\infty - T_c)}{2\rho_l L_v \left(\frac{1}{2a_i} + \frac{\rho\theta_c(1-\cos\theta_c)}{4\lambda_l\sin\theta_c} + \frac{\delta_c(1-\cos\theta_c)}{\lambda_c\sin^2\theta_c}\right)} \tag{4.12}$$

Here T_c is the substrate temperature, ρ^* is the critical nucleation radius (Eq. 3.14), λ_l is the liquid thermal conductivity, λ_c is the substrate coating thermal conductivity, whose thickness is δ_c. In Eq. 4.12, it is implicitly assumed that growth proceeds at constant contact angle. Another differential equation can be obtained when the contact line is pinned and drop evolution proceeds at constant base radius (Rykaczewski 2012). The differential equation, Eq. 4.12, is solved numerically. Figure 4.2a reports drop evolutions at different subcooling and the corresponding evolution calculated from Eq. 4.8.

Anand and Son (2010) performed the same kind of calculation without coating conduction to lead eventually to an analytical expression which mixes linear and exponential terms. With ρ_0 the initial drop curvature radius:

$$\rho^2 - \rho_0^2 + \left[\frac{4.6(1+\cos\theta_c)}{\theta_c \sin\theta_c a_i L_v^2}\right]\left[\rho - \rho_0 + \rho^* \ln\left(\frac{\rho - \rho^*}{\rho_0 - \rho^*}\right)\right] = \left[\frac{9.2\sin\theta_c \lambda_l (T_\infty - T_c)}{(2+\cos\theta_c)(1-\cos\theta_c)^2 \theta_c}\right]t$$

(4.13)

This evolution indeed accounts well with the observed drop growth behavior (Fig. 4.2b). One notes that both evaluations Eqs. 4.12 and 4.13 result in $d\rho/dt \sim \rho$, that is, to a behavior approximately as

$$\rho \sim t^{1/2}$$

(4.14)

4.2 Growth in Vapor with Non-Condensable Gases

Let us now consider the growth of an isolated droplet in vapor with non-condensable gases (e.g. air) (Figs. 4.1 and 4.3). Drop can be cooled by contact by means of a substrate of high thermal conductivity, itself in contact with a thermostat, or by radiative cooling of a substrate and drop, noting that for water it is mainly liquid emissivity that matters during condensation (Beysens 2018; Trosseille 2019; Trosseille et al. 2021a). Similarities and differences between conductive and radiative modes are also discussed in Sect. 13.6.

In the following, drop temperature is considered to be kept nearly homogeneous thanks to its small size and/or the presence of convections (see Sect. 13.5.4), which means that its growth will not be limited by the thermal resistance of drop plus substrate but by the resistance due to the diffusive gradient in the surrounding gas.

The sessile drop grows by incorporation of the diffusing water vapor molecules (monomers) around it that hits the interface and is incorporated with the same process as observed with pure vapor, see Sect. 4.1. However, the limiting process now is not any more condensation/evaporation in the Knudsen layer (except for very small drops smaller than this layer, see Hinds, 1999) but the diffusion of molecules in the surrounding gas.

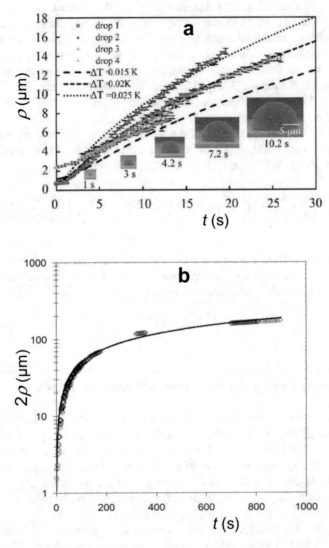

Fig. 4.2 Radius evolution observed in an environmental scanning electron microscope during water vapor condensation on a flat silicon substrate. **a** Constant contact angle $\theta_c = 106° \pm 2°$. Curves are the calculated radius from Eq. 4.12 with $\theta_c = 100°$ and different surface subcooling $T_\infty - T_c = 0.015, 0.02, 0.025$ K. (From Rykaczewski, 2012, with permission). **b** Constant contact angle $\theta_c = 62° \pm 2°$ and subcooling 0.006 K. The curve is Eq. 4.13. (Adapted from Anand and Son, 2010, with permission)

The concentration of monomers, $c(r,t)$, counted in mass per volume, where r is the distance from the drop center, obeys the following equations (Fick's laws). Let us define $\vec{j}$ as the diffusive flux of monomers (mass per unit surface and unit

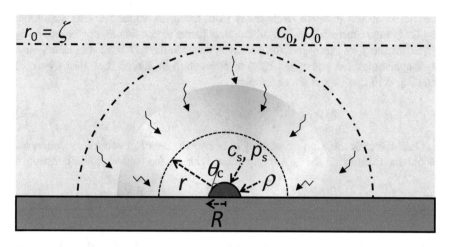

Fig. 4.3 Schematics of a 3D drop with contact angle θ_c in a 3D space (side view)

time) resulting from a concentration gradient. With D the diffusion coefficient of the water monomers in air is

$$\vec{j} = -D\,\vec{\nabla}c \tag{4.15}$$

The conservation law implies that the evolution of concentration results from the gradient of flux, that is,

$$\frac{\partial c}{\partial t} + \vec{\nabla}.\vec{j} = 0, \tag{4.16}$$

The problem governed by Eqs. 4.15–4.16 is a Stefan problem with a moving boundary at $r = R(t)$, the spherical cap perimeter radius. Analytical solutions are rare. For the present problem, one will thus assume a growth that is slow enough such that the time dependence of c can be neglected in Eq. 4.16. This is a quasi-static approximation. Thus Eqs. 4.15–4.16 reduce to the Laplace equation:

$$\Delta c = 0 \tag{4.17}$$

Its solution has to fulfill the following boundary conditions:

$$c(\text{dropsurface}) = c_S \tag{4.18}$$

c_S corresponding to the water saturation pressure at the drop temperature, p_s, and

$$c(r = r_0) = c_0 \tag{4.19}$$

meaning that at a distance r_0 far away from the drop, vapor takes its bulk value c_0. This length corresponds in mathematical terms to ∞, that is, $c(r \to \infty) = c_0$. Experimentally, r_0 represents the boundary layer value ζ. Above this value, air is no longer quiescent but mixed by free convection or forced flow (see Chap. 3), thus Eq. 4.19 can be written as follows:

$$c(r = \zeta) = c_0 \tag{4.20}$$

Once the concentration of monomers is known, the drop volume evolution can be obtained following the growth equation, with ρ_w the (water) liquid density:

$$\frac{dV}{dt} = \frac{1}{\rho_w} \int_S \vec{j} \, (r = R). \vec{n} \, dS = \frac{1}{\rho_w} D \int_S \left(-\vec{\nabla} c \right)_R . \vec{n} \, dS \tag{4.21}$$

Here $\vec{j} \, (r = R)$ is the flux of monomers at the drop surface, S is the surface of drop-air interface and $\vec{n}$ is the unit vector locally normal to the drop surface. For a drop cap, the volume is (Appendix D, Eq. D.8).

$$V = \pi G(\theta_c) R^3 \tag{4.22}$$

Here R is the drop contact radius and $G(\theta_c)$ is the function that expresses the drop cap volume with respect to the drop contact angle θ_c (Eq. D.8):

$$G(\theta_c) = \frac{2 - 3\cos\theta_c + \cos\theta_c{}^3}{3\sin\theta_c{}^3} \tag{4.23}$$

4.2.1 2D Concentration Profile

Let us consider a 2D droplet (a disk) or a 3D droplet growing in a 2D space (a plane) by incorporating at its perimeter monomers diffusing on the plane (Fig. 4.4), with surface concentration c^s (in mass per unit surface). The Laplacian of concentration is

$$\Delta c^s = \frac{\partial^2 c^s}{\partial r^2} + \frac{1}{r} \frac{\partial c^s}{\partial r} + \frac{1}{r^2} \frac{\partial^2 c^s}{\partial \theta^2} \tag{4.24}$$

The solution to Eq. 4.24 is symmetrical with respect to the drop center; it is not θ dependent and is of logarithmic form with respect to r:

$$c^s = A \ln r + B \tag{4.25}$$

Fig. 4.4 Schematics of 2D or 3D drop evolution in a 2D space (top view)

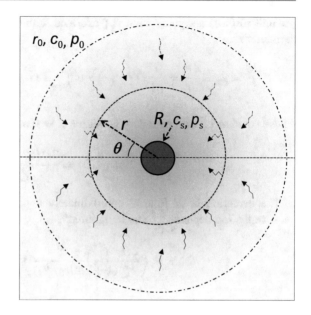

where A and B are constants that are determined from the boundary conditions Eqs. 4.18 and 4.19, using surface concentration c_0^s for c_0 and c_s^s for c_s. Equation 4.25 thus becomes

$$c^s = \frac{c_0^s - c_s^s}{\ln(r_0/R)} \ln(r/R) + c_s^s \tag{4.26}$$

Another solution can be found when the boundary conditions are different. Following Steyer et al. (1991), one can assume, instead of constant concentration at r_0, Eq. 4.19, a constant flux ϕ_0 (mass per unit time), corresponding to

$$\phi_0 = -2\pi D \left(r \nabla c^s \right)_{r_0} \tag{4.27}$$

Going back to Eq. 4.25, and still assuming boundary conditions at the drop perimeter Eq. 4.18, one finds

$$c^s = -\frac{\phi_0}{2\pi D} \ln(r/R) + c_s^s \tag{4.28}$$

4.2.2 2D Drop Growth

Growth of a 2D droplet (a disk) from its perimeter follows Eq. 4.21, where volume V is now disk surface S, ρ_w becomes (water) liquid surface density ρ_w^s (mass per unit surface) and integration is performed with curvilinear abscissa l on disk perimeter P:

$$\frac{dS}{dt} = 2\pi R \frac{dR}{dt} = \frac{1}{\rho_w^s} D \int_P \left(-\vec{\nabla} c^s \right)_R \cdot \vec{n} \, dl = \frac{2\pi D}{\rho_w^s} \frac{c_0^s - c_s^s}{\ln(r_0/R)} \tag{4.29}$$

whose solution is, with boundary condition $R = 0$ at $t = 0$,

$$R^2(1 + 2\ln(r_0/R)) = \frac{2D(c_0^s - c_s^s)}{\rho_w^s} t \tag{4.30}$$

The evolution of R follows approximately a power law with exponent 1/2 because the log dependence can be ignored:

$$R = \left(\frac{2D}{\rho_w^s} \frac{c_0^s - c_s^s}{(1 + 2\ln(r_0/R))} \right)^{1/2} t^{1/2} \tag{4.31}$$

4.2.3 3D Drop Growth

For the growth of a 3D droplet (a cap with contact angle θ_c) from its perimeter, the mixing of dimensions leads to difficulties. From Eq. 4.21, it becomes, with φ the azimuthal angle as defined in Fig. 4.3,

$$\frac{dV}{dt} = 3\pi G(\theta_c) R^2 \frac{dR}{dt} = \frac{1}{\rho_w} D \int_0^{2\pi} \left(\frac{c_0^s - c_s^s}{\ln(r_0/R)} \frac{1}{R} \right) R \, d\varphi = \frac{2\pi D}{\rho_w} \frac{c_0^s - c_s^s}{\ln(r_0/R)} \tag{4.32}$$

With boundary condition $R = 0$ at $t = 0$, one obtains

$$R^3(1 + 3\ln(r_0/R)) = \frac{6D(c_0^s - c_s^s)}{\rho_w G(\theta_c)} t \tag{4.33}$$

The evolution of R follows approximately a power law with exponent 1/3 because of the log dependence:

$$R = \left(\frac{6D}{\rho_w G(\theta_c)} \frac{c_0^s - c_s^s}{(1 + 3\ln(r_0/R))} \right)^{1/3} t^{1/3} \tag{4.34}$$

With constant flux ϕ_0 at r_0 as a different boundary condition (Eq. 4.27), the volume evolution of the drop obeys the equation:

$$\frac{dV}{dt} = 3\pi G(\theta_c) R^2 \frac{dR}{dt} = \frac{D}{\rho_w} \int_0^{2\pi} \frac{\phi_0}{2\pi D} d\varphi = \frac{\phi_0}{\rho_w} \tag{4.35}$$

giving the growth law:

$$R = \left(\frac{\phi_0}{\pi G(\theta_c)\rho_w}\right)^{1/3} t^{1/3} \tag{4.36}$$

It is interesting to note that the quasi-static assumption $\frac{\partial c}{\partial t} = 0$ in Eq. 4.16 is only asymptotic because of the time dependence of R:

$$\frac{\partial c}{\partial t} = \frac{\partial c}{\partial R}\frac{dR}{dt} = \frac{\phi_0^2}{6\pi^2 G(\theta_c)\rho_w D}\frac{1}{R^3} \tag{4.37}$$

Equation 4.37 shows that at long times (large R), $\partial c/\partial t$ can indeed be neglected.

4.2.4   3D Drop and 3D Concentration Profile

For a single sessile drop condensing from vapor with non-condensable gases on a surface kept at constant temperature (for thermal effects, see Chap. 13), a simple way to solve the problem is to assume an inverse process to evaporation (Picknett et al. 1977; Sokuler et al. 2010a), a process that has been much studied. It is implicitly assumed that the probability of incorporating the monomers is uniform on the drop surface, which means that the latent heat of condensation is uniformly removed.

Because of drop symmetry, one can use polar coordinates $(r, \theta, \varphi$; see Fig. 4.5):

$$\Delta c = \frac{\partial^2 c}{\partial r^2} + \frac{2}{r}\frac{\partial c}{\partial r} + \frac{1}{r^2}\frac{\partial^2 c}{\partial \theta^2} + \frac{1}{r^2}\frac{\partial^2 c}{\partial \theta^2} + \frac{1}{r^2 \tan\theta}\frac{\partial c}{\partial \theta} + + \frac{1}{r^2 \sin^2\theta}\frac{\partial^2 c}{\partial \varphi^2} \tag{4.38}$$

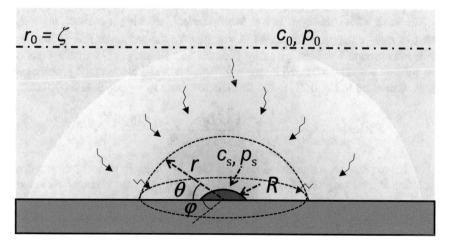

Fig. 4.5 Schematics of 3D drop evolution in a 3D space

Let us consider first a *hemispherical* drop. The concentration profile is invariant with respect to angles θ, φ. Equation 4.32 thus reduces to

$$\frac{\partial^2 c}{\partial r^2} + \frac{2}{r}\frac{\partial c}{\partial r} = 0 \tag{4.39}$$

whose solution, taking into account the boundary conditions Eqs. 4.12 and 4.13, is

$$c = \left(\frac{1}{1 - R/r_0}\right)\left[\left(c_0 - c_s\frac{R}{r_0}\right) - \frac{R(c_0 - c_s)}{r}\right] \tag{4.40}$$

A simplified formulation is found when considering the limit $r_0 \rightarrow \infty, c_0 = c_\infty$:

$$c = c_\infty - (c_\infty - c_s)\frac{R}{r} \tag{4.41}$$

Volume evolution follows from Eq. 4.15 with $\left(-\vec{\nabla}c\right)_R \cdot \vec{n} = (c_\infty - c_s)/R$. It thus comes, expressing the drop radius from the volume expression $V = (2\pi/3)R^3$:

$$R = (A_s t)^{\frac{1}{2}} \tag{4.42}$$

with

$$A_s = 2\frac{D(c_\infty - c_s)}{\rho_w} \tag{4.43}$$

The case where the drop is *not hemispherical*, showing a contact angle θ_c, leads to more complex calculations. A solution for the monomer flux $\vec{j}$ leading to droplet growth has been given by Picknett and Bexon (1977), making use of the analogy noted by Maxwell between diffusive flux and electrostatic potential. It leads, according to Eq. 4.21, to the formulation (case $r_0 \rightarrow \infty$; Sokuler 2010a):

$$V = \left(\frac{1}{\pi \sin^3\theta_c G(\theta_c)}\right)^{1/2}\left[\frac{4\pi D(c_\infty - c_s)f_0(\theta_c)}{3\rho_w}\right]^{3/2} t^{3/2} \tag{4.44}$$

In this expression, the function $G(\theta_c)$ is from Eq. 4.23. In terms of drop radius evolution and making use of Eqs. 4.23 and 4.44, it follows the classical evolution:

$$R = (A_s t)^{1/2} \tag{4.45}$$

Fig. 4.6 Typical volume evolution for an isolated drop with contact angle 95° condensing on a thin silanized cantilever cooled by 4–8 °C from room temperature (air relative humidity: 50–60%). The fitted line is the function $V = Bt^{3/2}$. The coefficient $B = 14.4 \times 10^{-18}$ m³.s$^{-3/2}$ compares favorably with the value predicted by Eq. 4.44. The double arrow gives the scale of the pictures. (From Sokuler 2010a, with permission)

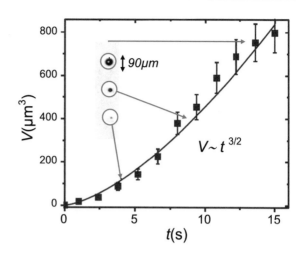

with:

$$A_s = \frac{4D(c_\infty - c_s)f_0(\theta_c)}{3\rho_w \sin\theta_c G(\theta_c)} \tag{4.46}$$

The function $f_0(\theta_c)$ results from a series expansion, which can be approximated with good precision (less than $0.1 - 0.2\%$) by the following polynomial functions:

$$f_0(\theta_c) = \frac{1}{2}(0.6366\theta_c + 0.09591\theta_c^2 - 0.614403\theta_c^3)$$
$$\text{for } \theta < 0.175 rd \ (10°)$$

$$f_0(\theta_c) = \frac{1}{2}(0.00008957 + 0.6333\theta_c + 0.116\theta_c^2 - 0.08878\theta_c^3 + 0.01033\theta_c^4)$$
$$\text{for } 0.175 < \theta_c < \pi$$

$$\tag{4.47}$$

For a hemispherical drop where $f_0(\theta_c) = 1$ and $G(\theta_c) = 2/3$, one recovers Eq. 4.43

$$A_s = 2\frac{D(c_\infty - c_s)}{\rho_w} \tag{4.48}$$

This result can be readily obtained by estimating droplet growth from Eq. 4.21 in considering the concentration gradient dc/dr of a hemispherical drop, Eq. 4.41.

Sokuler (2010a) carried out careful experiments on isolated drops placed on a thin (900 nm thick, 90 μm wide, 950 μm long) silanized cantilever cooled by a Peltier element. The expected $t^{3/2}$ growth for the droplet volume (Eq. 4.44) was indeed observed (Fig. 4.6).

Drop Coalescence

5

Droplets growing on a substrate unavoidably touch and coalesce, that is, fusion. Droplet coalescence appears a key process in condensation as it is the source of scaling in the growth of a drop pattern (Chap. 6). Below I address the coalescence process of free standing and sessile drops. The former mechanism implies one singularity when the surfaces of the drops approach sufficiently close to each other to nucleate a bridge between them. The coalescence process associated to sessile drops with contact angle less than 90° is concerned with two singularities, at the contact point and at the three-phase contact line.

Sessile drops can be considered on a smooth or patterned solid and also on liquid or plastic surfaces. The relaxation of the composite drop resulting from the coalescence of two drops is due to inertial and viscous effects, to which the contact line relaxation must be added when the substrate is a solid. Below I present only free standing drops and sessile drops on a smooth solid surface. Patterned solid surfaces will increase the pinning of the contact line when the drops are in the Wenzel state (see Chap. 10). On liquid surfaces the contact line is free to move (see Chap. 11) and relaxation will be alike free standing drops with, however, a different geometry when drops start to coalesce at their contact line. Differences due to flow specificities are found when thin droplets coalesce on quasi-2D liquid smectic films (Klopp and Eremin 2020).

5.1 Free Drops

5.1.1 Bridge Nucleation

When two droplets approach each other, the first stage is concerned with the nucleation of a bridge between them. The nucleation process has been only studied through molecular dynamics simulations (Pothier and Lewis 2012) or functional density (Niu et al. 2020) for free standing, nanoscale droplets. Coalescence starts

© Springer Nature Switzerland AG 2022
D. Beysens, *The Physics of Dew, Breath Figures and Dropwise Condensation*,
Lecture Notes in Physics 994, https://doi.org/10.1007/978-3-030-90442-5_5

(Fig. 5.1) when the droplets are placed at a distance d_m slightly shorter than the interaction range of the potentials (Pothier and Lewis 2012). Then the process is dominated by liquid–vapor nucleation and ordinary diffusion mechanisms. Molecular attraction is the driving force (Niu et al. 2020), counterbalanced by the resistance to bridge growth.

Another point of view would be to consider the thermal fluctuations of droplets interface, which, according to Buff and Lovett (1965), corresponds to capillary waves. The amplitude $\langle z^2 \rangle$ of these fluctuations is related to the liquid–vapor interfacial tension σ through

$$\langle z^2 \rangle = \frac{k_B T}{2\sigma} \ln \left(\frac{1 + q_M^2 l_c}{1 + q_m^2 l_c} \right) \tag{5.1}$$

In the above relation, $l_c = \sqrt{\sigma / \rho_w g}$ is the capillary length (see Appendix E, Eq. E.6), q_M is the maximum wave vector corresponding to the smallest wavelength of allowable capillary waves, i.e. the interfacial thickness itself. The minimum wave number q_m is determined by the drop dimension. Far from the liquid–vapor critical point (where σ tends to zero, see, e.g. Zappoli et al. 2015), the interface thickness is in the order of the intermolecular distance and corresponds to the interaction range of the molecular potentials. The definition of the maximum distance between drops to obtain coalescence:

$$d_m = \langle z^2 \rangle \tag{5.2}$$

therefore also corresponds to what is observed with molecular dynamics (Pothier and Lewis 2012).

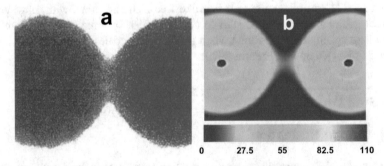

Fig. 5.1 Onset of coalescence between nanodroplets. **a** Two liquid Si nanodroplets from molecular dynamics study. (Adapted from Pothier and Lewis, 2012, with permission). **b** Two nanodroplets from density functional study. (Adapted from Niu et al. 2020, with permission)

5.1.2 Bridge Evolution

Once a liquid bridge has nucleated, the negative curvature at the liquid bridge drives the bridge expansion. In other words, because of the curvature difference between the liquid bridge radius and the droplet radius, Laplace pressure moves the liquid mass to the center, leading to the expansion of the liquid bridge. The geometry of the problem is drawn in Fig. 5.2.

The process has been comprehensively studied (see, e.g. Lafaurie et al. (1994); Eggers 1997; Eggers et al. 1999; Thorodsen 2005; Aarts et al. 2005; Paulsen et al. 2012; Pothier and Lewiw 2012; Khodabocus et al. 2018; Niu et al. 2020). It can be summarized as follows (Eggers et al. 1999; Aarts et al. 2005).

The coalescence process consists of three regimes, called viscous, viscous-inertial and inertial (Paulsen et al. 2012; Khodabocus et al. 2018, see Fig. 5.3), depending on the Ohnesorge number, Oh, which relates the viscous forces to

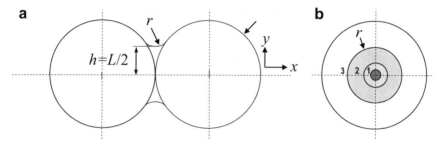

Fig. 5.2 Geometry of coalescence for free standing drops. **a** View from above. **b** Side view. The numbers 1, 2, 3 refer to the development of the bridge

Fig. 5.3 Phase diagram for 3D coalescence. Open circles and triangles: experiments; solid and dashed lines: theory. (Adapted from Paulsen et al. 2012, with permission)

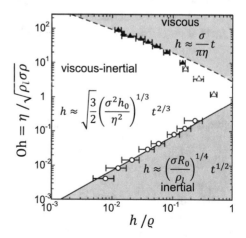

inertial and surface tension forces:

$$Oh = \frac{\eta}{\sqrt{\varrho_l \sigma \varrho}} \tag{5.3}$$

Here ϱ is the drop radius (see Fig. 5.2.), supposedly equal for each drop, σ is the air–liquid surface tension, ϱ_l is the liquid mass density and η is the liquid dynamic or shear viscosity.

5.1.2.1 Viscous Regime

When considering pure viscous flow, one must define the viscous length, that is, the distance l_v from the walls to a point where the flow velocity has reached the free stream velocity, or the Reynolds number exhibits a value on order unity:

$$Re = \frac{\rho_l u l_v}{\eta} = 1 \tag{5.4}$$

The flow velocity in the laminar flow approximation (see Eq. 5.30) being in the order of

$$u = \frac{\sigma}{\eta} \tag{5.5}$$

one readily obtains

$$l_v = \frac{\eta^2}{\rho_l \sigma} \tag{5.6}$$

When $h < l_v$, viscosity effects dominate the process and a linear evolution is expected for small h. Then, for any viscosity value, viscous flow occurs at the very early moments of coalescence. The complete treatment of the problem is given by Eggers et al. (1999) and leads to the evolution:

$$h = L/2 \approx \left[-\frac{1}{\pi} \mathrm{Ln}\left(\frac{\sigma}{\rho \eta} t \right) \right] \left(\frac{\sigma}{\pi \eta} \right) t \tag{5.7}$$

The weak logarithmic correction is often ignored. It therefore comes in agreement with the scaling approach of Khodabocus et al. (2018):

$$h = L/2 \sim \left(\frac{\sigma}{\pi \eta} \right) t \tag{5.8}$$

One can come to the same result from simple scaling arguments. The viscous flow is driven by the Laplace capillary pressure corresponding to the negative bridge curvature (see Fig. 5.2), expressed as

$$p_L = \frac{\sigma}{r} \tag{5.9}$$

Simple geometry (see Appendix F, Eq. F.2 with $\theta_c = 90°$) gives, noting that $h/\rho \ll 1$:

$$r = \frac{h}{2}\frac{2\rho\cos\theta_c + h}{\rho(1 - \cos\theta_c) - h} = \frac{h^2}{2\rho} \tag{5.10}$$

The Laplace pressure is balanced by the viscous stress, of order $\eta u \rho / \pi h^2$, where $u \sim dh/dt$ is the fluid velocity near the bridge:

$$p_L \sim \frac{\rho\sigma}{h^2} \sim \frac{\eta\rho}{h^2}\left(\frac{dh}{dt}\right) \tag{5.11}$$

The integration of which gives Eq. 5.8.

5.1.2.2 Inertial Regime
It occurs when h is larger than the viscous length. The Laplace capillary pressure, $p_L = \rho\sigma/h^2$ as given by Eq. 5.11, thus equals the dynamical, inertial pressure:

$$p_{iner} \sim \rho_l u^2 \sim \rho_l (dh/dt)^2 \tag{5.12}$$

it readily follows:

$$h\frac{dh}{dt} \sim \left(\frac{2\rho\sigma}{\rho_l}\right)^{1/2} \tag{5.13}$$

giving

$$h = L/2 = A\left(\frac{\sigma\rho}{\rho_l}\right)^{1/4} t^{1/2} \tag{5.14}$$

The prefactor is near unity ($A \approx 1.1$ according to Aarts et al. (2005)).

5.1.2.3 Visco-Inertial Regime
In this regime, both inertia and viscosity play a role in the dynamics. In addition to the viscous flow, the inertia of drops must be taken into account, the drops being considered as moving solid objects (Paulsen et al. 2011; Paulsen et al. 2012). From scaling mathematical arguments, Khodabocus et al. (2018) proposed for the bridge radius evolution (calculated for a sessile drop with non-pinned contact line):

$$h = L/2 = \sqrt{\frac{3}{2}}\left(\frac{\sigma^2 h_0}{\eta^2}\right)^{1/3} t^{2/3} \tag{5.15}$$

Here h_0 is the bridge radius at the beginning of coalescence (time $t = 0^+$). The inertia associated with each drop moving as a rigid object prevents the system from being in the Stokes regime.

Figure 5.4 provides the transition observed from the inertial regime to the visco-inertial and viscous regimes when viscosity is increased. One notes that, in agreement with Fig. 5.3, all regimes collapse at large h/ρ (rounding of evolutions).

Fig. 5.4 Coalescence of free-standing drops showing the effect of viscosity on the power law scaling, Eqs. 5.8, 5.15). From left to right the viscosities are (1.00, 2.17, 5.15, 42.6, 220, 493) × 10^{-3} Pa.s. (From Thoroddsen et al. 2005, with permission)

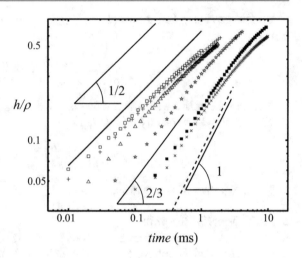

5.2 Coalescence of Sessile Drops

The description of the coalescence process of sessile drops requires an explanation of two singularities. The first singularity is concerned with the contact point of the merging drops and is the same as above with free-standing drops. The second singularity corresponds to a three-phase contact line and is particular to sessile drops when the contact angle is below 90°. In addition to bulk dissipation by viscous effects, the relaxation process also includes the motion of the contact line. The description of the coalescence process therefore appears quite complex.

The formation and growth of the bridge should obey the same general conditions as encountered with free-standing drops: interdroplet distance lower than the merging droplet interfaces molecular inter-distance for bridge nucleation and further expansion thanks to different capillary pressures in the bridge and the drops. However, there exists a supplementary factor in the process, the existence of a three-phase contact line whose motion must be considered. The general problem of the moving contact line can itself be split into four different sub-problems (Fig. 5.5), depending on whether the contact angle is (i) zero (perfect or complete wetting), (ii) finite below 90° (partial wetting with coalescence at the contact line), (iii) finite above 90° (partial wetting with coalescence above the contact line)

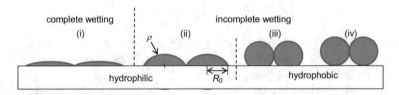

Fig. 5.5 Various coalescence cases of sessile droplets according to the wetting properties

or (iv) 180° (perfect drying, the same case as free-standing drops). Each situation requires a specific theoretical approach because of the obvious differences between the fluid mechanical approaches in the moving edge.

5.2.1 Perfect Wetting Liquid ($\theta_c = 0$)

The case (i) of Fig. 5.5 is concerned with the merging of two droplets of a perfectly wetting fluid after they have been deposited on a substrate. This situation has been discussed by Ristenpart et al. 2006 and is only briefly discussed here as such a configuration does not occur during condensation.

Coalescence corresponds to Fig. 5.6 geometry. For perfect wetting, there is no contact line pinning and dynamics are governed by the spreading of the droplets in the lubrication limit, using volume conservation and Tanner's law (Tanner 1979). The latter is written as

$$\frac{dR}{dt} = \frac{\sigma}{\eta}\left(\frac{h}{R}\right)^3 \tag{5.16}$$

where $R \approx R_0$ and $h \approx h_0$, the initial drop cap radius and height, respectively.

Mass conservation leads to the following relationship, with $V = lLh$ the bridge volume and $u = dR/dt$ the fluid velocity feeding the bridge:

$$\frac{dV}{dt} = Lhu = Lh\left(\frac{dR}{dt}\right) \tag{5.17}$$

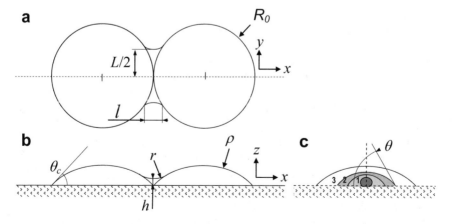

Fig. 5.6 Geometry of sessile drop coalescence. Bridge height is h, length is L and width is l. **a** View from above, (**b**, **c**) side views. The numbers 1, 2, 3 refer to the development of the bridge

For small contact angles, valid here, geometry gives (see Appendix F, Eq. F.6):

$$l \approx \frac{L^2}{4R_0} \tag{5.18}$$

The bridge volume variation becomes

$$\frac{dV}{dt} = \frac{d(lLh)}{dt} \sim h_0 \left(\frac{L^2}{R_0}\right)\left(\frac{dL}{dt}\right) \tag{5.19}$$

Expressing dR/dt from Eq. 5.16 and equaling Eqs. 5.17 and 5.19, it becomes, neglecting the weak $h \sim h_0$ variation,

$$\frac{LdL}{dt} \sim \frac{\sigma}{\eta} \frac{h_0^3}{R_0^2} \tag{5.20}$$

or

$$L \sim \left(\frac{\sigma}{\eta} \frac{h_0^3}{R_0^2}\right)^{1/2} t^{1/2} \tag{5.21}$$

and

$$l \sim \left(\frac{\sigma}{\eta} \frac{h_0^3}{R_0^3}\right) t \tag{5.22}$$

5.2.2 Hydrophobic Liquid ($\theta_c > 90°$)

The perfect hydrophobic case (iv) is similar to the free-standing drop process of Sect. 5.1 since there is no contact line pinning. Case (iii) where coalescence does not occur at the contact line location (see Fig. 5.5) presents similarities with the free-standing case (the bridge nucleates between two free interfaces) and differences (there is contact line pinning). In particular, during the early times (iii) when the connecting bridge radius, $h = L/2$ as in Fig. 5.1, is typically smaller than the drops radius, ρ, the drops contact line does not move due to pinning forces. As a matter of fact, several experiments mimicking free-standing drops have been performed with a pendant drop fusioning with the top of a sessile drop (see, e.g. Thorodsen et al. 2005; Aarts et al. 2005).

However, inertial effects as discussed above for free-standing drops can be more or less affected by the contact line pinning. Growth laws can thus exhibit some differences with Eqs. 5.8, 5.14, 5.15. It is also the case for late stages where the contact line motion interferes with the process, as discussed below for case (ii), $\theta_c < 90°$.

5.2.3 Hydrophilic Liquid ($\theta_c < 90°$)

Case (ii) in Fig. 5.5 is the most common case encountered with dew and breath figures. The difficulty in describing the coalescence process is mainly related to the description of the contact line motion. Most studies (Hernández-Sánchez et al. 2012; Lee et al. 2012a; Zheng et al. 2016; Khodabocus et al. 2018; Somwanshi 2018) do not consider in the interpretation the contact line motion and corresponding dissipation. The important role of contact line dissipation was recognized by Andrieu et al. (2002), Narhe et al. (2004) and Jiang et al. (2019), who studied the late stages of coalescence. Here the contact line motion can slow down the process by as much as seven orders of magnitude when compared to viscous flow dissipation. Narhe et al. (2008) also considered the early stages, where the large dissipation at the contact line leads to a dynamic drying of the connecting bridge.

5.2.3.1 Contact Line Motion

The motion of the contact line is determined by a mobility relation (see Pomeau 2000; Pomeau 2002; Ben Amar et al. 2003). The driving force f for contact line motion being related to the difference between the equilibrium and instantaneous values of the contact angle, it corresponds to

$$f = \sigma F(\theta, \theta_e) \tag{5.23}$$

$F(\theta, \theta_e)$ is a non-dimensional function of the contact angle, of order unity and equal to zero at equilibrium, when $\theta = \theta_e$. Classically:

$$F(\theta, \theta_e) = \cos\theta - \cos\theta_e \tag{5.24}$$

The driving force is balanced by a friction force, which can be expressed as

$$f = \left(\frac{\eta}{K}\right)v_n \tag{5.25}$$

Here v_n is the velocity of displacement of the line normal to its local orientation and K a constant, which accounts for the particular friction process at the contact line and will be discussed below. Equations 5.23 and 5.25 lead to a mobility relation between v_n and the difference between the equilibrium contact angle θ_c and its actual value θ:

$$v_n = K\left(\frac{\sigma}{\eta}\right)F(\theta, \theta_e) \tag{5.26}$$

The value of the parameter K is related to the details of the process of contact line motion. The contact line, sticking on the solid, cannot move by hydrodynamic motion, since the fluid velocity is zero on the solid (no-slip condition). This condition, however, can be met during the contact line motion when evaporation (when the liquid is receding) or condensation (when the liquid is advancing) occurs in the vicinity of the contact line (see Seppecher 1996). Although the rate of this phase change can be large (it is proportional to the contact line speed), it does not result in a large mass change of the drop because of the small part of the area where the phase change occurs. When coalescence occurs during condensation, the rate of this phase change is therefore independent of the condensation rate.

In a coalescence process only the receding motion is concerned, thus involving evaporation at the contact line. This evaporation is a thermally activated process because molecules in the liquid are in the bottom of a potential well, due to the attraction of the other molecules, an attraction necessary to maintain the cohesion of the liquid against spontaneous self-evaporation. Therefore, the rate of evaporation should be proportional to a very small Arrhenius factor:

$$K = \exp\left(-\frac{W}{k_B T}\right) \tag{5.27}$$

where W is the difference of potential energy between liquid and vapor, taken as positive (in practice, this potential energy is zero in the dilute vapor, and $-W$ in the liquid). Supposing that the potential energy grows monotonically from a well in the liquid to its zero value in the vapor, one would obtain for W the latent heat per molecule. Using the molar latent heat L_v ($= 44$ kJ mol^{-1} for water at 20 °C) one obtains, with $R = 8.314$ J.mol^{-1}.K^{-1}, the ideal gas constant:

$$K = \exp\left(-\frac{L}{RT}\right) \tag{5.28}$$

whose value $\sim 10^{-8}$ for water. Note that the pinning of the contact line on weak defects cannot cause large effects, as discussed in Nikolayev and Beysens (2001), because the retardation due to pinning at the defect is counterbalanced by acceleration when depinning.

The mobility relation, Eq. 5.25, loses its meaning in two limits, when the speed is "very large" in any direction where the liquid moves back or moves forward (Pomeau 2002). In both cases, a dynamical transition should take place to either perfect wetting or perfect drying depending on the orientation of the velocity with respect to the contact line. This assumes that, beyond a certain speed of advance, the droplet rolls over the solid surface, specifically when the speed is above a critical velocity for the transition to dynamical drying.

In order to characterize the flow, the Reynolds number (Eq. 3.1) can be evaluated as

$$\mathrm{Re} = \frac{u R_0 \rho_l}{\eta} \tag{5.29}$$

Using the water values at 20 °C (Table 2.2) and Eq. 5.36 $u = \frac{\sigma}{\eta}\theta_c^2 \approx 20$ m.s^{-1} with contact angle $\theta_c \approx 30°$, one finds Re $< 4 \times 10^4$ for drops of radius smaller than typically $R_0 = 2$ mm. This value of the Reynolds number is significantly lower than the critical value (5×10^5) above which turbulence typically occurs in an open planar structure (Rohsenow et al. 1998). Laminar flow approximation is thus valid.

The relevant number to determine the flow velocity will thus be the capillary number, which expresses the relative effect of viscous drag forces versus surface tension forces:

$$Ca = \frac{\eta u}{\sigma} \tag{5.30}$$

One can thus consider two distinct dynamical regimes in the merging process (Fig. 5.7). (i) A fast regime (Fig. 5.7a), during which the bridge between the two droplets is filled at a speed such that Ca is finite, although the rest of the droplet perimeter remains almost immobile. In this fast regime, there is an overhang in the bridge region, because there the advancing speed is above the transition value

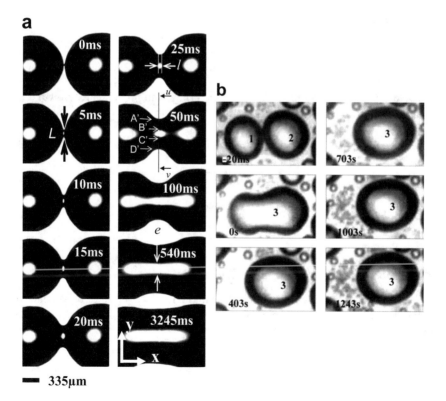

Fig. 5.7 Time sequence of two coalescing diethylene drops on silicon surface with advancing and receding contact angles 35° and 25°, resp. and mean contact angle $\theta_c = 30°$). **a** Early times (From Narhe and Beysens 2008). **b** Late times. The size of each image is 40×5 μm (From Beysens and Narhe 2006)

for dynamical drying. (ii) A slower regime (Fig. 5.7b), which is reached when the advancing speed becomes everywhere smaller than the critical value for dynamical dewetting. Afterwards, (iii), the dynamics is ruled by the low mobility of the contact line.

5.2.3.2 Fast Regime-Short Time Dynamics Ca ~ 1

The bridge is defined by its height h, width l, the dimension perpendicular to the line of the centers of the drops, length L, in the direction parallel to the line of drop centers (Fig. 5.6). Typical evolution is shown in Fig. 5.7a.

(i) *Low contact angles.* The Laplace capillary pressure is expressed as

$$p_L = \sigma\left(\frac{1}{r} + \frac{1}{L}\right) \approx \frac{\sigma}{r} \tag{5.31}$$

with (see Eq. F.5 in Appendix F):

$$r \approx \frac{h}{\theta_c^2} \tag{5.32}$$

The pressure p_L drives the fluid inside the bridge and provokes its growth. One obtains from simple geometric arguments:

$$L \approx 2(R_0 l)^{1/2} \tag{5.33}$$

see Appendix A4, Eq. A4.6 and Eq. A4.7:

$$h \sim \frac{\theta_c l}{2} \tag{5.34}$$

making $L \gg h$. Therefore, the curvature of the surface of the bridge is mostly in the direction normal to the line of the centers with radius r.

(ii) *Low contact angles and high viscous fluids.* The Laplace pressure p_L is balanced by the viscous stress, in the order of $\eta u/h$, where u is the fluid velocity near the bridge:

$$p_L \approx 2\frac{\sigma\theta_c^2}{h} \sim \frac{\eta u}{h} \tag{5.35}$$

One deduces the flow velocity value:

$$u = \frac{\sigma}{\eta}\theta_c^2 \tag{5.36}$$

With $u \sim \frac{dh}{dt}$, it comes the meniscus height evolution:

$$h \sim \left(\frac{\sigma}{\eta}\theta_c^2\right)t \tag{5.37}$$

Making use of Eq. 5.34:

$$l \sim 2\left(\frac{\sigma}{\eta}\theta_c\right)t \tag{5.38}$$

From Eq. 5.33:

$$L \sim 2\sqrt{2}\left(\frac{\sigma}{\eta}\theta_c R_0\right)^{1/2}t^{1/2} \tag{5.39}$$

The bridge height evolution $h \sim t$, in agreement with what is found for free-standing drops evolution for pure viscous flow (Eq. 5.8).

The above evolutions can be put in scaled form by defining a typical evolution time $t^* = \frac{t}{t_c}$, with:

$$t_c = \left(\frac{\eta}{\sigma}\right)R_0 \tag{5.40}$$

The time t_c corresponds to the capillary/viscosity balance or molecular velocity scale (σ/η) on typical length R_0.

One now defines dimensionless, scaled quantities. Height is scaled by the bridge height at the end of the coalescing process $h_\infty \approx \theta_c R_0$:

$$h^* = \frac{h}{\theta_c R_0} \tag{5.41}$$

Width is scaled by the width $l_\infty = 2R_0$ at the end of coalescence:

$$l^* = \frac{l}{2R_0} \tag{5.42}$$

Length is scaled by the length $L_\infty = 2R_0$ at the end of coalescence:

$$L^* = \frac{L}{2R_0} \tag{5.43}$$

From Eqs. 5.37–5.39, one thus deduces

$$h^* = l^* = \theta_c t^* \tag{5.44}$$

$$L^* = \sqrt{2}\left(\theta_c t^*\right)^{1/2} \tag{5.45}$$

The evolution of bridge height h^* and width l^* obeys the same power law with exponent unity while the exponent of the bridge length L evolution is 1/2. These different exponents are related to the viscous dissipation in sessile drops where the contact line remains pinned. Narhe et al. (2008) confirmed the above evolutions in viscous diethylene glycol from optical measurements (see Fig. 5.8).

They also observed the dynamic drying of the bridge during the early stages of evolution. Figure 5.9 reports on the bridge contact angle in the y-direction and the corresponding Ca number. Ca is always small, below 0.05. The measured angle corresponds to an apparent angle θ^* in the drying stage ($t < 30$ ms) and to the actual contact angle θ_c at later times.

(iii) *Large contact angles.* The approximation in the Laplace capillary pressure for the evaluation of the bridge negative curvature ($r \approx 2h/\theta_c^2$, Eq. 5.31) does not hold any more. Simple geometry (Eq. F.2 in Appendix F) gives

$$r = \frac{h}{2} \frac{2\rho\cos\theta_c + h}{\rho(1 - \cos\theta_c) - h} \tag{5.46}$$

Eddi et al. (2013) used a slightly different estimation, valid for θ_c near 90°, by approximating the circle rope to the bow (see Eq. F.4 in Appendix F):

$$r \approx \rho\sin\theta_c - \left[\rho^2 - (h + \rho\cos\theta_c)^2\right]^{1/2} \tag{5.47}$$

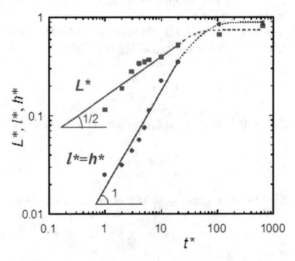

Fig. 5.8 Diethylene glycol drops of radius $R_0 = 770$ μm coalescing on silicon (advancing and receding contact angles 35° and 25°, resp., and mean contact angle $\theta_c = 30°$). Typical evolution with respect to scaled time $t^* = t/t_c$ ($t_c = 5.1$ ms) of the scaled bridge height $h^* = h/\theta_c R_0$, width $l^* = l/2R_0$, length $L^* = L/2R_0$, showing at early time $t^* < 10$ the expected 1 and 1/2 power law behavior (full lines, Eqs. 5.44, 5.45) and at late time $t^* > 10$ an exponential decay to unity (dotted and interrupted curves, fit to Eq. 5.60 with $a = 0.75 \pm 0.08$, $\tau^* = 16 \pm 7$ for L^* and $a = 0.9 \pm 0.08$, $\tau = 38 \pm 3$ for h^* and l^*). (From Narhe et al. 2008)

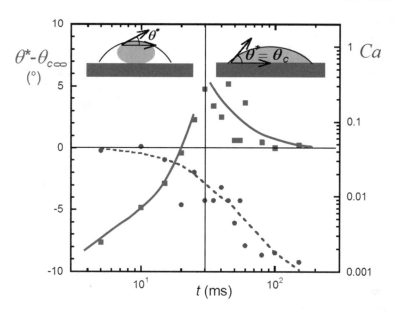

Fig. 5.9 Correlated evolution of the bridge contact angle $\theta^* - \theta_{c\infty}$ in the y-direction in Fig. 5.8a (left ordinate) and capillary number Ca (right ordinate). The limiting contact angle at the end of relaxation is $\theta_{c\infty}$. The drying stage region $t < 30$ ms corresponds to an apparent contact angle $\theta^* \neq \theta_c$. The interrupted and continuous lines are data smoothening. (From Narhe et al. 2008)

(iv) *Large contact angles and low viscous fluids.* For example, water, the capillary pressure is balanced by the dynamical pressure $p_{iner} = \rho_l u^2$ (Eq. 5.12). The growth of the bridge between low viscous, freely suspended drop with radius ρ obeys a power law evolution (inertial regime) with time, $h = A\left(\frac{\sigma\rho}{\rho_L}\right)^{1/4} t^{1/2}$, with exponent 1/2 (Eq. 5.14). Eddi et al. (2013) and Pawar et al. (2019) studied bridge evolution in sessile drops of water for contact angles of 90° and slightly less (84°, 81°, see Fig. 5.10). For $\theta_c = 90°$, the exponent 1/2 is recovered as in free-standing drops. This is understandable when the contact line friction is negligible; the substrate acts as a mirror and the geometry is identical to that of two freely suspended spherical drops. When $\theta_c \neq 90°$, the bridge evolution still obeys a power law but with exponent 2/3 at early times and 1/2 at late times (Fig. 5.10). This behavior stems from the θ_c dependence of the bridge radius of curvature r, Eq. 5.46. Equalizing Laplace and dynamic pressures as in Eq. 5.12, it comes, with B a constant of order unity:

$$\frac{h^3}{2} \frac{2\rho\cos\theta_c + h}{\rho(1 - \cos\theta_c) - h} = B\frac{\sigma}{\rho_l}t^2 \qquad (5.48)$$

Near $\theta_c = 90°$, Eq. 5.48 can be developed, since $h/\rho \ll 1$:

$$\frac{h^3}{2} \frac{2\rho\cos\theta_c + h}{\rho} \approx B\frac{\sigma}{\rho_l}t^2 \qquad (5.49)$$

Fig. 5.10 Evolution of the bridge height h/ρ between sessile drops for three different contact angles with respect to the scaled time t/τ_{ic}, where $\tau_{ic} = \sqrt{\rho_L \rho^3 / \sigma}$. The open circles are experimental data from Paulsen et al. (2012) for freely suspended water drops. (Adapted from Eddi et al. 2013, with permission)

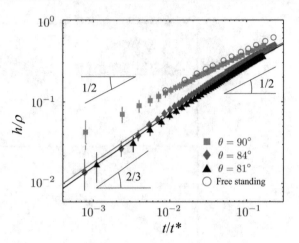

Depending whether h/ρ is larger or smaller than $\cos \theta_c$ ($\approx \frac{\pi}{2} - \theta_c$ near $\theta_c = \frac{\pi}{2}$), one obtains two different evolution for h:

Small time $h/2\rho \ll \frac{\pi}{2} - \theta_c$

$$h = \left(\frac{B\sigma}{\rho_l} \right)^{1/3} t^{2/3} \tag{5.50}$$

Large time $h/2\rho \gg \frac{\pi}{2} - \theta_c$

$$h = \left(\frac{2B\sigma\rho}{\rho_l} \right)^{1/4} t^{1/2} \tag{5.51}$$

The latter expression corresponds to Eq. 5.14 for free-standing drops with $B = A^4/2$.

The domain where the exponent 2/3 applies thus vanishes when $\theta_c \rightarrow 90°$, see Fig. 5.10. In this figure, the scaled bridge height h/ρ is plotted with respect to the scaled time $t^* = t/\tau_{ic}$, with $\tau_{ic} = \sqrt{\rho_L \rho^3 / \sigma}$ the inertial-capillary timescale of single drops (see Eq. 5.54). Note that the exponent 2/3 observed with such low viscous fluids, although equal to the exponent found in the visco-inertial regime in free-standing drops (Eq. 5.15), corresponds to a pure inertial regime.

5.2.4 Short Time Oscillations of Bridge and Drop

During the initial times of coalescence, oscillations of the bridge and composite drops can be observed in low viscous fluids. These oscillations can be triggered by the mode used to induce coalescence, e.g. by syringing or by condensation. Experimental studies indeed have used different means to induce coalescence. For pendant or sessile drops, growth of one or both drops is ensured by liquid injection (see, e.g. Somwanshi et al. 2018). For sessile drops, growth of one drop can be made by depositing with a microsyringe a very small drop (see, e.g. Narhe et al. 2004b). Condensation (see, e.g. Andrieu 2002; Narhe et al. 2004b) seems to be the smoothest way to induce coalescence. Narhe et al. 2004b noticed that oscillations of a composite water drop can be observed when coalescence is induced by depositing a drop on one of the drops to coalesce but not when coalescence is induced by condensation.

For low viscous liquids like water, the initial flow induced by coalescence is able to induce capillary waves at the surface of the early developing bridge, as noticed by Eddi et al. (2013) (see Fig. 5.11). Observation and 2D numerical simulations by Keller et al. (2000) and Billingham and King (2005) are also shown in Fig. 5.11. Simulations show more pronounced oscillations, which can be attributed to their 2D nature ignoring the saddle shape of the meniscus in 3D. Here the positive curvature along y-axis (perpendicular to the plane of Fig. 5.11) should lower the effect in 3D. Oscillations of the bridge and the composite drop are also observed (Narhe et al. 2004b; Zheng et al. 2016; Somwanshi et al. 2018). For viscous enough fluids, the oscillations are damped and are not visible, as discussed below.

Some general features can be seen in the bridge and composite drop oscillations. The amplitude H of liquid–air surface can be described by periodic oscillations whose amplitude decreases exponentially:

$$H = H_0 + H_m \exp\left(-\frac{t}{\tau}\right)\cos\left(2\pi\frac{t}{T} + \varphi\right) \tag{5.52}$$

Here H_m is the mean value of oscillations, C is the amplitude of the oscillation, τ is the damping time, T is the oscillation period and φ is the phase.

Fig. 5.11 Capillary waves at the surface of two merging conical drop of equal size. The comparison of the experimental profile (background photo) and 2D numerical solutions from Keller et al. (2000) and Billingham and King (2005) shows that the 2D simulation amplifies the effect (see text). The drop contact angle is ≈ 75°. (Adapted from Eddi et al. 2013, with permission)

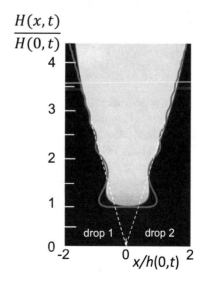

With ρ_1, ρ_2 the radius of curvature of coalescing droplets, one defines a volume-averaged radius of curvature of droplets, ρ, by

$$\rho = \left(\frac{\rho_1^3 + \rho_2^3}{2}\right)^{1/3} \tag{5.53}$$

Ignoring the motion of the contact line, three different timescales should enter in the problem:

(i) *Inertial-capillary timescale τ_{ic}*

The pressure gradient (σ/ρ) induced by the surface tension balances the inertial term $(\rho_L \partial u/\partial t \sim \rho_L \rho/\tau_{ic}^2 \sim \sigma/\rho)$:

$$\tau_{ic} = \sqrt{\frac{\rho_L \rho^3}{\sigma}} \tag{5.54}$$

(ii) *Viscous-capillary timescale τ_{vc}*

The capillary pressure (σ/ρ^2) is balanced by the viscous stress, of order $\eta u/\rho \sim \eta/\tau_{vc} \sim \sigma/\rho^2$:

$$\tau_{vc} = \frac{\eta \rho}{\sigma} \tag{5.55}$$

(iii) The inertial-viscous timescale τ_{iv}

The inertial force per unit area $(\rho_L \rho^2/\tau_{iv}^2)$ is counterbalanced by shear stress $(\eta \rho/\tau_{iv} \rho \sim \rho_L \rho^2/\tau_{iv}^2))$:

$$\tau_{iv} = \rho_L \frac{R^2}{\eta} \tag{5.56}$$

For water, the times are in the order of $\tau_{vc} \ll \tau_{ic} \ll \tau_{iv}$ (see Table 5.1).

Figure 5.12 presents the relaxation of two sessile drops of water-glycerine mixtures where coalescence is induced by syringing, as reported by Zheng et al. (2016). Shear viscosity η is varied from 1 to 10 mPa·s by changing the glycerine concentration (from 0 to 70 weight %). Obvious oscillations are observed until viscosity reaches 6 mPa·s (50% glycerine). Both the period T of oscillations and the damping time τ are seen to correspond to the unique inertial-capillary timescale (Fig. 5.13):

$$T \approx 2.3\tau_{ic}; \tau \approx 1.6 \times 10^3 \tau_{ic} \tag{5.57}$$

Table 5.1 Time scales estimated for the coalescence of water drops under ambient conditions; *ic*: inertial surface tension, *vc*: viscous surface tension, *iv*: inertia viscous. (Adapted from Somwanshi et al. 2018)

Drop base R(mm)	τ_{ic} (ms)	τ_{vc} (us)	τ_{iv} (ms)
0.11	0.13	1.21	14
0.23	0.41	2.62	66
0.29	0.59	3.30	104
0.36	0.83	4.15	166
0.46	1.17	5.23	263
0.58	1.66	6.59	417
0.73	2.36	8.30	663

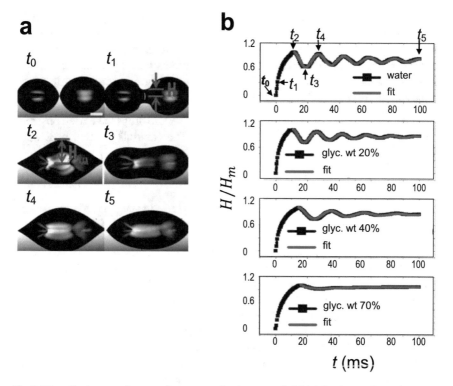

Fig. 5.12 **a** Coalescence images of two water droplets. $t_0 = 0$: initial droplet configuration. At $t_1 = 1$ ms, a bridge forms, with height H. At $t_2 = 11$ ms, H reaches its maximum value, H_m. Then it starts to decrease until $t_3 = 20$ ms. At $t_4 = 27$ ms, H reaches a second peak. After a few cycles of damping oscillation, the system approaches a static equilibrium state around $t_5 = 170$ ms. Bar is 500 μm. **b** Evolution of the normalized height in the middle of the bridge, H/H_m. Black dotted line corresponds to experimental measurements and the red line shows the fitting by Eq. 5.52. The typical times $t_{1,2,3,4,5}$ are the times noted in (**a**). Pure water and three aqueous glycerol solution are shown, with increasing viscosity. (Adapted from Zheng et al. 2016, with the permission of AIP Publishing)

Damped oscillation during the relaxation after the merging of droplets is indeed mainly dominated by inertial and capillary forces. It corresponds to a low value of the Ohnesorge number (Eq. 5.3):

$$Oh = \frac{\tau_{vc}}{\tau_{ic}} = \frac{\eta}{\sqrt{\varrho_L R \sigma}} \tag{5.58}$$

Such low values are observed in Fig. 5.13 with proportionality between τ_{vc}/τ and Oh. This property leads to also proportionality between τ and τ_{ic}. Viscous resistance thus does not apply as it can only decide when obvious oscillation can be observed. Once it happens, the influence of viscosity is marginal. The two-dimensional simulations in Zheng et al. (2016) show excellent agreement with the above analysis. Note that when viscous effects are weak, damped oscillation occurs and the liquid bridge experiences a damped oscillation process before reaching its equilibrium shape. However, if the viscous effects become significant, under-damped relaxation shows up. In this case, the liquid bridge relaxes to its stable shape in a non-periodic decay mode.

Centroid and interface oscillations in pendant and sessile drops have been observed after coalescence by Somwanshi et al. (2018). The centroid position of a water composite drop is defined by

$$\vec{R} = \sum_{1}^{N} \vec{r_i} \tag{5.59}$$

with $\vec{r_i}$, $i = [1-N]$, the position of N drop elements I. Coalescence was induced by syringing. The main results are in agreement with the above discussion, however more information is available concerning long time relaxation of oscillations, in particular, (i) the inertial-viscous time scale is seen to relate to the time required by the velocity amplitudes to decay monotonically in the long term by viscous dissipation and (ii) fluid oscillations diminish faster at the interface region as compared to the fluid at the core. Especially, interfacial oscillations diminish more rapidly for sessile drops than for pendant drops due to the fact that shear stresses are momentarily developed at the wall at the short time scale. These stresses are thus smaller in the pendant mode compared to the sessile mode.

5.2.5 Late Time Slow Regime

5.2.5.1 Bridge Relaxation

At late time, the dynamics are much slowed down and Ca $<<$ 1 (Eq. 5.30) and become ruled by the low mobility of the contact line. The bridge dimensions h, l, L, progressively reach their limit dimensions $h_\infty, l_\infty, L_\infty$ such as the composite drop becomes near elliptical (see Figs. 5.7 and 5.14). The corresponding evolutions

Fig. 5.13 **a** Scaled oscillation amplitude H_m/ρ as a function of shear viscosity η for seven liquid mixtures. The line is a guide for the eye. **b** Oscillation period T versus the inertial-capillary time scale τ_{ic} for seven liquid mixtures. The line slope is 2.3. **c** Plot of τ_{vc}/τ as a function of the Ohnesorge number Oh (Eq. 5.58) for seven kinds of liquids, showing proportionality to Oh. (Adapted from Zheng et al. 2016, with the permission of AIP Publishing)

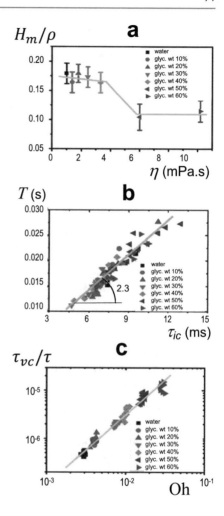

can be fitted by exponential decaying functions (Narhe et al. 2008) such as, with the notations of Eqs. 5.41–5.43,

$$h^* \sim l^* \sim L^* \sim a\left[1 - \exp(-t^*/\tau^*)\right] \tag{5.60}$$

The reduced damping τ^* (see Fig. 5.8) is found more than an order of magnitude larger than unity, corresponding to a damping time much larger than the time t_c ($= (\eta/\sigma)R_0$ from Eq. 5.40) as obtained from the typical molecular velocity scale (σ/η). It corresponds to a mix-up of hydrodynamics and contact line motion.

5.2.5.2 Composite Drop Relaxation

The composite drop of near-elliptical shape then relaxes to the equilibrium, near-circular shape (see Fig. 5.7b). With R_x and R_y the ellipsis major and minor axes,

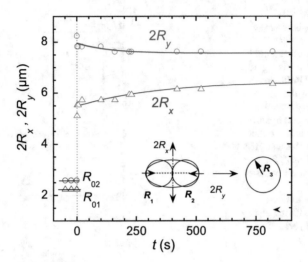

Fig. 5.14 Late time coalescence of DEG drops with radius $R_{01} = 2.6$ μm and $R_{02} = 2.3$ μm. Typical time evolution of the composite drop large axis $2R_y$ and small axis $2R_x$. (From Beysens and Narhe 2006). ($t_c = 16.5$ μs)

respectively, along x and y (Fig. 5.14), the relaxation time τ can be obtained by fitting the relaxation data by an equation of the form (Fig. 5.15), where $R_{0x,y}$ and $R_{1x,y}$ are amplitudes:

$$R_{x,y} = \left(R_{0x,y} + k_{x,y}t^*\right) + R_{1x,y}\exp(-t^*/\tau^*) \qquad (5.61)$$

Growth by condensation is accounted for the term $k_{x,y}$ and is quite often negligibly small. The data in Fig. 5.14 give $k_{x,y} = 0$ and $T^* \sim 1.5 \times 10^5$. Considering the contact line dissipation through the Arrhenius factor K of Eq. 5.25, it follows that

$$\tau^* = 1/K \qquad (5.62)$$

*corresponding, as in Eq. 5.26, to a contact line velocity:

$$u_{cl} = K\left(\frac{\sigma}{\eta}\right) \qquad (5.63)$$

From the T^* value, one deduces $K \approx 7 \times 10^{-8}$. A more refined treatment of the contact angle (Nikolayev and Beysens 2002) gives $K \approx 6 \times 10^{-7}$, which agrees (within the large error providing from the exponential function) with the expected value 2×10^{-8} from Eq. 5.28 using the latent heat for DEG, $L = 52.3$ kJ/mol (Lide 1999). Data for water (Andrieu et al. 2002) gives values $K \approx 10^{-8}$. In any case, such very small K values clearly show that the dynamics of viscous sessile drops in the regime of partial wetting is limited by the very large dissipation at the region of the drop close to the contact line.

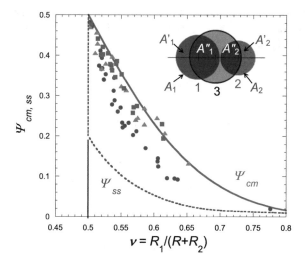

Fig. 5.15 Position of the center of mass of drop (3) resulting from the coalescence of droplets (1), (2) as a function of the relative radius $v = R_1/(R_1 + R_2)$. $A_{1,2}$: Drops (1), (2) area; $A'_{1,2}$: Drop (1), (2) swept area; $A''_{1,2}$: Drops (1), (2) and (3) overlap areas. ψ_{cm}, ψ_{ss}: non-dimensional position (see Eq. 5.68) assuming momentum conservation (negligible or strong contact line pinning, subscript cm) or minimizing contact line friction (moderate contact line pinning, subscript ss). Data points are experiments from Bouverot-Dupuis et al. (2021). Full circles: Bare Silicon (moderate pinning: $\theta_c = 41°$, CAH = 21°); squares: TiO$_2$-coated silicon (strong pinning: $\theta_c = 41°$, CAH = 38°); triangles: PTFE-coated silicon (strong pinning: $\theta_c = 74°$, CAH = 59°)

5.2.5.3 Position of the New Drop

The position of the new drop (see Fig. 5.15) is difficult to predict as it corresponds to a mix-up of inertial and viscous effects and contact line dissipation, whose combined effects vary during relaxation. However, following Viovy et al. (1988), it is possible to calculate the center of mass of the new drop in two limiting cases, when dissipation can be neglected and when dissipation occurs only at the contact line. A third case where the pinning of the contact line is strong can be treated as the case where dissipation is negligible.

(i) *Negligible or strong contact line pinning.* For negligible pinning, the position of the center of the new droplet (3) resulting from the coalescence of drops (1) and (2) is imposed by momentum conservation. The center of mass of drop (3) ($\overrightarrow{R_3}$, coordinates x_3, y_3) will be the same as the coalescing drops (1) and (2) (coordinates $x_{1,2}$, $y_{1,2}$, respectively). This is also the case of strong contact line pinning because the drops that coalesce hardly move. For drops showing little deviation from spherical cap with same contact angle θ_c, one obtains

$$\overrightarrow{R_3} = \frac{\overrightarrow{R_1} R_1^3 + \overrightarrow{R_2} R_2^3}{R_1^3 + R_2^3} \tag{5.64}$$

The contact angle dependence of the drop volume, $\pi G(\theta_c)$, (see Appendix D, Eq. D.8) does not enter in the equations because it is the same for all drops.

Taking drop (1) as the origin:

$$\vec{R_3'} = \vec{R_2} - \vec{R_1} \tag{5.65}$$

and defining the dimensionless radius asymmetry:

$$v = \frac{R_1}{R_1 + R_2} \tag{5.66}$$

it becomes

$$\vec{R_3'} = \Psi_{cm}(v)\left(\vec{R_2} - \vec{R_1}\right) \tag{5.67}$$

where the non-dimensional center-of-mass function is

$$\Psi_{cm}(v) = \frac{(1-v)^3}{v^3 + \left(1 - v^3\right)} \tag{5.68}$$

The corresponding variation of $\Psi_{cm}(v)$ is shown in Fig. 5.15.

(ii) *Moderate contact line pinning.* What matters is the dissipation due to the contact line displacement as discussed above in this section. The contact line displacement can be estimated by minimizing the area swept by these lines, or equivalently by maximizing the overlap between the surface contacts of the two "old" droplets and of the "new" droplet since the sum $A'_{1,2}$ (swept) + $A''_{1,2}$ (overlap) = $A_{1,2}$ = constant (Fig. 5.15). This purely geometrical condition leads to a more complicated, but still scale-invariant, relative displacement,

$$\vec{R_3'} = \Psi_{ss}(v)\left(\vec{R_2} - \vec{R_1}\right) \tag{5.69}$$

The corresponding function $\Psi_{ss}(v)$ has been numerically evaluated by Viovy et al. (1988) (see Fig. 5.15).

The difference between Ψ_{ss} and Ψ_{cm} is significant, except when $R_1 \approx R_2$ ($v = 0.5$) and can be tested experimentally. Published data are scarce (the study by Somwanshi (2018) was concerned with the centroid fast dynamics of the composite drop after coalescence and did not address the position of the center of mass). In Fig. 5.16, it is reported the evolution of the droplet center of mass after coalescence (Andrieu et al. 2002) where unfortunately drops are of near-equal radius ($v = 0.495$) and Ψ_{ss} and Ψ_{cm} take the same value 1/2. In this figure are reported the coordinates $(x_i; y_i)$ of the center of mass of the drops $i = 1, 2, 3$ as calculated from their contact area on the substrate. Drops (1) and (2) and at late times drop (3) are

Fig. 5.16 Evolution of the coordinates of water drop centers before and after coalescence (coalescence time is indicated by a vertical interrupted line). Parent drops are (1): diameter 45 μm and (2): diameter 55 μm. The composite drop after coalescence is (3). **a** Abscissa X, **b** ordinate Y. (From Andrieu et al. 2002)

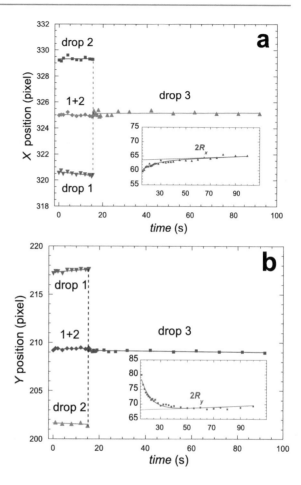

nearly hemispherical and this approximation is thus justified. Minor and major axes of the relaxing composite drop are also shown. During the relaxation process the center of mass does not appreciably move during (i) drop $(1 + 2)$ growth and (ii) (drop) 3 relaxation towards a spherical cap. An unpublished study using different substrates by Bouverot-Dupuis et al. (2021) is also reported in Fig. 5.15. It is observed that strong pinning data are only slightly below the Ψ_{cm} behavior when moderate pinning gives data well below Ψ_{cm}, closer to Ψ_{ss}, in agreement with the above analysis.

Drop Pattern Evolution

6

Droplets in a pattern have a growth different from the situation where a drop is alone. In vapor with non-condensable gases, droplets have indeed to compete to collect water molecules from the environment. Expressed differently, the water vapor concentration gradients of each drop whose spatial extent is in order of the diffusion boundary layer, ζ, overlap and correspond to the condition $\langle d \rangle < 2\zeta$, where $\langle d \rangle$ is the average interdroplet distance. In addition, droplet growth leads to the drop surface or drop contact linse to touch, making them coalesce with each other. Such coalescence events appear to be the key process in rescaling the growth of the drop pattern.

When vapor alone is present, the fact that single drop growth is limited by the available heat flux and that water pressure is uniform, allowing multiple drop nucleation, changes the growth laws. In particular. scaling is no longer valid although drop coalescence still occurs.

6.1 Individual Drop Growth in a 2D Pattern

6.1.1 3D Drop Growth by 2D Surface Diffusion

The case where the growth of three-dimensional (3D) droplets is dominated by dynamics on the surface has been considered by Rogers et al. (1988). When the surface mobility is large, the two-dimensional (2D) motion of molecules on the substrate determines the growth dynamics. Upon contact with a drop, monomers are assumed to be instantaneously absorbed at the drop perimeter.

This situation is similar to the case studied in Sect. 4.2.3 for a single drop but with different boundary conditions. In addition, the presence of many droplets in the pattern modifies the concentration gradient around each droplet. Rogers et al.

© Springer Nature Switzerland AG 2022
D. Beysens, *The Physics of Dew, Breath Figures and Dropwise Condensation*,
Lecture Notes in Physics 994, https://doi.org/10.1007/978-3-030-90442-5_6

(1988) linearized the 2D surface concentration profile $c^s(r)$ around each droplet, to become

$$c^s = c^s_\infty \quad r < R$$

$$c^s = c^s_s + \frac{c^s_\infty - c^s_s}{\xi}(r - R)\xi > r > R$$

$$c^s = c^s_\infty \quad r > \xi \tag{6.1}$$

The length ξ is a boundary layer extent and is an average property of the effective medium. It varies with time as droplets grow (Fig.6.1).

From droplet volume evolution, using Eq. 4.21 adapted to surface concentration and surface flux as in Sect. 4.2

$$\frac{dV}{dt} = \frac{1}{\rho_w} D \int_P \left(-\vec{\nabla} c\right)_R \cdot \vec{n}\, dl \tag{6.2}$$

Here, P is the drop perimeter and l is the curvilinear abscissa. Using the concentration profile Eq. 6.1 and the relation drop volume-cap radius (Appendix D, Eq. D.8), it comes, with ρ_w the density of condensed liquid (water)

$$\frac{dR}{dt} = \frac{2D}{3\rho_w G(\theta_c)} \frac{c^s_\infty - c^s_s}{R\xi} \tag{6.3}$$

Using the fact that the number of molecules entering the growing drops equals the amount leaving the boundary layers, Rogers et al. (1988) demonstrated that the boundary layer thickness ξ varies with time as

$$\xi \sim 2[Dt]^{1/2} \tag{6.4}$$

Substituting ξ in Eq. 6.3, one obtains

$$R \sim 2\left[\frac{D^{1/2}}{3\rho_w f(\theta_c)}\left(c^s_\infty - c^s_s\right)\right]^{1/2} t^{1/4} \tag{6.5}$$

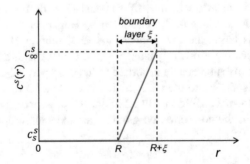

Fig. 6.1 Concentration profile around a growing droplet. The abscissa is the radial distance from the droplet center. (Adapted from Rogers et al. 1988)

It is interesting to compare the above droplet growth in a pattern with the growth of an isolated droplet with the same boundary conditions (Sect. 4.2.3, Eq. 4.34). A single drop exponent is 1/3, larger than the pattern drop exponent 1/4. The smaller exponent in the pattern reflects the competition between drops to collect diffusing monomers.

6.1.2 3D Drop Growth by 3D Diffusion. Equivalent Film

Let us consider an array of drops separated by a mean distance $\langle d \rangle < 2\zeta$ (Fig. 6.2). The individual water vapor concentration profiles then overlap to form a concentration profile parallel to the substrate.

Since the water vapor concentration profiles merge in a single profile parallel to the substrate, the drop pattern can thus be treated as a homogeneous film (Picknett and Bexon 1977; Briscoe and Galvin 1991; Beysens 2006; Sokuler et al. 2010a; Guadarrama-Cetina et al. 2014), with average thickness

$$h = \frac{V_T}{S_c} = \frac{V_i}{A_i} \tag{6.6}$$

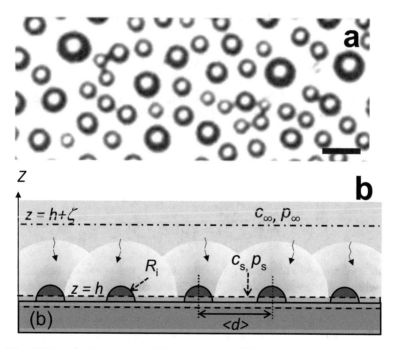

Fig. 6.2 **a** Picture of a droplet pattern. The bar represents 150 μm. **b** Schematics of droplet growth in a pattern with overlapping water vapor concentration profile and equivalent film of average thickness h. Notations: see text

Fig. 6.3 Construction of Voronoi-Dichlet polygons from the bisectors of segments connecting the drop centers to determine the area A_i of water vapor collection for drop i

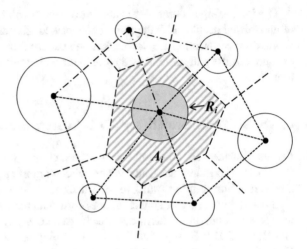

where A_i is the surface of influence of drop i where water molecules can be collected. It corresponds (Fig. 6.3) to the area determined by Voronoi polygons or Dirichlet tessellation (see e.g. Okabe et al., 1992 and Bormashenko et al., 2018). One can define a mean interdroplet distance $\langle d \rangle$ from the surface area S_c or the Voronoi polygons area A_i (see Fig. 6.3) such as

$$\langle d \rangle^2 = \frac{\sum_{i=1}^{N} A_i}{N} = \frac{S_c}{N} \tag{6.7}$$

with N *as* the number of drops. A relation between the mean interdrop distance $\langle d \rangle$ and the number of drops N can be found since the total condensation surface $S_c = N A_i \approx N \langle d \rangle^2$)

$$\langle d \rangle \approx S_c^{1/2} N^{-1/2} \tag{6.8}$$

In Eq. 6.6,

$$V_T = \sum V_i = \pi G(\theta_c) \sum R_i{}^3 \tag{6.9}$$

is the total condensed volume and S_c is the surface area of the condensation substrate, with $G(\theta_c)$ the function Eq. D.8 in Appendix D and θ_c the drop contact angle. Each individual drop (i) with volume $V_i = \pi G(\theta_c) R_i{}^3$ (Eq. 6.9) is assumed to grow like a thin film, where the vapor concentration profile depends only on the normal to the substrate (z axis). Equation 4.17 then becomes

$$\Delta c = \frac{d^2 c}{dz^2} = 0 \tag{6.10}$$

whose solution is a linear variation for the water vapor profile. The boundary conditions in Eqs. 4.18 and 4.19 becomes $c_s = c(z = h)$ and $c_0 = c_\infty = c(z = h + \varsigma)$, where ς is the diffusion boundary layer. One eventually finds

$$c = c_s + (c_\infty - c_s)\left(\frac{z - h}{\varsigma}\right) \tag{6.11}$$

It follows from the volume evolution, Eq. 4.21

$$\frac{dV_T}{dt} = S_c\frac{dh}{dt} = \frac{D}{\rho_w}\left(\frac{dc}{dz}\right)_h S_c \tag{6.12}$$

leading to

$$h = \frac{D(c_\infty - c_s)}{\rho_w \varsigma}t \tag{6.13}$$

In this equation,

$$\frac{dh}{dt} = \frac{D(c_\infty - c_s)}{\rho_w \varsigma} \tag{6.14}$$

is the condensation rate per unit surface. The relationship Eq. 6.6 implies that V_i is proportional to h, and hence to t. It follows that

$$R_i = (A_{pi}t)^\beta \text{ with } \beta = 1/3 \tag{6.15}$$

In this relation,

$$A_{pi} = \frac{D(c_\infty - c_s)}{\pi G(\theta_c)\rho_w \varsigma}A_i \tag{6.16}$$

The same results as already noted above in Sect. 6.1.1 is found (Fig. 6.4): growth exponent in a pattern (1/3) is lower than the growth exponent of an isolated drop (1/2) with the same boundary conditions (Sect. 4.2.4, Eq. 4.45). The smaller exponent in the pattern also reflects the competition between drops to collect water monomers.

Relation Eq. 6.15 holds as long as $\langle d \rangle$ remains constant, that is, as long as the drop growth does not lead to coalescence events. Such coalescence events occur for $2\langle R \rangle > \langle d \rangle$. Drop coalescence then modifies in depth the growth laws, as described in the next Sect. 6.3.

Fig. 6.4 Drop volume
($V \sim R^{1/3}$) evolution (log–log
plot in arbitrary units)
showing single drop growth
with exponent 3/2 (1/2 for
radius, Eq. 4.45) and pattern
drop growth with exponent 1
(1/3 for radius, Eq. 6.15).
Insets: analyzed drops.
(Adapted from Sokuler et al.
2010a, with permission)

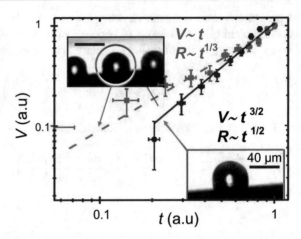

6.2 Individual 3D Drop Growth in a 1D Pattern

This situation is concerned with, e.g. drops growing around a thread (length d_c),
very thin with respect to the drop radius. Drops are spherical. The individual concentration profiles overlap to form a cylindrical profile (Fig. 6.5). Similar to the
situation above in Sect. 6.1.2 and adapting Eq. 6.6, one can define an equivalent
cylindrical film layer of thickness h that relates to individual drop radius R_i such
as

$$h^2 = \frac{V_T}{\pi d_c} = \frac{V_i}{\pi d_i} = \frac{4R_i^{\,3}}{3d_i} \tag{6.17}$$

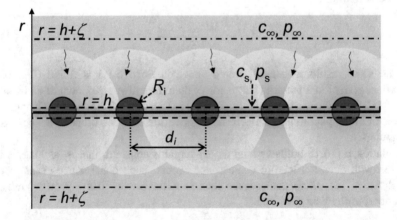

Fig. 6.5 Schematics of droplet growth in a 1D pattern on a thread with overlapping water vapor
concentration profile forming a cylindrical profile and equivalent film of average thickness h.
Notations: see text

In the equivalent film approximation, Eq. 4.38 can be written in cylindrical coordinates. The profile being symmetrical around the thread at $r = 0$ and invariant along the axis, only remains the r-dependence:

$$\Delta c = \frac{1}{r}\left(r\frac{\partial c}{\partial r}\right) = 0 \tag{6.18}$$

The solution is a logarithmic variation for the water vapor profile. The boundary conditions are similar to Eqs. 4.18 and 4.19, $c_s = c(r = h)$ and $c_\infty = c(r = h + \varsigma)$. One eventually finds

$$c = c_\infty - \frac{c_\infty - c_s}{\ln(1 + \varsigma/h)}\ln(r/h) \tag{6.19}$$

It follows from the volume evolution, Eq. 4.21

$$\frac{dV_T}{dt} = 2\pi d_c h\frac{dh}{dt} = 2\pi h\frac{D}{\rho_w}\left(\frac{dc}{dr}\right)_h d_c \tag{6.20}$$

leading to

$$h\ln(1 + \varsigma/h)\frac{dh}{dt} = \frac{D}{\rho_w}(c_\infty - c_s) \tag{6.21}$$

This equation can be solved exactly, however, if one neglects the log dependence of h, it becomes

$$h \approx \left[\frac{2D}{\rho_w}(c_\infty - c_s)\right]^{1/2} t^{1/2} \tag{6.22}$$

From Eq. 6.16–1, one finds for the droplet radius

$$R_i \approx \left(\frac{3d_i}{4\pi}\right)^{1/3}\left[\frac{2D}{\rho_w}(c_\infty - c_s)\right]^{1/3} t^{1/3} \tag{6.23}$$

As a result of the concentration gradient at the film surface in $\approx 1/h$, the film growth rate decreases with time. Similar to equivalent film growth on a 2D substrate, the total condensed volume $V_t \sim h^2 \sim t$ and is thus proportional to time. In contrast to other surfaces, note that the condensation surface $S_c = 2\pi h d_c$ is not constant and increases with time.

6.3   3D Drop Pattern Evolution with Coalescence

In this next growth stage, drops eventually touch each other and coalesce. It corresponds to drops having a critical radius R_c such as $2R_c \geq \langle d_0 \rangle$, the mean spacing between the nucleation sites. One can roughly estimate R_c by assuming that the nucleation sites are distributed on a square lattice. With n_s the number of nucleation sites per unit surface, $R_c = 1/2\sqrt{n_s}$.

Coalescence is the key process that leads to a self-similar growth of the drop pattern and acceleration of its evolution. Most descriptions of drop pattern evolution when coalescences rescale the individual drop growth, dated from the late 1980s and early 1990s. A common result is the following relation between the growth laws concerning a single drop where cap radius $R \sim t^\beta$ and the mean droplet in a pattern with mean cap radius $\langle R \rangle \sim t^\gamma$:

$$\gamma = 3\beta \tag{6.24}$$

However, some theoretical and/or numerical approaches do not correspond to the 3D diffusion growth process as encountered here. For instance, Rogers et al. (1988), assumed a 2D diffusion process, while Family and Meakin (1988) assumed a rain-like process. Family and Meakin (1989), Meakin and Family (1989) and Briscoe and Galvin (1990) assumed an ad hoc growth law exponent that can be varied in the model in order to test the theoretical predictions.

A simplified approach is proposed in Sect. 6.3.2, which captures the main features of the process. A more rigorous approach for any drop and substrate dimensionality is given in Sect. 6.4.

6.3.1   Constancy of Surface Coverage

Let us consider a drop (i), whose surface coverage is $S_i = \pi R_i^2$. This coverage increases with drop growth, leading unavoidably to a coalescence event with another drop. It appears, however, that coalescence lowers the drop surface coverage. The drop that results from the coalescence of the two "parent" drops indeed exhibits a surface coverage lower than the sum of the parents' coverage because the drop grow above the surface, in the third dimension.

Let us consider simplifying two parent drops of same radius R and volume V, and neglect the contact angle hysteresis effects. The new drop resulting from coalescence has the same geometry than the parents. The new drop thus exhibits a surface coverage S' lower than the parent drop surface coverage $(2S)$.

$$S' = \pi R'^2 = 2^{2/3}\pi R^2 < 2 \times \pi R^2 = 2S \tag{6.25}$$

For drops i and j with different radii R_i and R_j and surface coverages S_i and S_j, the new radius after coalescence becomes

$$R' = \left(R_i^3 + R_j^3 \right)^{\frac{1}{3}} \tag{6.26}$$

Equation 6.25 transforms into

$$S' = \pi R'^2 = \pi \left(R_i^3 + R_j^3 \right)^{\frac{2}{3}} = \pi \left(R_i^2 - 2R_i^{-1} R_j^3 - \cdots + R_j^2 \right)$$

$$< \pi \left(R_i^2 + R_j^2 \right) = S_i + S_j \qquad (6.27)$$

The total surface area S_T covered by the drops on the substrate

$$S_T = \pi \sum R_i^2 \qquad (6.28)$$

It is thus the balance of droplet growth (which increases surface coverage) and droplet coalescences (which decrease surface coverage). This eventually leads to constant surface coverage during the pattern evolution. The 2D surface coverage fraction

$$\varepsilon_2 = \frac{S_T}{S_c} = \frac{\pi \sum R_i^2}{S_c} \qquad (6.29)$$

therefore becomes constant ($\varepsilon_2 = \varepsilon_{2,\infty}$), see Fig. 6.6. The limiting value for hemispherical drops is close to the random packing limit at 2D (e.g. throwing randomly disks on a plane and removing disks which overlap) whose value is about 55% (Solomon, 1967; Hinrichsen et al., 1986). In both experiments and simulations, the value $\varepsilon_{2,\infty} \approx 0.57$ is found (Figs. 6.6 and 6.8b). The weak deviations from random packing limit value is due to the existence of correlations between drops, see Figs. 6.10 and 6.11.

Constant surface coverage (from which follows scaling in the growth) therefore comes from the fact that, after coalescence, the surface wetted by the new drop on the substrate is smaller than the surface wetted by the coalescing drops. Equations 6.25 and 6.27 can be generalized for two drops with dimensionality D_d coalescing on a substrate with dimensionality D_s. Assuming for simplicity that the

Fig. 6.6 Experimental evolution of surface coverage ε_2 and polydispersity $g = \sigma/\langle R \rangle$ (see Eq. 6.36). The limiting value is $\varepsilon_{2,\infty} \approx 0.57$. (Adapted from Fritter et al. 1991)

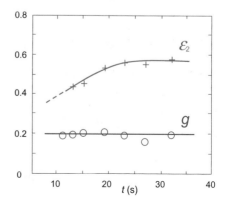

drops have same radius, the straightforward generalization of Eq. 6.25 gives for
the new wetted surface S' compared to the initial drop wetted surface S:

$$\frac{S'}{2S} = 2^{\frac{D_s}{D_d}-1}$$

(6.30)

The condition $S'/2S < 1$ needed for constant surface coverage and scaling in
growth therefore corresponds to

$$D_s < D_d$$

(6.31)

In the time regime with no drop coalescence, the number of drops is constant
and equal to the number of nucleation sites, N_s. Expressing (see Appendix G, Eq.
G.4) the mean drop radius from length-weighted moment (see Appendix G, Eq.
G.4),

$$\langle R \rangle = \frac{\sum_{i=0}^{N_s} R_i^2}{\sum_{i=0}^{N_s} R_i} = \left(\frac{\sum_{i=0}^{N_s} R_i^2}{n_s} \right) \left(\frac{N_s}{\sum_{i=0}^{n_s} R_i} \right) = \left(\frac{\sum_{i=0}^{N_s} R_i^2}{N_s} \right) \frac{1}{\langle R \rangle}$$

(6.32)

This expression is valid for weak drop polydispersity, as it is the case here (see
Figs. 6.6 and 6.8b) one obtains (from Eqs. 6.29 and 6.30)

$$\varepsilon_2 = \frac{\pi N_s \langle R \rangle^2}{S_c} \sim t^{2\beta}$$

(6.33)

Such behavior has been observed in numerical simulations, see Fig. 6.8b.

When the contact angle hysteresis is not negligible, the new drop remains
pinned at some place during the coalescence process. A detailed description of
the process is given in Sect. 6.5. It considers the new drop as being homomorphic from the parent drops, although the drops shape is no more a spherical
cap (see Fig. 6.18). Surface coverage, although showing value different from
$\varepsilon_2 \approx 55\%$, remains also constant but dependent on the drop contact angle
$(\epsilon_2 \approx 1 - \theta_c(\text{deg.})/180)$, see Eq. 6.78).

6.3.2 Mean Radius Growth Law

6.3.2.1 Simple Approach

This approach leads to the main results concerning drop law exponents. When
accounting for such coalescence events, the mean radius evolution of the droplet
pattern can be obtained in terms of area-weighted moments (see Appendix G, Eq.
G.5):

$$\langle R^{(3)} \rangle = \frac{\langle R^3 \rangle}{\langle R^2 \rangle}$$

(6.34)

and (Eq. G.8)

$$\langle R^{2(4)} \rangle = \frac{\langle R^4 \rangle}{\langle R^2 \rangle} \tag{6.35}$$

assuming that the droplet polydispersity, defined as

$$g = \frac{\sigma}{\langle R \rangle} \tag{6.36}$$

remains small (as is the case, see Fig. 6.6, experiments, and Fig. 6.8b, simulation) where $g \approx$ constant ≈ 0.2 or less). Here σ is the standard deviation

$$\sigma = \langle (R - \langle R \rangle)^2 \rangle^{1/2} = \left(\langle R^2 \rangle - \langle R \rangle^2 \right)^{1/2} \tag{6.37}$$

In this case,

$$\langle R^{(p)} \rangle = \frac{\langle R^p \rangle}{\langle R^{p-1} \rangle} \approx \langle R \rangle \tag{6.38}$$

and

$$\langle R^{2(p)} \rangle = \frac{\langle R^p \rangle}{\langle R^{p-2} \rangle} \approx \langle R^2 \rangle \tag{6.39}$$

Going back to the mean radius as written in terms of area-weighted moments Eq. 6.34, the numerator can be expressed in terms of total volume $V_T = \pi G(\theta_c \sum R_i{}^3$ (Appendix D, Eq. D.8), which itself can be written as a function of equivalent film thickness $h = \frac{V_T}{S_c}$, Eq. 6.6. The denominator of Eq. 6.34 can in turn be expressed in terms of surface coverage fraction ε_2 from Eq. 6.33. It eventually gives the following paradoxical relation, where the mean radius is proportional to the total condensed volume per unit surface

$$\langle R \rangle = \frac{1}{\varepsilon_2 S_c} \frac{V_T}{G(\theta_c)} = \frac{h}{\varepsilon_2 G(\theta_c)} \tag{6.40}$$

This proportionality is due to the fact that, during growth the mean distance $\langle d \rangle$ between drops, as defined in Eq. 6.7, remains proportional to the mean radius

$$\langle R \rangle = \sqrt{\frac{\varepsilon_2}{\pi}} \langle d \rangle \tag{6.41}$$

This relation results from the constant drop surface coverage. Let us consider the expression of the mean radius according to Eqs. G.1 and G.4 in Appendix 22:

$$\langle R \rangle = \frac{\sum R_i^2}{\sum R_i} = \frac{\sum R_i}{N} \tag{6.42}$$

Equation 6.33, which defines the surface coverage fraction, can then be written as

$$\varepsilon_2 = \frac{\pi \sum R_i^2}{S_c} = \frac{\pi \left(\sum R_i\right)^2}{N\langle d\rangle^2} = \pi \left(\frac{\langle R\rangle}{\langle d\rangle}\right)^2 \tag{6.43}$$

The $< R >$ growth law from Eqs. 6.40 can be written as

$$\langle R\rangle = \frac{h}{\varepsilon_2 G(\theta_c)} = k_P t \tag{6.44}$$

where the radius growth rate k_P, using Eqs. 6.13 and 6.14, is

$$k_P = \frac{1}{\varepsilon_2 G(\theta_c)} \frac{D(c_\infty - c_s)}{\rho_w \zeta} \tag{6.45}$$

When writing Eq. 6.44 in terms of growth law exponent γ

$$\langle R\rangle = k_P t^\gamma \text{ with} \gamma = 1 \tag{6.46}$$

and comparing with the single drop growth in the pattern Eq. 6.15, one deduces

$$\gamma = 3\beta \tag{6.47}$$

Linear growth ($\gamma = 1$) of mean radius is indeed well observed (Fig. 6.7).

Let us now consider the drop radius distribution polydispersity g (Eq. 6.36). In the self-similar regime, g is slightly less than 20% (see experiments in Fig. 6.7 and simulation in Fig. 6.8).

6.3.2.2 Simulations

Dropwise condensation including coalescence can be simulated. Many studies have been performed (Leach et al. (2006) and see the review by Singh et al. (2019)). The first simulation is attributed to Gose et al. (1967). Some authors (Tanasawa and Tachibana,1970); Glicksman and Hunt,1972; Rose and Glicksman, 1973) have explored the last, steady state of condensation when drops depart by gravity from the substrate (see Sect. 6.6.3) but did not describe the self-similar growth regime.

Typically, in the simulations, a computational grid is filled with randomly placed nucleated drops and their size is increased according to a dedicated growth law. When adjacent drops touch, a new drop is formed, generally at the center of mass (Rose and Glicksman, 1973; Briscoe and Galvin, 1991; Fritter et al., 1988; 1991; Steyer et al., 1990; Leach et al., 2006). In the experiments (see Andrieu et al., 2002 and Fig. 5.16), the new droplet is found to be centered at the center of mass of the coalescing pair. This is the location indeed chosen in most of the simulations but studies can be carried out with a computationally simpler procedure in which the new droplet is centered on the site of the larger coalescing

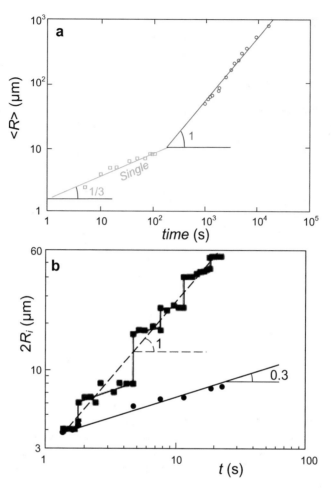

Fig. 6.7 Experimental evolution of droplet size showing **a** the single droplet power law $R_i \sim t^{1/3}$ (single, Eq. 6.15) and mean droplet power law $<R> \sim t$ (planar substrate, $D_s = 2$, Eq. 6.46). (Adapted from Steyer et al. 1990). **b** Effect of coalescence (steps) for a planar substrate; between coalescence growth is in $t^{1/3}$ (single, Eq. 6.15), while the mean growth is in t (Eq. 6.46). (Adapted from Beysens and Knobler 1986)

droplet following Gose et al. (1967). However, this situation increases the ability for the coalescing drops to coalesce with others as discussed by Fritter et al. (1988). Continuous nucleation of drops can be envisaged (Ulrich et al., 2004; Family and Meakin, 1988; Meakin, 1992), occasionally with dynamical lattice rescaling (Glicksman and Hunt, 1972). To simplify computational work, simulations with more complex growth laws can limit the number of nucleation sites (Gose et al., 1967; Tanasawa and Tachibana, 1970) or the simulation duration (Burnside and Hadi, 1999).

Fig. 6.8 Simulation of the
evolution of a pattern of
hemispherical drops with
respect to reduced time τ^*
(Eq. 6.49). **a** Mean reduced
droplet radius ρ^* (Eq. 6.50)
showing the cross-over
between exponents $\beta(=1/3)$
and $3\beta(=1)$. **b** Surface
coverage fraction ε_2
(Eq. 6.22) and radius
polydispersity g (Eq. 6.25)
with late time asymptotic
values $\varepsilon_{2,\infty}= 0.57$ and $g_\alpha =$
0.17 (interrupted lines). The
continuous line corresponds
to 2β $(= 2/3)$ (see text).
(Adapted from Fritter et al.
1991)

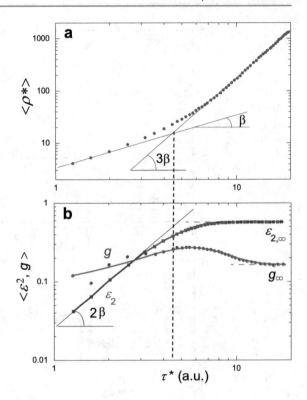

The effect of drop coalescence on a substrate is usually performed by considering initially disks randomly placed on a planar surface, as done by Fritter et al. (1988; 1991). The disks are sequentially placed at random positions while rejecting coordinates that would result in overlaps. The time is then incremented and the radii of the disks increase at each time step according to a power law with exponent μ

$$R_i^{new} = R_i^{old}\left(\frac{t_{new}}{t_{old}}\right)^\beta = R_i^{old}\left(1 + \frac{\delta t}{t_{old}}\right)^\beta \qquad (6.48)$$

Here, $\delta t = t_{new} - t_{old}$. It is generally convenient to work with a reduced time

$$\tau^* = (t_{new}/t_{old})^\beta \qquad (6.49)$$

In this case, the simulation is independent of the choice of the β value. It is also convenient to define a reduced radius

$$\rho^* = \langle R\rangle/\langle R\rangle_0 \qquad (6.50)$$

where $< R >$ is the area-weighted average radius defined by $< R >= \frac{\sum R_i^3}{\sum R_i^2}$ and $<R>_0$ is its value at the initial time. When two disks (radii R_{i-1} and R'_{i-1})

touch, they coalesce and are replaced by a new disk centered at their mutual center of mass, whose radius R_i is computed on the assumption of mass conservation

$$R_i = \left(R_{i-1}^{D_d} + R_{i-1}'^{D_d} \right)^{1/D_d} \tag{6.51}$$

The search for coalescences is carried out by calculating the distance between the centers of two disks and comparing it to the sum of the radii. After the coalescence takes place, a check is made for contacts with other droplets that occur as the result of the coalescence. When these have been dealt with, the time is again incremented and single droplet growth continues until the next contact and coalescence. Periodic boundary conditions are utilized to eliminate edge effects.

Note that some simulations simplify the process by considering drops of equal length and/or mean-field approximation. In the latter, the correlations between drops are neglected and the iterating configurations are random (see Derrida et al., 1990; 1991 and Sect. 9.2).

Figure 6.8a shows the evolution of the reduced radius ρ^*. The exponent β representing the single drop growth between coalescences is taken to be $\beta = 1$. The cross-over between the growth regimes without coalescences (exponent β) and with coalescences (exponent 3β) occurs for $\varepsilon_2 \sim 0.35$ when comparing with Fig. 6.8b. The cross-over with the single drop growth regime with exponent $1/2$ and pattern growth without coalescences with exponent $1/3$ is dependent on the number density of nucleation sites (see Sect. 6.3).

Figure 6.8b presents the evolution of the surface coverage fraction ε_2 with the reduced time τ^* (random initial condition of 10^5 monodisperse drops with 3% coverage). The initial behavior follows Eq. 6.33 with exponent 2β. The limiting ε_2 value (0.57) is slightly larger than the 2D random packing limit (0.55) because some correlations are present between drops (see Fig. 6.8).

Figure 6.9a is concerned with the *radius drop distribution*, which is bimodal. The small contribution at low R results from a few droplets that do not coalesce and therefore grow more slowly than the mean radius $< R >$. Due to coalescence, the distribution is broadened into a bell-shaped curve. The most characteristic feature of the distribution is that the position of the peak moves to larger sizes with increasing time. At the same time, the amplitude of the peak moves down indicating that the density of large droplets decreases. Figure 6.9b corresponds to the same information but is concerned with the droplet volume $V \sim R^3$ and is in log–log scale to evidence the power law decrease of peak amplitude with time (exponent θ) and scaling of the distribution peak with $< V >$

$$n(V,t) = V^{-\theta} f\left(\frac{V}{\langle V \rangle} \right) \tag{6.52}$$

The mean drop volume evolution with coalescence follows the power law evolution $\langle V \rangle \sim \langle R^{D_d} \rangle \sim t^{D_d}$ corresponding to steady condensation. This scaling

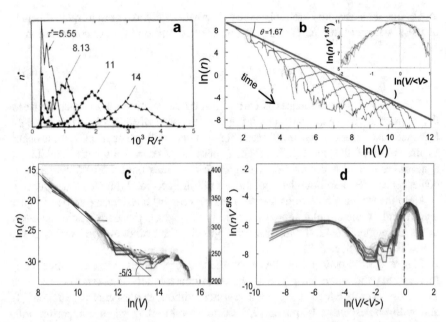

Fig. 6.9 Drop size distribution. (**a–b**) Simulations. **a** Drop radius distribution at different time τ^* (Eq. 6.49). Abscissa has been divided by τ^* to factor out growth not due to coalescence, and the ordinate n^* has been normalized to give an area of unity under each curve. Drop radius and time are in arbitrary units. (Adapted from Fritter et al., 1991). **b** Droplet volume V distribution at different times showing power law decrease with exponent $\theta = 5/3 \approx 1.67$ (log–log plot). Inset shows scaling. (Adapted from Meakin and Family, 1989, with permission). (**c–d**) Experimental data. **c** Self-similar evolution of the droplet volume (V) distribution (n). The triangles mark the estimates for average drop sizes (peaks in (b)). The colors correspond to different times (in s). **b** Distribution scaled obtained using Eq. 6.52. (Adapted from Baratian et al. 2018, with permission)

function has been proposed by Meakin and Family (1989) as a generalization of scaling for aggregation of clusters (Vicsek and Family, 1984).

One notices that $n(V, t)$, representing the number of droplets per droplet volume and surface area, has a dimension of length to the power $-(D_d + D_s)$ and must be equal to the dimension of the right-hand side of Eq. 6.52, that is, $D_d + D_s = \theta D_d$, giving eventually

$$\theta = 1 + \frac{D_s}{D_d} \tag{6.53}$$

For the usual cases where $D_d = 3$ and $D_s = 2$, $\theta = 5/3 \approx 1.67$ (see Fig. 6.9b, c and d). More rigorous treatments are given by Family and Meakin (1988; 1989), Meakin and Family (1989), and Meakin (1992). Blashke et al. (2012) improved the scaling description by considering a lower cutoff. The cutoff enables the weak distortion to scaling at small V to be accounted for.

A necessary condition for self-similarity is the scaling behavior of the structure factor or equivalently of the pair correlation function. The average distance

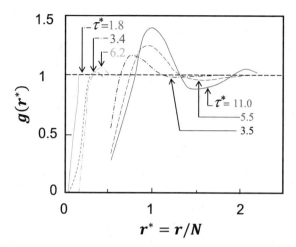

Fig. 6.10 Pair distribution function $g(r^*)$ against the reduced interdroplet distance $r^* = r/N^{-1/2}$ (Eq. 6.54) at several times during simulation (initial random condition of 10^5 monodisperse drops with 3% coverage $\tau^* = 3.5, 5.5, 11$ with τ^* from Eq. 6.49). Results of the simulation for an initial random condition of 10^4 monodisperse disk with 0.3% coverage are also shown ($\tau^* = 1.8, 3.4, 6.2$). (Times are not directly comparable since the initial coverages at $\tau^* = 1$ are different). (Adapted from Fritter et al. 1991)

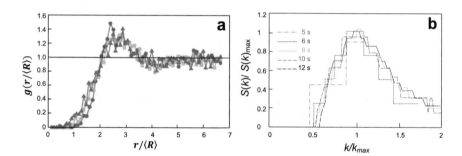

Fig. 6.11 **a** Experimental pair function $g(r/\langle R \rangle)$ vs $r/\langle R \rangle$ showing scaling for different surface coverages (circles: coverage 39%; square: coverage 42%; triangles: coverage 45%). (Polymerized nanodroplets growing from an oversaturated liquid phase, adapted from Xu et al., 2017). **b** Experimental scaled structure factors $S(k)/S(k)_{max}$ vs ratios of wavevectors k/k_{max} at several times showing scaling. (water condensation, adapted from Fritter et al. 1991)

between disk centers follows (Eq. 6.8) $\langle d \rangle \sim N^{-1/2}$, where N is the number of droplets. This gives a characteristic length to define a reduced distance between drop centers

$$r^* = r/N^{-\frac{1}{2}} \sim r/\langle d \rangle \tag{6.54}$$

The radial or pair distribution function $g(r^*)$ of the center-to-center distances is shown as a function of interdroplet distance in Fig. 6.10 for different times. The initial disks are of negligible radius and randomly placed. At the start of the simulation, there is no correlation between the droplets and $g(r^*)$ is unity. As coalescences proceeds, the short center-to-center distances are preferentially eliminated and a short-range cutoff appears in the distribution function (at $r^* \sim 1/2$). At the same time, local order develops, as evidenced by the peak that arises in $g(r^*)$. As the system enters the regime in which the 3 β growth law applies, the maximum in the peak increases and there is evidence of a second peak, indicating that the correlation is increasing. At longer times, the distribution function approaches the $g(r^*)$ for systems in which the coalescence mechanism is well developed, with two peaks close to $r = N^{1/2}$ and $r = 2N^{1/2}$. Once the 3 β growth is well established, $g(r^*)$ does not change and the structure is therefore self-similar. The shape of $g(r^*)$ at small r^* indicates that the development of this local order seems to be due to an interaction which, in contrast to the hard-sphere model, should be infinitely attractive at the point of contact.

6.3.3 Droplet Number

From Eq. 6.33, which defines the surface coverage fraction, it comes with $\langle d \rangle$ the mean interdrop distance as defined in Eq. 6.7 and assuming the drop to be on a square lattice,

$$\varepsilon_2 \approx \pi \left(\frac{\langle R \rangle}{\langle d \rangle} \right)^2 \tag{6.55}$$

It follows a relationship between the number of drops, N, and the mean radius $\langle R \rangle$ since, from Eq. 6.8, $\langle d \rangle = S_c^{1/2} N^{-1/2}$

$$N = \frac{\varepsilon_2 S_c}{\pi} \langle R \rangle^{-2} = A_N t^{-2} \tag{6.56}$$

where

$$A_N = \frac{(\varepsilon_2)^3 S_c}{\pi} \left[\frac{G(\theta_c) \rho_w \zeta}{D(c_\infty - c_s)} \right]^2 \tag{6.57}$$

6.4 Self-Similar Regime for Any Drop and Substrate Dimensionalities

6.4.1 Scaling

It is interesting to note that Eq. 6.23 obtained for 3D drops condensing on a 2D surface can be generalized to any drop dimensionality D_d and surface dimensionality D_s. According to Appendix G, Eq. G.6, the mean radius in the drop pattern can be written as

$$\langle R \rangle = \frac{\sum R_i^{D_d}}{\sum R_i^{D_d-1}} \tag{6.58}$$

The above equation can be developed as

$$\begin{aligned}
\langle R \rangle &= \sum R_i^{D_d} \cdot \frac{\sum R_i^{D_d-2}}{\sum R_i^{D_d-1}} \cdot \frac{\sum R_i^{D_d-3}}{\sum R_i^{D_d-2}} \cdot \frac{\sum R_i^{D_d-4}}{\sum R_i^{D_d-3}} \cdots \frac{\sum R_i^{D_s}}{\sum R_i^{D_s+1}} \cdot \frac{1}{\sum R_i^{D_s}} \\
&= \frac{\sum R_i^{D_d}}{\sum R_i^{D_s}} \cdot \frac{1}{\langle R \rangle^{D_d-D_s-1}}
\end{aligned} \tag{6.59}$$

which leads to

$$\langle R \rangle^{D_d-D_s} = \frac{\sum R_i^{D_d}}{\sum R_i^{D_s}} \tag{6.60}$$

The numerator is proportional to the total condensed volume V_t and the numerator to the drop surface coverage ε_{D_s}. Assuming the steady state of condensation where $V_i \sim V_t \sim t$ and self-similar growth where ε_{D_s} remains constant, it becomes.

$$R_i \sim V_i^{1/D_d} \sim t^\beta \text{ with } \beta = 1/D_d \tag{6.61}$$

$$\langle R \rangle \sim t^\gamma \text{ with } \gamma = \frac{1}{D_d - D_s} \tag{6.62}$$

The relation between γ and β can thus be expressed as

$$\gamma = \frac{D_d}{D_d - D_s} \beta \tag{6.63}$$

Another approach can be followed according to Viovy et al. (1988). It is based on the recognition that individual droplet growth laws and the position of the drop after coalescence follow renormalization.

Since the law for the evolution of individual droplets is known (Eq. 6.15), the evolution of an assembly of droplets can be predicted. Let us consider a particular realization at time $t = 0$

$$\left\{ \vec{R}_i, R_i \right\} t = 0 \tag{6.64}$$

and another homomorphic realization with factor λ at same time

$$\left\{ \lambda \vec{R}_i, \lambda R_i \right\} t = 0 \tag{6.65}$$

From droplet growth law Eqs. 6.15 and 6.16, the following renormalization

$$\vec{R}_i \rightarrow \lambda \vec{R}_i$$

$$A_i \rightarrow \lambda^2 A_i$$

$$R_i \rightarrow \lambda R_i$$

$$t \rightarrow \lambda t \tag{6.66}$$

leaves unchanged the evolution equation for growth and condensation, since the equation for the occurrence of the first coalescence involving droplet i

$$\min_j \left\{ \left| \vec{R}_i - \vec{R}_j \right| - [R_i(t) + R_i(t)] \right\} = 0 \tag{6.67}$$

Equation 5.66 for inertia and Eq. 5.68 for dissipation-limited coalescence are also invariant upon the renormalization Eq. 6.66. The evolution of the entire pattern is thus invariant for at least these two limiting coalescence cases. Individual realizations of the system can therefore be separated into homomorphic time-dependent series of patterns:

$$\left\{ \lambda \vec{R}_i, \lambda R_i, \lambda t \right\} \tag{6.68}$$

If we assume that the pattern of nucleation sites on the substrate is random and "fine grained" (i.e. smaller than the mean distance between droplets at the beginning of the coalescence-dominated growth phase), then the range of correlation is in the order of the droplet size (as observed in the correlation function Figs. 6.10 and 6.11a), and all realizations are statistically independent on the "large system limit". In other words, the pattern of a realization at time t_2 can always be considered as homomorphic with a subdomain of a statistically equivalent realization at any given time t_1 [in regime (iv)]. This ensures self-similarity of the pattern, as indeed observed in the simulations of pair correlation function (Fig. 6.10) and in the experiments (Fig. 6.11a) or equivalently its Fourier transform, the drop pattern structure factor $S(k)$ (Fig. 6.11b). Comparisons of pair correlation functions in the simulations and experiments show good agreement. Note that this self-similarity is the direct consequence of the scale invariance of the growth processes involving one droplet.

Let us now consider a mean-field approach by using the spatial average values of drop volume and drop radius. A general relation can be drawn between the exponent β of droplet growth between coalescence $\langle R \rangle \sim t^\beta$ and the exponent γ of the mean droplet radius including coalescence $\langle R \rangle \sim t^\gamma$.

The condensed volume between coalescences obeys the growth law

$$\langle V \rangle \sim \langle R \rangle^{D_d} \sim t^{\beta D_d} \tag{6.69}$$

This volume is also, when considering coalescence, related to the N droplets on the substrate through

$$\langle V \rangle \sim N \langle R \rangle^{D_d} \tag{6.70}$$

Coalescence makes N to scale as

$$N \sim \langle R \rangle^{-D_s} \tag{6.71}$$

It results from Eqs. 6.70 and 6.71:

$$\langle V \rangle \sim \langle R \rangle^{D_d - D_s} \sim t^{\gamma(D_d - D_s)} \tag{6.72}$$

Since the condensed volume is the same, including or not the coalescence events, it follows, from comparing Eqs. 6.69 and 6.72:

$$\gamma = \beta \frac{D_d}{D_d - D_s} \tag{6.73}$$

This relation is the same as Eq. 6.63. The same results were obtained by Meakin and Family (1988) according to a scaling approach derived from cluster aggregation.

6.4.2   Various Drop and Substrate Dimensionalities

The scaling relationship Eqs. 6.63 or 6.73 between growth exponents β (without coalescences) and γ (with coalescences) depends solely on drop and substrate dimensionalities. The validity of the relationship can therefore be tested by making substrate and drop dimensionality to vary in experiments and/or simulations.

6.4.2.1   3D Drops on 1D and 2D Substrates

There are no examples of experiments on threads or fibers, because contact cooling cannot ensure uniform and stable thermal conditions. One notes an interesting study of non-uniform condensate morphologies on a cantilever due to the coupling between vapor and heat diffusion by Zhu et al. (2020).

Radiative cooling could be a good alternative (see e.g. Trosseille, 2019; Trosseille et al. 2021a). As a matter of fact, natural dew occurs on spider webs.

Condensation has been studied by Steyer et al. (1990) on biological materials (iris leaf) naturally made of parallel fibers where condensation preferably occurs in lines (Fig. 6.12a). Linear scratches on silanized glass also offered 1D substrates (Fig. 6.12b). In these latter cases, individual drop growth laws therefore correspond to a 2D substrate and only drop coalescence is constrained on a 1D substrate (Fig. 6.12).

Figure 6.12cd presents the individual growth of droplets without coalescence (exponent 1/3, time $t < t_1$), growth with 1D coalescences after time t_1 on scratches (exponent 1/2), and growth with 2D coalescences after time t_2 (exponent 1). After time t_3, 1D growth becomes 2D. This is because the substrates are 2D in nature and, when the drop size exceeds a typical distance $d_0 \approx 25$ μm corresponding to the line mean spacing, coalescence occurs between lines and growth becomes 2D everywhere. The transition from no coalescences to 1D coalescences ($t_1 \approx 50$ s) corresponds to line coverage $\varepsilon_1 \approx 68\%$. The exponent values are in agreement with what is expected from Eq. 6.61: no coalescence, ($\beta = 1/D_d = 1/3$) and Eqs. 6.61, 6.73: 1D coalescences, $\gamma = D_d/(D_d - D_s)\beta$ 1/2; 2D coalescences, ($\gamma = 3\beta = 1$).

The limiting line surface coverage value on a line of length L is defined as

$$\varepsilon_1 = \frac{2 \sum R_i}{L} \tag{6.75}$$

The measured value found $\varepsilon_1 \approx 0.795$ is only slightly larger than the value expected from the 2D measurements $\sqrt{\varepsilon_2 = 0.57} = 0.754$ (Sect. 6.3.1) and the 1D random packing limit $\sqrt{0.55} = 0.742$.

Simulations from Steyer et al. (1990) are shown in Fig. 6.13. The simulation was performed as explained in Sect. 6.3.2.2, using 3D drops and a 1D substrate. Reduced time τ^* (Eq. 6.49) and drop radius ρ^* (Eq. 6.50) were used. The transition no-coalescences (exponent β) and 1D coalescence (exponent γ) occurs for surface coverage $\varepsilon_1 \approx 50\%$, as observed in the experiments (Fig. 6.12). Exponent $\gamma = 3\beta/2$ is found, as expected. Drop polydispersity increases with surface coverage due to enhanced coalescences and reaches a maximum at the transition between the β and γ behavior, as also observed with 3D droplets on a 2D substrate (Fig. 6.8). The limiting line surface coverage value is the same as in the experiment above, with $\varepsilon_1 \approx 0.82$, slightly larger than the expected value $\sqrt{\varepsilon_2 = 0.57} = 0.754$ (Sect. 6.3.1) and the 1D random packing limit $\sqrt{0.55} = 0.742$.

6.4.2.2   2D Drops on a 1D Substrate

Condensation on grooved substrates (see also Sects. 10.3.1 and 10.3.2) exhibits an intermediate stage where drops growing on the plateaus are constrained at the

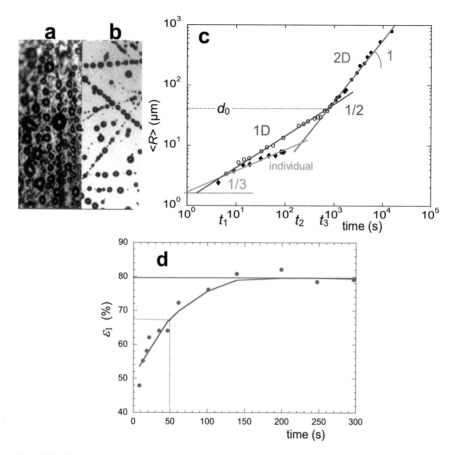

Fig. 6.12 Photos of water droplets growing with 1D coalescences. The largest dimension of the photos corresponds to 285 μm. **a** Iris leaf. **b** Silanized glass with linear scratches. (Adapted from Steyer et al., 1990). (**c–d**) Droplets growing on silanized slide **b** with linear scratches (Iris leaf gives the same results with a larger uncertainty). **c** Mean droplet radius < R > evolution, with no coalescences (individual) and with 1D and 2D coalescences. (Log–log plot). **b** Evolution of the line surface coverage ε_1 (Eq. 6.75). (Semi-log plot). Experimental conditions: $T_c = 15.7\,°C$, $T_d = 22.3\,°C$, air flux 18.5 $cm^3 s^{-1}$. (Adapted from Steyer et al. 1990)

plateau-groove edge (Fig. 6.14). Drops become elongated along the groove direction (y). Drops can grow only along y and in the direction z perpendicular to it and can thus be considered as 2D. Coalescence is constrained along the groove direction and is 1D. The drop characteristic size is taken as the length size $2R_\parallel$ along the groove direction. Before coalescence becomes important (surface coverage $\varepsilon_1 < 0.4$, see inset in Fig. 6.14b), the mean size $\langle R_\parallel \rangle \sim t^{1/D_d} = t^{1/2}$. When surface coverage increases above 0.4, growth accelerates to $\langle R_y \rangle \sim t^{1/D_d - D_s} = t^1$. The limiting line surface coverage value is $\varepsilon_1 \approx 0.83$, slightly larger than the

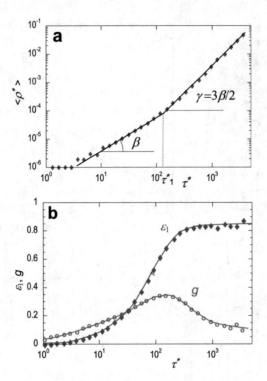

Fig. 6.13 Numerical simulation of 3D droplets growing on a 1D substrate. **a** Evolution of the mean radius. **b** Evolution of the line coverage (ε_1) and droplet polydispersity (g). (Adapted from Steyer et al. 1990)

expected value $\sqrt{\varepsilon_2 = 0.57} = 0.754$ (Sect. 6.3.1) and the 1D random packing limit $\sqrt{0.55} = 0.742$.

6.4.2.3 3D Drops on a 2.14D Grooved Substrate

Patterned substrates offer the opportunity to observe dropwise condensation on substrates with apparent dimensionality different from 1 or 2. On grooved substrates (Narhe et al. 2004a), individual drops in the steady, and final stage grow with radius:

$$R \sim t^{1/2} \tag{6.76}$$

This law is due to the fact that water vapor condenses both on drop surfaces and on groove and plateau surfaces to which the drop is connected (see Sects. 10.3.1 and 13.3.2 and Fig. 6.15). Coalescence between drops is long range from the connections by channels along the groove direction and short ranged by direct contact perpendicular to grooves.

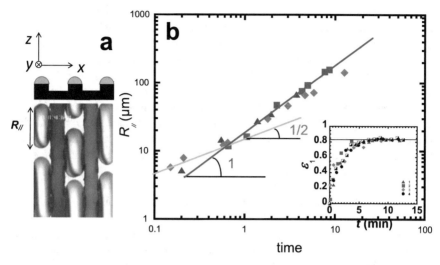

Fig. 6.14 2D droplets in the plane (y, z) growing along y and z on a 1D, grooved substrate (thickness $a = 22$ μm, spacing $b = 25$ μm, depth $c = 52$ μm). **a** Picture. **b** Evolution of the mean radius drop size parallel to the grooves, $R_{//}$. **b** Evolution of the line coverage (ε_1) and droplet polydispersity (g). (From Narhe et al., 2004a)

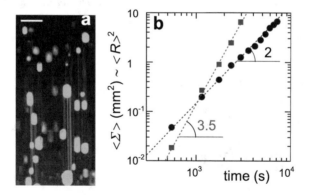

Fig. 6.15 **a** Picture of fluorinated water drops condensing on a grooved substrate (steady state, see Sect. 10.31). Growth proceeds from condensation on drops and substrate area connected to drops. Coalescence between drops is due to long-range interactions by the channels along groove direction and direct contact perpendicular to grooves. **b** Evolution of the mean drop surface in contact with the substrate, $\langle \Sigma \rangle$, proportional to the mean radius $\langle R \rangle^2$. Circles: Smooth surface $\langle R \rangle^2 \sim t^2$; squares: Grooved surface $\langle R \rangle^2 \sim t^{3.5}$. (Adapted from Bintein et al. 2019)

Figure 6.15 presents the evolution of the contact surface of drops on substrate, Σ, a quantity proportional to the square drop mean radius $\langle R \rangle^2$. On a smooth surface where $D_s = 2$ and D_d, $= 3$, the exponent $\gamma = 1/(D_d\text{-}D_s) = 1$ for the mean drop radius (Eq. 6.63), corresponding for the contact surface to $\langle \Sigma \rangle \sim t^{2\gamma} = t^2$.

On a grooved surface, it is $\langle R \rangle^2 \sim t^{3.5}$. From the relation between growth exponents with and without drop coalescences (Eq. 6.63), it is possible to deduce the effective substrate dimensionality. The growth law of a 2D drop without coalescences on grooves is indeed $R_i \sim t^{\beta=1/2}$ (Eq. 6.76). When coalescence proceeds, the growth law becomes $\langle R \rangle \sim t^{\gamma=1.75}$ (Fig. 6.15). Simple algebra using Eq. 6.63 with $D_d = 3$, $\beta = 1/2$ and $\gamma = 1.75$ thus gives $D_s = D_d(1 - \beta/\gamma) = 2.14$, a value indeed larger than $D_s = 2$. The origin of such an apparent substrate dimensionality is found in the particular long-ranged coalescence process in the groove direction, where drops are connected through the grooves. In contrast, perpendicular to the groove direction, coalescence occurs only at drop contact as on planar surfaces.

6.5 Contact Angle Dependence of Surface Coverage

A sessile drop can be characterized by the radius of its spherical cap, R, and its contact angle with the substrate, θ_c. When two drops coalesce with negligible contact angle hysteresis (see Appendix E), the resulting drop exhibits the same contact angle as the parent drops and its radius can be obtained by volume additivity, as discussed in Chap. 5. Coalescence occurs at the triple contact line of the drops when $\theta_a < 90°$ (or at a contact point when $\theta_a > 90°$, see Fig. 5.5). A bridge between drops firstly forms and very rapidly a composite drop is shaped, of nearly ellipsoidal form. Then the composite drop slowly relaxes to an equilibrium shape by moving its contact line (see Sect. 5.3.5 and Fig. 6.16).

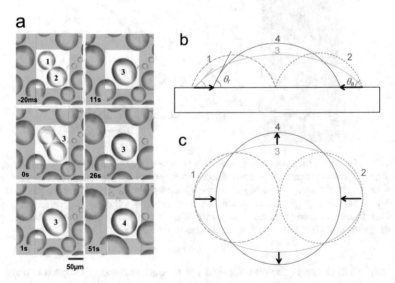

Fig. 6.16 **a** Drop coalescence. (Adapted from Narhe et al., 2004b). (**b-c**) Schematics. **b**: Side view. **c**: Front view. 1, 2: Coalescing circular parent drops. 3: Composite drop after parent coalescence, relaxing to 4, circular drop

The force F_c that moves the contact line is of capillary nature and depends on the difference in contact angle of the composite drop, $\theta_c(t)$ and the receding angle θ_r (Sect. 5.3.3.1). The contact line moves if F_c is larger than the pinning force F_s, such as

$$F_c \sim \sigma R(\cos\theta_c(t) - \cos\theta_r) > F_s \qquad (6.77)$$

Relaxation stops when θ_c reaches the receding angle θ_r. The dynamics of relaxation to a circular drop is related to the dissipation in the motion of the contact line and is 10^5–10^6 times larger than the viscous dissipation of the liquid internal flow line (see Sects. 5.3.3.1 and 5.3.5). When the capillary force is much larger than the pinning force, the final shape of the composite drop is circular. This generally happens for large θ_c values (larger than 70°, see Fig. 6.17). For smaller θ_c values, the contact line remains pinned on substrate defects during the relaxation process. The final drop after coalescence then becomes rather elliptical or triangular ($\theta_c = 40°$) or present even more complicated shapes ($\theta_c = 20°, 10°$, see Fig. 6.17).

The important feature for these small contact angles is that coalescence always leaves free some substrate area, whatever the complexity of the resulting drop is. Then ε_2 remains constant as the result of competing tendencies, growth-induced drop surface increase and growth-induced drop coalescence decrease. The surface

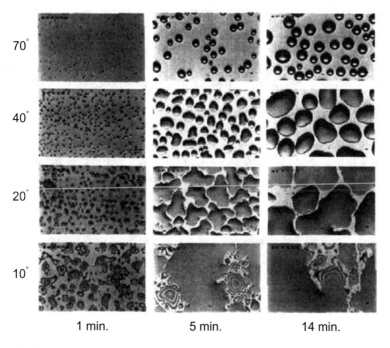

Fig. 6.17 Growth patterns of water condensing on cold silicon wafer with a coating providing different contact angle. The width of the photo is 385 µm. (Adapted from Zhao and Beysens 1995)

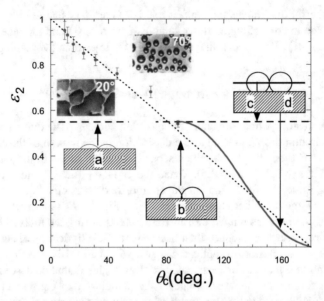

Fig. 6.18 Surface coverage variations with respect to drop contact angle (full circles). Capillary forces larger than pinning forces lead to near 2D random packing limit 0.55 surface coverage ($\theta_c \sim 90°$ and horizontal interrupted line). For $\theta_c < 90°$ (**a**), pinning increase the surface coverage. For $\theta_c > 90°$, drops coalesce by their diameter leading to 0.55 surface coverage in the plane (**c**) of the drop centers. The wetted surface (**d**) thus decreases as $0.55 \times \sin^2\theta_c$ (curve). The dotted line is Eq. 5.56. Experimental data: red full circles. (From Zhao and Beysens 1995)

coverage value, however, is larger than the value for hemispherical drops where capillary forces are much larger than pinning forces. The latter corresponds to near random packing limit case $\varepsilon_2 \approx 0.55$ of Sect. 6.3.1 where the composite drop shape is homothetic from the parent drops.

Surface coverage variations with respect to mean contact angle $\theta_c = (\theta_a - \theta_r)/2$ is reported in Fig. 6.18 from the experiments by Zhao and Beysens (1995). When $\theta_c \geq \sim 90°$, capillary forces are much larger than pinning forces. The actual surface coverage should decrease as the area of the spherical cup in contact with the substrate, that is as $0.55 \times \sin^2\theta$.

Eventually, the reduced surface coverage should follow a continuous variation from the points $\theta_c = 0°$, $\varepsilon_2 = 1$ to $\theta_c = 180°$, $\varepsilon_2 = 0$, crossing the point $\theta_c = 180°$, $\varepsilon_2 \approx 0.55$. The simplest continuous curve between these two points is linear, as was experimentally found by Zhao and Beysens (1995)

$$\epsilon_2 \approx 1 - \frac{\theta_c(\text{deg.})}{180} \tag{6.78}$$

Fig. 6.19 Different coexisting homogeneous families appearing thanks to large droplet inter-distance. (Photo Briscoe and Galvin 1989)

6.6 New Drops Generation

6.6.1 Surface Coverage

In the above stage of growth where drops grow and coalesce, drop surface coverage remains constant. Then the average droplet radius and the mean distance between the drop centers increase linearly with time $< R > \sim < d > \sim t$ (Eqs. 6.7, 6.41 and 6.44). Assuming for the sake of simplicity that the drops are set on a square lattice and making use of Eq. 6.78 to evaluate ε^2, the following relationship can be obtained:

$$\langle d \rangle = \left(\frac{\pi}{\varepsilon_2} \right)^{1/2} \langle R \rangle = \left(\frac{\pi}{1 - \frac{\theta_c}{180}} \right)^{1/2} \langle R \rangle = \left(\frac{\pi}{1 - \frac{\theta_c}{180}} \right)^{1/2} k_P t \qquad (6.79)$$

It means that at some time t_c, the interdroplet distance will be larger than the extent of the concentration profile, that is, with ζ the boundary layer thickness:

$$\langle d \rangle > 2\zeta; t > t_c = \frac{\zeta}{k_P} \left(1 - \frac{\theta_c}{180} \right)^{1/2} \qquad (6.80)$$

In this situation, nucleation of a new "family" of droplets can proceed between such remote droplets (see Fig. 6.19). This renucleation phenomenon corresponds to the failure of the assumption of a mean linear vapor concentration profile directed perpendicular to the surface for drop growth (see Sect. 6.1.2). In other words, the vapor profiles around each drop no longer overlap and nucleation of new droplets can occur between them. Equation 6.80 above gives the condition for the onset of this late stage of growth. Note that droplet nucleation on the free surface left free after coalescence proceeds from the same argument where droplets can nucleate only if the concentration profiles of coalescing drops do not overlap.

The same cascade of growth as already experienced by the former drop pattern (isolated, non-isolated, non-isolated with coalescence) is then observed. Over time, other families can again nucleate within the second family, and so on.

Although the surface coverage exhibits the same value ε_2 for each family, the total surface coverage increases. A simple calculation (Beysens et al., 1991) leads to a surface coverage $\varepsilon_{2,n}$ that increases in the order (n) of the generation as

$$\varepsilon_{2,n} = \varepsilon_{2,n-1} + \left(1 - \varepsilon_{2,n-1}\right)\varepsilon_{2,0} \tag{6.81}$$

which leads to

$$\varepsilon_{2,n} = 1 - \left(1 - \varepsilon_{2,0}\right)^n \tag{6.82}$$

When n tends to infinity, $\varepsilon_{2,n}$ tends to unity. However, the value unity, corresponding to a film, cannot be reached even approximately because (i) the drops are never at contact and (ii) gravity effects will occur (see next Sect. 6.6.2 and Chap. 14).

6.6.2 Effect of Gravity

Gravity affects drop behavior in two different manners. It depends whether the substrate is horizontal or tilted.

6.6.2.1 Horizontal Substrates

The shape of horizontal drop is modified, a hemispherical cap becoming a pancake. This is due to the non-negligible effect of hydrostatic pressure on droplet height as compared to the capillary pressure (see Appendix E, Section E.1.4). The length scale of the deformation is the capillary length $l_c = \sqrt{\sigma/\rho_w g}$ (Eq. E.6).

6.6.2.2 Inclined Substrates

On substrates making an angle α with horizontal, gravity makes the drops slide down when the drop weight becomes larger than the capillary pinning forces. It corresponds to a departure volume V_0 or radius R_0 ($\approx$ mm for usual substrates) above which all drop detach while coalescing with smaller droplets on its path. This process is detailed in Chap. 14 and Sect. 14.1.2 for a smooth surface. It leads (see Eq. 14.17) to

$$R_0 = l_c \left(\frac{k^*}{\pi G(\theta)}\right)^{1/2} \left(\frac{\cos\theta_r - \cos\theta_a}{\sin\alpha}\right)^{1/2} \tag{6.83}$$

Here, k^* is a numerical constant which depends on the precise shape of the drop. When the drop contour can be approximated by a circle, $k^* = \frac{48}{\pi^3} = 1.548$ (ElSherbini and Jacobi, 2006). It must be noted that drops generally start to slide

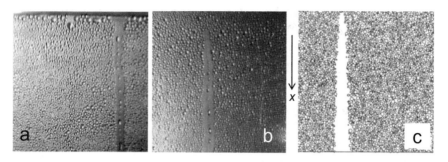

Fig. 6.20 Drop pattern with first sliding drop. Note the V-shape of the tracks in (**a, b**). **a** Experimental picture with edge effects. The drop slides from the upper edge where drops grow faster. **b** Same as (**a**) but edge effects have been suppressed on the top by a piece of blotting paper. In (**a**) and (**b**), the sliding drop leaves droplets on the substrate as resulting from the pearling transition (see text and Appendix H). **c** Simulation. (Adapted from Meakin 1992, with permission)

during a coalescence event because the contact line moves during this process, decreasing the pinning forces (Gao et al., 2018).

Meakin (1992) extended the scaling approach of Sect. 6.4.1 to gravity effects. The first sliding event, corresponding to a drop with radius R_0 and volume V_0, is assumed to occur with equal probability anywhere in the system and wipe out on its path all drops, leaving the substrate dry (see Fig. 6.20c). In reality, as can be seen in Fig. 6.20a–b, this is not exactly the case as (i) edge effects will make the drops grow faster on the top and detach sooner from this location (Fig. 6.20a and Chap. 8), (ii) a sliding drop can leave on its path smaller drops. The shape of the sliding drop is indeed deformed during their motion because the contact line cannot follow the imposed speed. With increasing velocity (e.g. due to a larger substrate tilt angle), the drop exhibits a transition from ellipsis-like shape to corner shape (see details in Appendix H). Above a critical capillary number Ca or critical velocity, the corner develops into a rivulet that, thanks to a Rayleigh-Plateau—like instability, breaks into tiny droplets (see Figs. 6.20a–b). This is the so-called "pearling transition". The radius of these droplets is a fraction of the rivulet width, itself dependent on the distance to the critical velocity at which the pearling transition occurs (see Appendix 4). The closer to the critical value, the smaller the drop diameter (Eqs. H.6 and H.8). While edge effects do not happen in all real situations and can be removed, as done in Fig. 6.20b, this is not the case for the droplets remaining on the sliding drop path coming from pearling.

If one neglects these drops, then the Meakin (1992) approach can be used. The sliding droplet collects volume from a system showing a uniform condensation volume per unit surface $h_s(V_0)$ (the equivalent condensation layer, see Sect. 6.1.2), where $h_s(V_0) \sim R_0 \sim V_0^{1/D}$ according to Eq. 6.40. Here, the drop volumes are in reality slightly less than V_0 because those drops did not yet slide. The volume increase of the sliding droplet by coalescence with drops on its path along axis x parallel to gravity (Fig. 6.20) then follows:

$$\frac{dV_0}{dx} \sim h_s R_0^{D_s-1} \sim h_s V_0^{\frac{D_s-1}{D_d}} \tag{6.84}$$

From the above equation, one readily obtains, as $V_0 \sim R_0^{D_d}$

$$\frac{dR_0}{dx} \sim h_s R_0^{D_s-D_d} \tag{6.85}$$

After integration, it comes the following increase, with x_0 the initial position of the sliding drop:

$$R_0 \sim [h_s(x - x_0)]^{\frac{1}{1+D_d-D_s}} \tag{6.86}$$

It is assumed that sliding drops are uniformly distributed over the sample length $0-L$. From this consideration and scaling equation Eq. 6.52, it follows that the distribution, n', of sliding drop volumes along x, is given by

$$n_s^1(V_0) = V_0^{-\tau} f\left(\frac{V_0}{V_{0c}}\right) \tag{6.87}$$

The cutoff V_{0c} corresponds to the maximum volume of the sliding drop, that is, from Eq. 6.84, to

$$V_{0c} \sim (h_s L)^{\frac{D_d}{1+D_d-D_s}} \tag{6.88}$$

The exponent τ obeys to dimension arguments. Since n' is in unit of (D_s-1) length, it readily comes $D_s - 1 = D_d \tau$ or

$$\tau = \frac{D_s - 1}{D_d} \tag{6.89}$$

6.6.2.3 Diffuse Boundary Layer Perturbation

An interesting feature is the fact that sliding droplets may disturb the boundary layer where drops grow (see Chap. 4 and Sect. 6.2). Droplet growth can be affected. This effect has been investigated by Wen et al. (2019) in the particular case of a hydrophilic surface coated with hydrophobic bumps.

6.6.3   Late Stage Drop Radius Distribution

The steady state of condensation consists therefore in the coexistence of many families whose maximum radius corresponds to the departure radius R_0 above which drops slide from the surface. Such drops sweep the surface and a new generation of drops can nucleate and grow.

6.6.3.1   Geometry Approach

Rose and Glicksman (1973) adequately described this steady state for dropwise condensation of steam vapor (pure water vapor with no condensable gas like air). Their approach is not restricted to steam and can be used for dew and BFs. They simplify the description of the pattern by considering hemispherical drops and monodisperse families uniformly spaced, i.e. the drop centers form an equilateral triangular array. They describe the pattern without the regions of sweeping, which in general corresponds to a small fraction of the pattern. The different regions available to the different families were estimated through the fraction A of total area covered by drops having radius greater than R. With $n(R)$ the drop size distribution (the mean number of drops per area and per drop radius), it comes

$$A = \pi \int_R^{R_0} R^2 n(R) dR \qquad (6.90)$$

Lefevre and Rose (1966) used the following power law to estimate A:

$$A = 1 - \left(\frac{R}{R_0}\right)^m \qquad (6.91)$$

It follows the drop size distribution by differentiating Eq. 6.90 with respect to R (with a change of sign as A diminishes when R increases) and using the power law Eq. 6.91 to estimate A:

$$n_c(R) = \frac{m}{\pi R_0^3}\left(\frac{R}{R_0}\right)^{m-3} \qquad (6.92)$$

$n_c(R)dR$ represents the number of drops per surface area whose radius is between R and $R + dR$. When compared with the distribution in droplet volume (Eq. 6.52), once expressed as a function of R, the parameter $m = 1/3$. From a critical review of experiments and simulations, including Lefevre and Rose (1966), Barati et al. (2017) indeed estimated the exponent in the range $m = [0.33 - 0.35]$. This distribution well fits the experimental data as seen in Fig. 6.21a. The departure radius R_0 comes from observations and corresponds to about 0.45 mm. In Kim and Kim (2011), the departure radius is calculated from the balance of weight and pinning forces from Eq. 6.83 (Eq. 14.17).

Figure 6.21b presents the evolution of the drop size distribution from $t = 0$ to 40 s as calculated by Tanaka (1975). It is interesting to note how the curves exhibit similar shapes after $t = 0.3$ s and simply move along the line of slope -2.7 to eventually reproduce the expected distribution Eq. 6.92 (interrupted line).

6.6.3.2   Fractals

Another approach due to Mei et al. (2009; 2015) uses the physical concept of *fractals* for determining the drop distribution. Since the drop evolution is self-similar (see Sect. 6.3), pictures taken at different scales exhibit similar structures and can be analyzed by the fractal theory (for a review of fractal properties, see e.g. Nurujjaman et al., 2017). One thus defines as follows the fractal dimension, d_f, of the droplets pattern (not to be confused with the fractal dimension D_f of the condensation substrate as used in Sect. 6.4.2, nor the fractal dimension d_f of the contour of substrate regions never visited by a drop in Sect. 9.2.2). With N, the number of drops whose radius is larger than R, and R_0, the radius of the biggest droplet

$$N = (R_0/R)^{d_f} \tag{6.93}$$

This relation defines the fractal dimension by the limiting value

$$d_f = \lim_{R \to 0} \frac{\ln(N)}{\ln(R_0/R)} \tag{6.94}$$

For 2D droplets growing on a planar surface, $d_f < 2$, while for 3D droplets, $d_f < 3$. The size distribution is obtained by differentiating Eq. 6.93 (with a negative sign as the number of droplets decreases with increasing R):

$$n_c(R) = \frac{d_f}{R_0} \left(\frac{R}{R_0} \right)^{-(d_f+1)} \tag{6.95}$$

This distribution is similar to Eq. 6.92 with $d_f = 2-m$. Using experimental data and simulations, Barati et al. (2017) estimated the exponent d_f between 1.79 and 1.99. It corresponds to exponent $1 + d_f = [2.79 - 2.99]$ in Eq. 6.95, to be compared to the exponent $3-m = [2.65 - 2.67]$ (Eq. 6.92). Both determinations are in good agreement with experimental data (Fig. 6.21).

6.7   Drop Pattern Evolution in Pure Vapor

Growth of droplets in pure vapor at saturation is limited by the heat transport through the drop and the substrate (see Sect. 4.1 and Chap. 13) and contrasts with droplets growing in humid air, where growth is limited by the vapor concentration gradient (see Sect. 4.2). In addition, the absence of vapor concentration gradient between drops authorizes constant nucleation between drops on the available

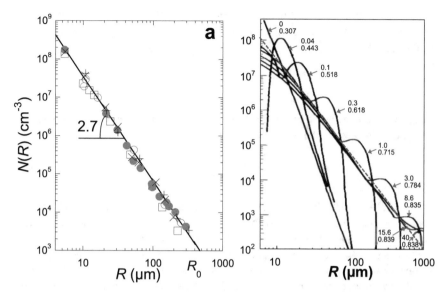

Fig. 6.21 **a** Drop size distribution in the steady state where gravity detaches drops above the departure radius R_0. The full line is Eq. 6.92, exhibiting the exponent $-8/3 = -2.7$, in agreement with Eqs. 6.92 and 6.95. Open circles and squares: From Graham (1969); full circles, squares and crosses: From Tanasawa and Ochiai (1973). (From Rose and Glicksman, 1973). **b** Calculated evolution of drop size distribution from $t = 0$ to 40 s. Time and surface coverage are shown for each distribution ($\rho^* = 6$ μm, distance between nucleation sites $\langle d_0 \rangle = 18$ μm, departure radius $\rho_0 = 1$ mm, $T_\infty = T_d = 100$ °C, subcooling $\Delta T = 1$ K. The interrupted line has a slope -2.7. (Adapted from Tanaka 1975, with permission)

nucleation sites. Although both process share some common features (e.g. drop coalescence), pure vapor growth exhibit therefore some particularities, in particular in the distribution of drop sizes.

Droplets initially form on nucleation sites, with surface density n_s. They are separated on average, for uniform distribution of nucleation sites, by the distance $\langle d_0 \rangle$. If one assumes that the sites are on a square lattice (Sect. 6.3),

$$\langle d_0 \rangle = \frac{1}{\sqrt{n_s}} \qquad (6.96)$$

The region where no coalescences occur between drops correspond to having a drop radius of curvature $\rho < \rho_c = \langle d_0 \rangle / 2$ (contact angle $\theta_c \geq \pi/2$) or cap radius $R = \rho \sin\theta_c < R_c = \langle d_0 \rangle / 2$ (contact angle $\theta_c < \pi/2$).

6.7.1 Drop Growth

There is no scaling in the growth of individual droplets because growth is limited by the available heat flux (Sect. 4.1). Then coalescences, although they accelerate the growth of the mean droplet radius, do not rescale the pattern as is the case for droplet growth in humid air (Sect. 6.3). However, the single drop growth law without coalescences $\rho < \rho_c$ (Eqs. 4.12 and 4.13) is generally close to

$$\rho \sim t^\beta \tag{6.97}$$

with $\beta = 1/2$ (Eq. 4.14, Fig. 6.22b). The effect of coalescence will thus rescale growth (see Sect. 6.3.2 above) such as

$$\langle \rho \rangle \sim t^\gamma \tag{6.98}$$

with $\gamma = 3\beta = 3/2$. However, new drops can easily nucleate between growing droplets because the vapor pressure is uniform (Fig. 6.22a—growth/coalescence). This situation contrasts with growth in vapor containing non-condensable gas where drops grow thanks to a vapor concentration gradient around them, preventing new drops nucleation before they are sufficiently apart, by a length on order the diffusive boundary layer (see Sect. 6.6.1). Then the rescaled growth can hardly be seen and multiple renucleated drop patterns coexist, increasing the surface coverage until the largest drop slides when its weight overcomes the pinning forces. The corresponding radius is evaluated in Sect. 14.1.2, Eq. 14.17. As a consequence of multiple nucleation, the surface coverage reaches large values ($\sim$0.7–0.8, see Fig. 6.22c) in the steady state where many renucleated patterns coexist.

6.7.2 Drop Size Distribution (No Coalescences)

In this region where $\rho < \rho_c$, an estimation of the overall drop distribution can be performed (Abu-Orabi, 1998); Kim and Kim, 2011) by accounting for the thermal resistances vapor—substrate listed in Sect. 13.5.5, Eq. 13.56. When expressed in temperature differences, they correspond to a drop interface with vapor, ΔT_i, drop conduction, ΔT_d, drop curvature, ΔT_c and substrate coating ΔT_{cc}. The heat flux through the drop is assumed by conduction since accounting for convections due to buoyancy and drop expansion leads to only weak differences (Sect. 13.5.6; note that Marangoni should not be present in pure vapor growth, see Sect. 13.5.4.2). With $T_\infty = T_d$, the vapor temperature at saturation and T_c the substrate temperature

$$\Delta T = T_\infty - T_c = \Delta T_i + \Delta T_d + \Delta T_c + \Delta T_{cc} \tag{6.99}$$

Let us now consider a given range of drop size characterized by radii between ρ and $\rho + d\rho$. To keep the number of drops constant in this range, the number of drops entering by growth must equal the number leaving by growth, plus the number removed by large drops sliding off the substrate. With S_c the substrate

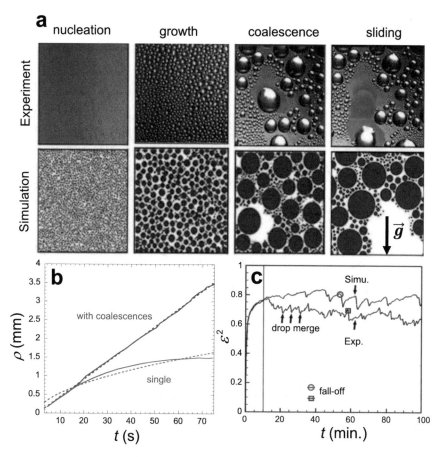

Fig. 6.22 a Different stages of growth for condensation in pure water vapor. The self-similar stage is hardly seen since nucleation can take place between packed droplets. (Adapted from Sikarwar et al. 2012, with permission **b** Hemispherical drop radius evolution from numerical simulation. Single drop evolution (continuous red curve) obeys an evolution similar to Eq. 4.14. Fit to Eq. 6.97 for single drop growth gives $\beta = 0.55 \pm 0.05$ (interrupted curve). When coalescences are taken into account (continuous blue line), growth is nearly linear. Fit to Eq. 6.98 gives $\gamma = 1 \pm 0.007$. This value is different from expected by scaling ($\gamma = 1.5$) because here scaling does not apply. (From Sikarwar et al., 2012, with permission). **c** Surface coverage ε^2 evolution. To the right of the vertical line corresponds the region where coalescence begin to be efficient. (Adapted from Sikarwar et al. 2011, with permission)

surface area and $\dot{\rho} = d\rho/dt$ the drop growth rate, the equality above is written as follows:

$$S_c n\dot{\rho}(\rho)dt = S_c n\dot{\rho}(\rho + d\rho)dt + B_s n d\rho dt \qquad (6.100)$$

Here, n is the number of drops per unit area per unit drop radius in the range $\rho, \rho + d\rho$ and B_s is the rate at which the substrate is renewed by drop sliding (surface area per unit time). Writing the sweeping period

$$\tau = \frac{S_c}{B} \tag{6.101}$$

Equation 6.100 becomes

$$-\frac{d(n\dot\rho)}{d\rho} = \frac{n}{\tau} \tag{6.102}$$

The growth rate $\dot\rho$ can be determined by considering the heat transfer vapor-interface-drop-coating as estimated above in Sect. 4.1. Since $\dot\rho$ is dependent on ρ only (see Eq. 4.12), Eq. 6.102 can be expanded as follows, where $\dot\rho$ and $d\dot\rho/d\rho$ derive from Eq. 4.12:

$$-\dot\rho\frac{dn}{d\rho} - n\frac{d\dot\rho}{d\rho} = \frac{n}{\tau} \tag{6.103}$$

The boundary conditions are the following. At the limit where drop coalescence starts

$$\rho = \rho_c n(\rho_c) = n_c(\rho_c) \tag{6.104}$$

$n_c(\rho_c)$ is the value of the distribution with coalescence (Eq. 6.116) at $\rho = \rho_c$. The radius $\rho_c = \langle d_0\rangle/2$, with $\langle d_0\rangle$ the mean spacing between the nucleation sites.

One can roughly estimate ρ_c by assuming that the nucleation sites are distributed on a square lattice. The number of nucleation sites per unit surface, n_s, relates to ρ_c as (Eq. 6.96):

$$\rho_c = \frac{1}{2\sqrt{n_s}} \tag{6.105}$$

The solution to Eq. 6.103 is obtained as follows (Abu-Orabi, 1998). One defines a overall reduced growth rate $\dot\rho_1 = \frac{d\dot\rho}{d\rho} + \frac{1}{\tau}$. Equation 6.103 becomes

$$\frac{dn}{n} = -\frac{\dot\rho_1}{\dot\rho}d\rho \tag{6.106}$$

Integration of the above Eq. 6.106 gives

$$n(\rho) = n_0\exp\left[-\int_{\rho*}^{\rho}\left(\frac{\dot\rho_1}{\dot\rho}d\rho\right)\right] \tag{6.107}$$

where ρ^* is the nucleation critical radius (Eqs. 3.14 and 3.36). The amplitude n_0 is evaluated from the boundary conditions: Eq. 6.104

$$n_0 = n_c(\rho_c)\exp\left[-\int_{\rho^*}^{\rho_c}\left(\frac{\dot{\rho}_1}{\dot{\rho}}d\rho\right)\right] \tag{6.108}$$

Equation 6.107 thus becomes

$$n(\rho) = n_c(\rho_c)\exp\left[\int_{\rho}^{\rho_c}\left(\frac{\dot{\rho}_1}{\dot{\rho}}d\rho\right)\right] \tag{6.109}$$

When substituting the $\dot{\rho}$ and $\dot{\rho}_1$ values from Eq. 4.12 in Eq. 6.109 above, it comes

$$n(\rho) = n_c(\rho_c)\frac{\rho(\rho_c - \rho^*)}{\rho_c(\rho_c - \rho^*)}\frac{A_2\rho + A_3}{A_2\rho_c + A_3}\exp(B_1 + B_2) \tag{6.110}$$

In this Equation,

$$B_1 = \frac{A_2}{\tau A_1}\left[\frac{\rho_c^2 - \rho^2}{2} + \rho^*(\rho_c - \rho) - \rho^{*2}\ln\left(\frac{\rho - \rho^*}{\rho_c - \rho^*}\right)\right] \tag{6.111}$$

$$B_2 = \frac{A_3}{\tau A_1}\left[\rho_c - \rho - \rho^*\ln\left(\frac{\rho - \rho^*}{\rho_c - \rho^*}\right)\right] \tag{6.112}$$

$$A_1 = \frac{T_\infty - T_c}{2\rho_l L_v},\ A_2 = \frac{\theta_c(1 - \cos\theta_c)}{4\lambda_l\sin\theta_c},\ A_3 = \frac{1}{2a_i} + \frac{1 - \cos\theta_c}{a_{cc}\sin^2\theta_c} \tag{6.113}$$

Here, ρ_l is the liquid density, L_v is the latent heat of condensation, λ_l is the liquid thermal conductivity, a_i is the coefficient of interfacial heat transfer Eq. 4.10 and a_{cc} is the coating heat transfer coefficient, Eqs. 13.8–13.8.

The sweeping period, τ (Eqs. 6.101–6.103), can be expressed as a function of ρ_c by considering a second boundary condition:

$$\frac{d\ln n(\rho)}{d\ln\rho} = \frac{d\ln N(\rho)}{d\ln\rho} = -\frac{8}{3} \tag{6.114}$$

It becomes

$$\tau = \frac{3\rho_c^2(A_2\rho_c + A_3)^2}{A_1\left(11A_2\rho_c^2 - 14A_2\rho_c\rho^* + 8A_3\rho_c - 11A_3\rho^*\right)} \tag{6.115}$$

6.7.3 Drop Size Distribution (Coalescences)

The region where $\rho > \rho_c$ is characterized by drop coalescence, drop sliding by gravity effects and renucleation of drops on the refreshed surface left free by coalescence and sliding. It is the distribution detailed in Sect. 6.6.3, satisfactorily described by Eq. 6.92 for hemispherical drops $(R = \rho)$

$$n_c(\rho) = \frac{1}{3\pi\rho_0^3}\left(\frac{\rho}{\rho_0(\theta_c)}\right)^{-8/3} \tag{6.116}$$

Here, ρ_0 is the radius of the departing drop by gravity, whose evaluation is given in Sect. 14.1.2, Eq. 14.17 (also Eq. 6.83 above). The dependence on the contact angle is included in the departure radius $\rho_0(\theta_c)$.

6.8 Complete Drop Distribution

The total drop distribution corresponds to the distribution Eq. 6.116 for $\rho \geq \rho_c$ and Eq. 6.110 for $\rho < \rho_c$. In Fig. 6.23, Kim and Kim (2011) provide the corresponding distribution functions for three values of the contact angle, $\theta_c = 90°, 120°, 150°$.

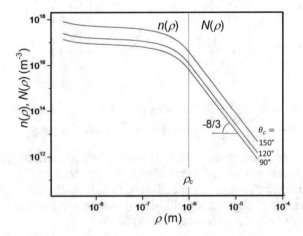

Fig. 6.23 Drop radius distribution for different contact angles θ_c with $T_\infty = 345$ K, $\Delta T = T_\infty - T_c = 5$ K, $a_i = 1.45 \times 10^7$ W.m^{-2}.K^{-1}, $a_{cc} = 2 \times 10^5$ W.m^{-2}.K^{-1}, $n_s = 2.5 \times 10^{11}$ m^{-2}. Notations: see text. (Adapted with permission from Kim and Kim 2011)

Humidity Sink and Inhibited Condensation

A drop grows at the expense of surrounding water molecules concentration. It thus creates around it a region of lower supersaturation (see Chaps. 4 and 6). In this region, condensation on the substrate can be hampered and new droplets prevented to nucleate. This effect can be magnified when considering condensation around hygroscopic materials showing a lower vapor saturation pressure than pure water as, e.g. ice or salty water, or by lowering the nucleation energy barrier by increasing wetting properties (e.g. hydrophilic coating or hydrogel). A dry region of inhibited condensation (RIC), indicative of the vapor concentration profile near the substrate, can then be observed around the hygroscopic materials (Fig. 7.1).

According to the arguments developed by Nath et al. (2018), one can determine two main processes at the origin of the RIC. Four different situations can occur depending on nucleation and flux RIC values and initial configuration (Fig. 7.2):

(i) Nucleation RIC, δ_N, where droplets grow in the concentration gradient created by a hygroscopic drop on an initial bare substrate. The vapor profile, as depicted in Fig. 7.3b, prevents the nucleation of droplets because locally the supersaturation is not large enough to overcome the nucleation barrier (case I in Fig. 7.1). This pathway is observed in condensation experiments around a lyophilic patch (Sect. 7.1) or a hydrogel (Sect. 7.2). Another situation concerns condensation around a salt crystal (case II in Fig. 7.2) with or without icing (Sects. 7.4 and 7.5).

(ii) Flux RIC, δ_F, where a large hygroscopic material is placed in a pre-existing droplet pattern and makes them evaporate around it. This case corresponds to ice crystals set in a pre-existing pattern of supercooled water droplets (cases III and IV in Fig. 7.2, Sect. 7.32).

© Springer Nature Switzerland AG 2022

D. Beysens, *The Physics of Dew, Breath Figures and Dropwise Condensation*, Lecture Notes in Physics 994, https://doi.org/10.1007/978-3-030-90442-5_7

Fig. 7.1 Condensation around a salty drop at $t = 2000$s (air temperature $T_a = 23$ °C, RH = 100%, flow rate 165 mL min^{-1}; substrate temperature $T_s = 10$ °C) showing the drop pattern around a region of inhibited condensation (RIC) of radius δ and thickness $\delta^* = \delta - R$ (Adapted from Guadarrama-Cetina et al. 2014)

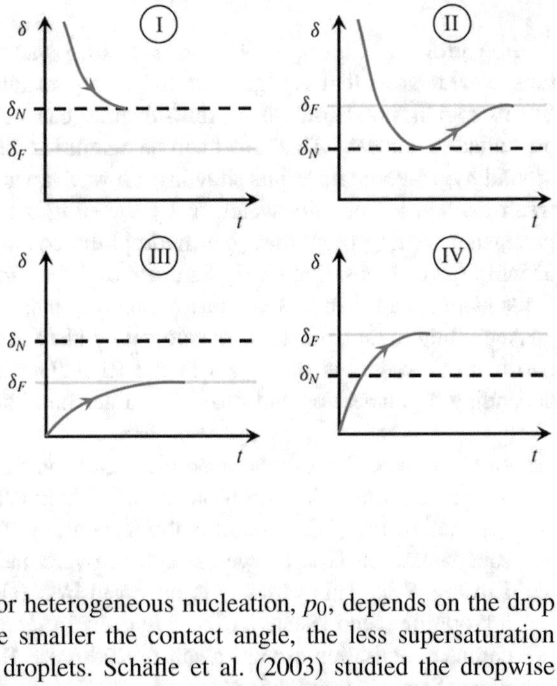

Fig. 7.2 Four different possible pathways for attaining a RIC over time. δ_N is the nucleation RIC, δ_F is the flux RIC. (Adapted from Nath et al. 2018, with permission)

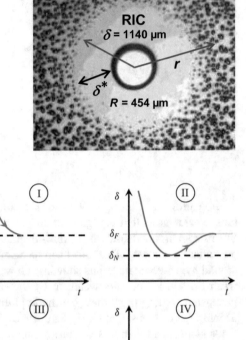

7.1 Lyophilic Spots

The vapor pressure needed for heterogeneous nucleation, p_0, depends on the drop contact angle (Chap. 3). The smaller the contact angle, the less supersaturation is needed to nucleate water droplets. Schäfle et al. (2003) studied the dropwise condensation of diethylene glycol on a substrate at $T_c = 25$ °C, corresponding to a saturation pressure p_s, around lyophilic patches (contact angle $\theta_c \approx 0°$) on a lyophobic substrate (contact angle $\theta_c = 69°$). Nitrogen saturated with diethylene glycol was sent at temperature $T_a = 68$ °C (vapor pressure, p_∞). Nucleation started on the lyophilic patch and a drop developed. Around it a concentration profile formed, which prevented droplet nucleation in a region of inhibited condensation (RIC, radius δ) (Fig. 7.3a).

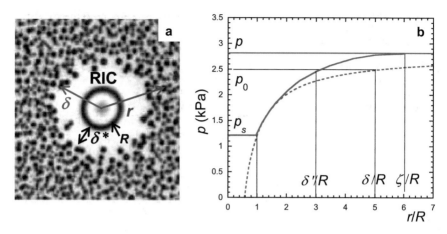

Fig. 7.3 Dropwise condensation of **a** diethylene glycol on a lyophobic substrate around a lyophilic substrates of 10 μmdiameter, showing a region of inhibited condensation (RIC). (Adapted from Schäfle et al. 2003, with permission). **b** Vapor concentration profile (schematics) without vapor flow (interrupted curve) or with flow (full curve) around the patch as a function of the reduced distance r/R. $p0$ is the saturation pressure corresponding to minimum supersaturation $(p_\infty - p_0)$ at which nucleation occurs, determining the RIC region of radius δ (no flow) or δ' (with flow). ζ is the boundary layer thickness

This situation thus corresponds to a nucleation RIC (I) in Fig. 7.2 since the patch pre-existed to condensation. One can consider that the concentration profile is similar to the hyperbolic profile developed around a hemispherical drop, Eq. (4.41). Making use of vapor pressure p rather that concentration by using Eqs. (1.45) and (1.46) (which shows that $c \propto p$), it comes (Fig. 7.3b):

$$p = p_\infty - \frac{R(p_\infty - p_s)}{r} \tag{7.1}$$

With p_0, the vapor pressure at nucleation and the RIC corresponds to the minimum supersaturation $(p_0 - p_s)$ at which nucleation can occur. This supersaturation depends on the droplet contact angle. On the lyophilic patch, $\theta_c \approx 0°$, $p_0 - p_s = 0$, on the lyophobic substrate, $\theta_c = 69°$, $p_0 - p_s > 0$ (see Sect. 3.2).

Schäfle et al. (2003) observed that the RIC diameter is dependent on the vapor flow rate. This is due to the fact that the hyperbolic profile is valid only within the boundary layer (see schematics of Fig. 7.3b). Following Eq. (2.5), increasing the volume flow rate F, proportional to the external flow velocity U_0, decreases the boundary layer thickness as

$$\zeta \sim U_0^{-1/2} \sim F^{-1/2} \tag{7.2}$$

Figure 7.4 reports the variation of the width of the RIC with the gas flow rate as

$$\delta^* = \delta - R \tag{7.3}$$

Fig. 7.4 Dependence of the RIC $\delta^* = \delta - R$ on the gas flow rate around the lyophilic patches with radius $R = 2.5\ \mu$m (triangles) and $R = 5\ \mu$m (squares). The line is a fit to Eq. (7.4) (data from Schaffle et al. 2003)

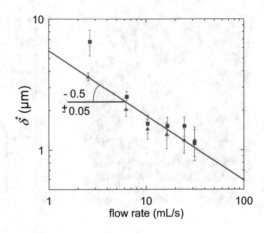

The data are indeed seen to follow a power law:

$$\delta^* \sim F^{-x} \tag{7.4}$$

with $x = 0.5 \pm 0.05$.

When the patches are set on a lyophilic lattice with periodicity p, the presence or the absence of interstitial droplets depends on the respecting values of RIC radius and periodicity. In particular, when $\delta > p$, no droplets can condense between the patches. When $\delta \approx p$, it is possible to make apparent a regularly ordered substructure.

Note that Yu et al. (2021) have observed that some droplets can evaporate during dropwise condensation on a smooth surface. This phenomenon was attributed to the local changes in the surface wettability due to the deposition of volatile organic compounds. Such compounds indeed created local nano-structures in the region where a droplet has previously been condensed and evaporated.

7.2 Hydrogels

Hydrogels are three-dimensional (3D) networks of hydrophilic polymers, whose properties are well known (see e.g. Wichterle and Lím 1960). There are many types of hydrogels (pH-sensitive, temperature-sensitive, electro-sensitive, light-responsive, etc.), with several applications related to their extraordinary swelling properties (volume can be multiplied by 1000) when immersed in liquid water (for a review, see Majee 2016). The osmotic pressure attributed to the polymer network is the driving force of swelling. An osmotic gradient indeed forms between the water solvent, low in ionic solute, towards the polymer, rich in ionic solute. The swelling process distends the network and is counterbalanced by the elastic contractility of the stretched polymer network.

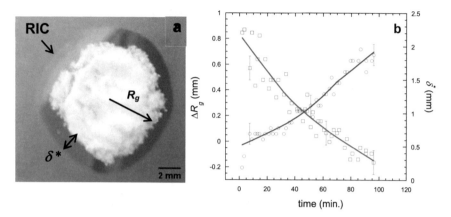

Fig. 7.5 **a** Region of inhibited condensation (RIC, thickness δ^*) around a thin (<1 mm) layer ≈ 0.2 mm diameter grain on a silicon surface after 25 s (T_a= 22 °C at 47% RH, $T_c = T_d - 6.7$ K). **b** Evolution of the gel layer radius variation $\Delta R_g = R_g(t) - R_g(0)$ (left ordinate, circles) and RIC thickness δ^* (T_a= 20 °C, $RH = 50\%$, $T_s = T_d - 4$ K) (right ordinate, squares) (Adapted from Urbina et al. 2020)

Due to this osmotic pressure, hydrogels can also exhibit high water adsorption from water vapor (they are hygroscopic), an adsorption which increases with relative humidity *RH* (see e.g. Delavoipière et al. 2018). Urbina et al. (2021) studied water condensation of Aquasorb 3005™ gel grains (gel used in agriculture) placed on a silicon wafer (contact angle $\theta_c \approx 60°$). When the gel temperature, T_c, is below the surrounding humid air dew point temperature, T_d, dropwise condensation proceeds on the substrate except around the gel, where a RIC is observed (Fig. 7.5). The RIC corresponds to pathway I in Fig. 7.2 since the water drops nucleate around the pre-existing hydrogel.

The hygroscopic nature of the gel due to the osmotic pressure build up by its hydrophilic sites induces enhanced adsorption of water vapor and thus lowers vapor pressure at the gel border, giving rise to the RIC. This RIC corresponds to case (I) in Fig. 7.2 and is similar to the RIC observed in Sect. 7.1 under diethylene glycol condensation.

Figure 7.5b shows that the gel swells as condensation proceeds, corresponding to the increase of gel layer radius, R_g. The RIC, which corresponds to vapor adsorption, decreases until reaching zero when gel adsorption reaches saturation. Gel mass evolution was determined by weighing in the Urbina et al. (2021) experiments. When looking at the water collected by the gel in units of volume per surface area (*h*, in mm, see Fig. 7.6), the initial condensation rate *dh/dt* is larger on gel than on the bare substrate, due to vapor adsorption which adds to condensation. As a matter of fact, water collected by the gel corresponds to the volume condensed on a bare substrate plus the adsorbed mass measured without condensation, at a temperature just above the dew point.

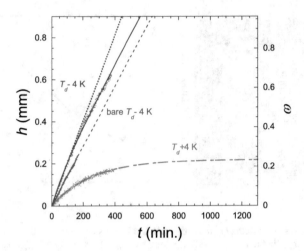

Fig. 7.6 Evolution of condensation in units of water volume per unit projected area, h, on gel grains (≈ 0.7 mm diameter, blue dots, left ordinate) and bare substrate (red dots, left ordinate) under same conditions ($T_a = 20\ °C$, $RH = 50\%$, $T_c = T_d - 4$ K). The interrupted line is a linear fit of bare substrate data, the continuous curve is the sum of the bare substrate linear fit and sorption isotherm. The dotted line is the initial slope for condensation on gel. Right ordinate corresponds to the sorption isotherm $\omega = m_w/m_i$ (m_w is adsorbed water mass, m_i is dry gel mass) just above the dew point at $T_a = 20\ °C$, $RH = 50\%$, $T_s = T_d + 4$ K (orange dots, fitted to exponential decay) (Adapted from Urbina et al. 2021)

7.3   Ice Crystals

When water condenses at temperature below 0 °C, metastable supercooled droplets first form. They freeze only after a lengthy time or if an ice crystal touches it by chance, like a crystal growing from the side of the sample. The process of ice invasion can proceed by slow growth of faceted crystals where supercooled droplets evaporate on ice or by fast propagation through dendritic growth when the supercooled droplet surface coverage is high, ensuring that dendrite propagation is faster than droplet evaporation (see also Appendix I9). Indeed, ice exhibits a lower water pressure p_{si} than saturated vapor pressure of supercooled (SC) water at the same temperature, p_s, ($p_{si} \approx 0.90\ p_s$, see Table 7.1). Therefore, SC water evaporates in the vicinity of an ice crystal, forming a RIC or dry zone.

Below are experiments made by Nath et al. (2018), corresponding to either an icy drop placed in a pattern of pre-existing supercooled droplets (case III or IV of Fig. 7.2) or condensation of supercooled droplets on a bare substrate around an icy drop (case I or II of Fig. 7.2). From the analysis of data, in all situations the flux RIC cases II, III and IV were found, validating flux as the dominant process. This was expected since the ice crystal appears in a pre-existing water droplet pattern.

Table 7.1 Water saturation pressures at two typical temperatures corresponding to dew and frost formation. Values are from Lide (1998), Sawamura et al. (2007) and Arias-González et al. (2010)

Conditions	Symbol	Temperature-1 (°C)	Saturation pressure (Pa; 10 °C)	Temperature-2 (°C)	Saturation pressure (Pa; − 12 °C)
Salty water (salt saturation)	p_{s0}	10	910	−12	180
Salty water	p_{ss}	10	910–1200	−12	180–245
Ice	p_{si}	–	–	−12	220
Water	p_s	10	1220	−12	245
Humid air	p_∞	23	2810	17	1950

7.3.1 Nucleation RIC

Case I corresponds to the vapor pressure profile around an icy drop, where p_s is replaced by p_{si} in Eq. (7.1)

$$p = p_\infty - \frac{R(p_\infty - p_{si})}{r} \qquad (7.5)$$

With p_0 the minimum vapor pressure for heterogeneous nucleation, condensation cannot proceed in a RIC, where $p < p_0 \leq p_s$. The RIC radius δ_N and thickness $\delta_N^* = \delta_N - R$ thus follows:

$$\delta_N = R\frac{p_\infty - p_{si}}{p_\infty - p_0}; \quad \delta_N^* = R\frac{p_0 - p_{si}}{p_\infty - p_0} \qquad (7.6)$$

In the Nath et al. (2018) experiments, the variation of the RIC thickness δ_N^* for different R obtained at different temperatures and then different values of the ratio $(p_\infty - p_i)/(p_\infty - p_0)$ did not follow Eq. (7.6), thus discarding nucleation as the limiting process.

7.3.2 Flux RIC

The assumption initially made by Guadarrama-Cetina et al. (2013b) is to consider diffusive flux inside the boundary layer ζ, with the underlying assumption that the icy drop radius $R < \zeta$. The border of the RIC is determined by a water molecules flux balance where the diffusive flux of condensing molecules, j_c, equilibrates the flux of evaporation, j_e, (see Fig. 7.7a). Using Eq. (4.15) to express the flux of monomers in a concentration gradient, it comes $\overrightarrow{j} = -D\overrightarrow{\nabla}c$, with c the concentration in mass per volume of water monomers. For the condensation flux on the water droplets on the substrate, one obtains, the subscripts for concentration

Fig. 7.7 Condensation around an icy drop. Drop radius $R < \zeta$, the diffusion boundary layer, with diffusive condensation and evaporation fluxes. **a** Schematics. **b** Axisymmetric simulation corresponding to (**a**), showing the lines of fluxes, the hyperbolic concentration profile around the icy drop and the linear profile above the BF droplets, schematized by the equivalent film of Sect. 6.1.2 (exaggerated ratio $p/p_s = 0.75$ was imposed at the icy drop to better visualize the streamlines). **c** Zoom of (**b**) near the film edge (Adapted from Nath et al. 2018, with permission)

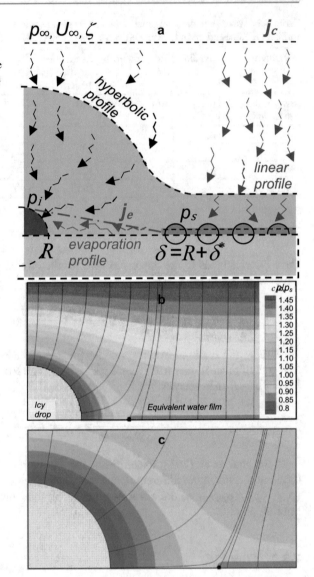

correspond to the subscripts (∞, s) used in Eq. (7.1) for vapor pressure:

$$j_c = D \frac{c_\infty - c_s}{\zeta} \tag{7.7}$$

Evaporation proceeds from the RIC border to the ice crystal. The flux can be then written as

$$j_e = D \frac{c_s - c_{si}}{\delta_D^*} \tag{7.8}$$

Here, δ_{F1}^* is the "diffusive" RIC thickness as based on mere diffusion of water molecules in the concentration gradients. Making $j_c = j_e$ at the RIC border, one obtains

$$\delta_{F1}^* \sim \zeta \left(\frac{c_s - c_{si}}{c_\infty - c_s} \right) = \zeta \left(\frac{p_s - p_{si}}{p_\infty - p_s} \right) \tag{7.9}$$

Here, the proportionality is used between p and c (Eqs. 1.45 and 1.46).

The data collected by Nath et al. (2018) correspond to an ice drop of a radius in order or larger than the boundary layer thickness ($R \geq \zeta \sim$ mm). Data were obtained at different temperatures and then different values of the relative super-saturation $(p_s - p_i)/(p_\infty - p_s)$. When the RIC thickness is compared to this supersaturation (Fig. 7.8a) in a log-log plot with the boundary layer thickness as a free parameter, the expected slope unity is observed for relative supersatura-tion typically lower than 0.1. The boundary layer thickness value that give best results corresponds to $\zeta = 1.3$ cm, however about 10 times the expected value

Fig. 7.8 RIC thickness δ^* as a function of different supersaturation (log-log plot). Slope unity is expected. The boundary layer thickness is set at $\zeta = 1.3$ cm. **a** According to Eq. (7.9). **b** According to Eq. (7.11). Crosses correspond to numerically obtained RIC (see text) for three different RH directly compared to their experimental analogues (circles and triangles) (Adapted from Nath et al. 2018, with permission)

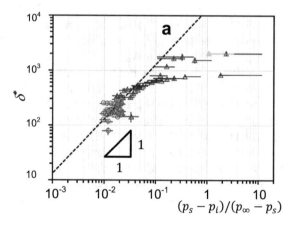

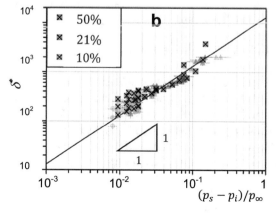

using Eq. (2.16). Equation (7.9) is nevertheless only a scaling relation, without precise values for the proportionality factor.

In order to try to improve scaling in the situation where $R \geq \zeta$, Nath et al. (2018) proposed a different evaluation of the condensation flux. They assumed the condensing flux to be proportional to the flux of molecules at the border of the boundary layer, that is,

$$j_c \sim p_\infty U_\infty \qquad (7.10)$$

Using Eqs. (7.8) and (7.10), the balance of evaporative and condensing flux gives

$$\delta_{F2}^* \sim \frac{D}{U_\infty} \left(\frac{p_s - p_{si}}{p_\infty} \right) = \zeta^* \left(\frac{p_s - p_{si}}{p_\infty} \right) \qquad (7.11)$$

Here, $\zeta^* = D/U_\infty$ corresponds to a typical length similar to the boundary layer thickness ζ (the latter, however, obeys a different velocity dependence, see Chap. 2). Figure 7.8b shows the comparison of the Nath et al. (2018) experiments with Eq. (7.11), using ζ^* as a free parameter. A good agreement is found over the whole range of relative supersaturation, however the same value as above for the expression Eq. (7.9) is found, $\zeta^* \approx 1.3$ cm, still about 10 times the expected value using Eq. (2.16). However, as already noted, Eq. (7.10) is a scaling relation without precise values for the proportionality factor.

Computational analysis was performed solving the Laplace equation Eq. (4.17) for the boundary conditions corresponding to Fig. 7.7a and using the experiment conditions. Figure 7.7b, c (zoomed) show the lines of fluxes and the expected hyperbolic concentration profile around the ice drop and linear profile above the BF droplets, schematized by the equivalent film of Sect. 6.1.2. Three different conditions of relative humidity were considered. The comparison with their experimental analogues in Fig. 7.8b show excellent agreement.

7.4　Salty Drop

7.4.1　General Features

The process of dropwise condensation in the vicinity of a salty water drop obeys features similar to encountered with lyophilic or hydrophilic materials and ice crystal as discussed in Sects. 7.1, 7.2 and 7.3. The situation corresponds in Fig. 7.2 to pathway I or transient pathway II, before evaporation, since the salty drop was already present before water drop condensation begins. Indeed, the RIC is always transient as condensation makes the salty drop vapor pressure decreasing with time. Figure 7.9c shows that the depletion zone, although shrinking with time in a non-continuous manner due to the presence of nucleation sites, does not exhibit increasing episodes, even at late times.

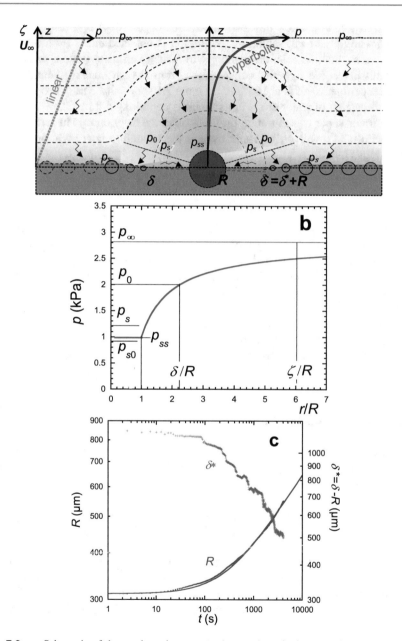

Fig. 7.9 **a** Schematic of the condensation process (see text). p_0 is the saturation pressure corresponding to minimum supersaturation ($p_\infty - p_0$) at which nucleation occurs, determining the RIC region with radius δ as observed in Fig. 7.1. The drawing shows droplets and the equivalent film layer (Sect. 6.1.2). **b** Water vapor concentration profile as a function of the reduced distance r/R. The salty drop water saturation pressure, p_{ss}, varies during growth from the value at salt saturation p_{s0} ($\approx 0.75\,p_s$) to the value for pure water, p_s. The vapor pressure values are from Table 7.1 for temperature-1 conditions. The vapor pressure profile Eq. (7.12) is, for $p_{ss} = 1$ kPa, $p(\text{kPa}) = 2.8 - 1.8(R/r)$. (Adapted from Guadarrama-Cetina et al. 2014). **c** Evolution (log-log plot) of salty drop radius R and RIC thickness $\delta^* = \delta - R$ for the typical experiment of Fig. 7.1. The solid line is a fit of R to Eq. (7.19). Stepwise evolution is due to nucleation on defects (Adapted from Guadarrama-Cetina et al. 2014)

A typical configuration (Guadarrama-Cetina et al. 2014) is a salty drop (cap radius R) initially at salt saturation. Saturation is obtained either from taking the drop from a saturated solution or simply by placing on a substrate a salt crystal, which dissolves due to water condensation from the surrounding humid air. In typical experiments, air is at 100% RH. Figure 7.1 shows that a salty drop prevents the condensation of water drops in its vicinity, delimiting a flux RIC.

The water vapor concentration profile around the salty drop is hyperbolic, following Eq. (7.1), with p_s replaced by p_{ss}, the saturation concentration of salty water (Figs. 7.9a, b) is

$$p = p_\infty - \frac{R(p_\infty - p_{ss})}{r} \tag{7.12}$$

Initially, $p_{ss} = p_{s0}$ ($\approx 0.75 ps$ from Lide 1998), corresponding to salt saturation. During the drop growth, p_{ss} increases due to water condensation. At late times the value for pure water is reached, $p_{ss} = p_s$ (Fig. 7.9b). As p_{ss} is always lower than the vapor pressure for nucleation of pure water on the substrate, p_0, the nucleation of water droplets is impeded inside a RIC ring at distances $r \leq \delta$ from the salty drop center (Figs. 7.1 and 7.9b). Vapor pressure $p_0 > p_s$ because of the presence of an energy barrier for nucleation (see Chap. 3). As time goes on and water condensation proceeds, the radius of the salty drop increases (Fig. 7.9c) and salt concentration decreases accordingly. It causes p_{ss} to increase and δ to decrease (Fig. 7.9c).

7.4.2   Salty Drop Evolution

The salty drop grows by diffusion in the concentration profile Eq. (7.12) and incorporation of monomers at its surface (Fig. 7.9a). The growth equation is deduced from Eq. (4.21):

$$R^2 \frac{dR}{dt} \propto R^2 \left(\frac{\partial p}{\partial r} \right)_R = R^2 \frac{(p_\infty - p_{ss})}{R} \tag{7.13}$$

During growth, the salt concentration (and thus the local supersaturation) decreases. This in turn increases the water pressure gradient. The variation of the saturation pressure with the salty drop radius can be evaluated by using the Raoult law. A detailed calculation is given in Appendix J and leads to the approximate variation (Eq. J.17):

$$p_s - p_{ss} \approx (p_s - p_{s0}) \left(\frac{R_0}{R} \right)^b \tag{7.14}$$

where the exponent $b \approx 3.3$. In the limit of small salt concentration, a simplified expression corresponds to $b = 3$

$$p_s - p_{ss} \approx (p_s - p_{s0}) \left(\frac{R_0}{R} \right)^3 \tag{7.15}$$

As b is not significantly different in both Eqs. (7.14) and (7.15), one makes, for sake of simplicity, $b = 3$. Equation (7.13) then becomes

$$R \frac{dR}{dt} \sim (p_\infty - p_s) + (p_s - p_{ss}) = (p_\infty - p_s) + (p_s - p_{s0}) \left(\frac{R_0}{R} \right)^3 \tag{7.16}$$

The usual way to solve this equation is to invert it and solve $t(R)$. It easily reduces to the integral of $R/(R^3 + \text{constant})$, which can be put in terms of an arctan and log of algebraic functions. The solution can be found in classical handbooks (see e.g. Lipschutz et al. 2008). Unfortunately, it is not possible to invert the result analytically to obtain the evolution $R(t)$. One can, however, consider the following asymptotic regimes:

(i) At long times, the salty drop becomes highly diluted and the first term in Eq. (7.16) is dominant. The depletion zone becomes of the same order as in pure water. Equation (7.16) can thus be written as

$$R \frac{dR}{dt} \sim (p_\infty - p_{ss}) \tag{7.17}$$

This equation corresponds to the classical evolution $R \sim t^{1/2}$ of Eq. (4.42).

(ii) At short times, when the drop is close to salt saturation, the second term in Eq. (7.16) is larger than the first term. Equation (7.16) becomes

$$R \frac{dR}{dt} \sim (p_s - p_{s0}) \left(\frac{R_0}{R} \right)^3 \tag{7.18}$$

whose integration gives

$$R^5 = R_0^5 + Bt \tag{7.19}$$

Here,

$$B \sim R_0^3 \tag{7.20}$$

In Fig. 7.10a, the variation of B with respect to R_0 indeed follows a power law with an exponent ~3, in accordance with Eq. (7.20). The different sets of data also fall on one single curve when properly scaled. Figure 7.10b shows such scaled data with the evolution of $\left(R^5 - R_0^5 \right) / B$.

Fig. 7.10 a Amplitude B (see Eqs. 7.19 and 7.20) for several initial salty drop radius R_0 (log-log plot) showing the power law dependence with exponent 3. Diamonds correspond to initial salt crystal experiments. **b** Evolution of $R^5 - R_0^5$ rescaled by the corresponding B values (Eq. 7.19). Stars in the values of R_0 (right column) correspond to initial salt crystal experiments (Adapted from Guadarrama-Cetina et al. 2014)

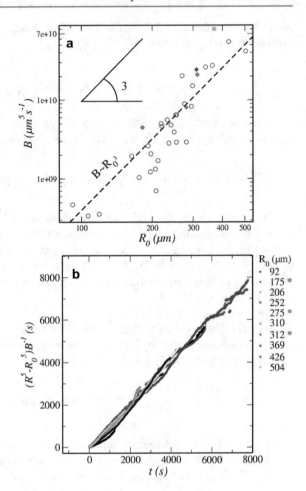

7.4.3 Region of Inhibited Condensation

As noted above at the beginning of the present section, the situation corresponds in Fig. 7.2 to pathway I or transient pathway II, before evaporation. One must therefore deal with a nucleation RIC, with radius δ_N. The RIC border corresponds to the place where supersaturation reaches the limit where nucleation just occurs, that is, where $p = p_0$. From the hyperbolic profile Eq. (7.1) it means that

$$p_0 = p_\infty - \frac{R(p_\infty - p_{ss})}{\delta_N} \tag{7.21}$$

Using Eq. (7.15) to evaluate p_{ss}, one obtains

$$\left(\frac{\delta_N}{R}\right) = \left(\frac{p_\infty - p_s}{p_\infty - p_0}\right) + \left(\frac{p_s - p_{s0}}{p_\infty - p_0}\right)\left(\frac{R}{R_0}\right)^{-3} \tag{7.22}$$

Fig. 7.11 Variation of δ_N/R with respect to R/R_0 fitted to Eq. (7.10) (log-log plot) (Adapted from Guadarrama-Cetina et al. 2014)

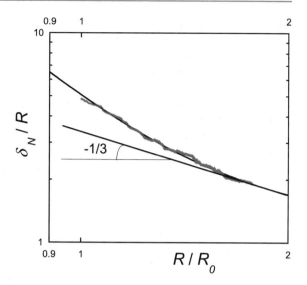

An experimental variation of δ_N/R with respect to R/R_0 was reported by Guadarrama-Cetina et al. (2014) in the conditions of Table 7.1—temperature-1 (Fig. 7.11). The data could be successfully fitted to Eq. (7.22) as $\delta_N/R = A + B(R/R_0)^{-3}$, with A and B as free parameters. The value of p_0 deduced from B is not accurate because the value of R_0 enters with the power 3. The value found for A (≈ 1.2) corresponds to a nucleation pressure limit $p_0 \approx 1.48$ kPa. This is a somewhat small value as one expects nucleation to start within a few degrees below air temperature when it is saturated. However, the concentration profile near the substrate is likely to deviate from a hyperbolic profile due to the interactions between the different profiles (see the simulation in Fig. 7.7c). In addition, as noted by Guadarrama-Cetina et al. (2014), the presence of non-random, uncontrolled nucleation sites add uncertainties in the determination of the RIC.

When considering the vapor concentration profile and the expression of δ_N, Eq. (7.21), one obtains

$$p = p_\infty - (p_\infty - p_0)\frac{\delta_N}{r} \tag{7.23}$$

As a result, the profile around the salty drop scales with δ_N. This result will be used in the Sect. 7.4.4 to determine experimentally the vapor concentration profile from the growth rate of droplets in the region $r > \delta_N$.

7.4.4 Concentration Profile

The BF pattern which forms beyond the ring exhibits the typical different stages of growth (see Sect. 3.3): nucleation at the RIC perimeter; initial growth with low surface coverage in $t^{1/2}$; further growth at higher surface coverage with individual droplets growing as $t^{1/3}$ corresponding to a linear vapor profile perpendicular to the surface; self-similar growth with constant high surface coverage; and mean droplet radius growing as t.

The vapor pressure depends on time t and distance r due to the time-dependent vapor gradient around the salty drop. The growth rate of water droplet (i) at distance r_i will be thus different at different times t_i and distances r_i. Since droplet (i) growth depends on local supersaturations $\Delta p_i = p_i - p_s$, all droplet growth can be rescaled on a unique curve R_i (t/t_i) with $t_i \propto 1/\Delta p_i$, as noticed by Fritter et al. (1991). The droplet evolution in the drop pattern around the salty drop can then be used to determine the local supersaturation profile near the substrate:

$$1/t_i \sim p_i - p_s = p(r_i) - p_s \qquad (7.24)$$

A difficulty arises because the growth laws correspond to a supersaturation that remains constant over time. In contrast, in the present configuration, the droplets evolve according to a time-dependent local supersaturation. In other words, supersaturation varies over time at constant r. However, the difficulty can be overcome if considering the vapor concentration profile Eq. (7.23) where the profile around the salty drop scales with r/δ_N. The droplet growth laws at given r/δ_N then correspond to constant supersaturation levels. The mean radius of water droplets, $<R_i>$, can be measured around the salty drop at distances r_i in the range $r_i > \delta_N$ inside concentric annuli of small thickness (ε) with respect to r_t (Fig. 7.12a). In Fig. 7.12b, the evolution at constant r/δ_N corresponds to what is currently observed for BF growth at constant supersaturation. An initial $t^{1/3}$ growth law followed, when droplet coalescence dominates the process, by a t^1 growth (Chap. 6). The various growth laws can be rescaled by times (t_i) to collapse on the same (t/t_i) growth curves (inset of Fig. 7.12b). From the rescaling by t_i, one eventually obtains the vapor pressure profile in units of r/δ_N:

$$p(r/\delta_N) - p_s \sim 1/t_i \qquad (7.25)$$

Using Eq. (7.23), the local supersaturation $p(r/\delta_N) - p_s$ can be written, with E a proportionality constant, as

$$\frac{1}{t_i} = E\left[1 - \left(\frac{p_\infty - p_0}{p_\infty - p_s}\right)\left(\frac{\delta_N}{r}\right)\right] \qquad (7.26)$$

This formulation with the hyperbolic variation implicitly assumes that the growth of the droplets does not appreciably modify the water vapor profile around the salty drop. Using the values p_∞, p_s from Table 7.1, the fit with Eq. (7.26) of data in Fig. 7.13 with p_0 and E as adjustable parameters gives $E \approx 5.1 \times 10^{-3}$ s^{-1} and $p_0 \approx 1.44$ kPa. The latter value is in accord with the value $p_0 = 1.48$ kPa found in Sect. 7.4.3.

Fig. 7.12 a Determination of droplet radius averaged at distance r_i averaged in an annulus of thickness ε. **b** Evolution of mean droplet radius $<R>$ for several ratios r/δ_N. The initial $t^{1/3}$ and final t growth laws are clearly visible (black lines). The inset show the same data with time rescaled by the onset of nucleation time, t_i. Data correspond to Fig. 7.9c (Adapted from Guadarrama-Cetina et al. 2014)

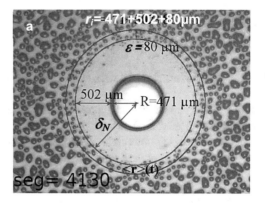

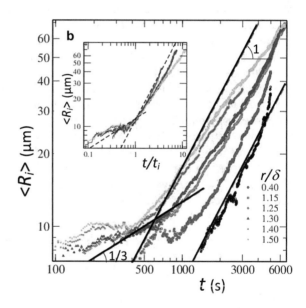

7.5 Salty Drops and Ice

Let us now address the process by which salt acts during frost formation. Before melting ice crystals by lowering the freezing temperature, the first action of salt is found to prevent nucleation of water and ice on the substrate (Guadarrama-Cetina et al. 2013b). Both ice and salty drop have indeed different water pressures, lower than the water pressure of pure water (Fig. 7.14). However, the fact that the vapor pressure of the salty drop increases with time due to condensation makes an interesting time-dependent interplay between water droplets, ice crystals and salty drop. In addition, there is simultaneously the presence of nucleation RIC (water and ice nucleation around the pre-existing salty drop) and flux RIC (water and solidified ice).

7.5.1 Ice-Water, Salt-Water and Salt-Ice RICs

Different RICs can be present, depending whether condensation proceeds by SC water droplets (water RIC) or ice crystals (ice RIC) or both. Figure 7.15a describes the different possible configurations according to the value of the salty drop saturation pressure p_{ss} with respect to ice saturation pressure p_i and pure water saturation pressure p_s. Using the data of Table 7.1 temperature-2 in Eqs. (7.15) and (7.19) with $R_0 = 103$ μm and $B = 1.3 \times 10^9$ μm^5 s^{-1}, one obtains p_{ss} corresponding to the study of Guadarrama-Cetina et al. (2013b).

Three kind of RICs can be observed in Fig. 7.15. (i) Ice-water RIC (δ_{iw}^*, Fig. 7.15b) and (ii) water-salty water RIC δ_{ws}^*, Fig. 7.15c) are present all along the salty drop evolution because $p_{ss} < p_i < p_s$. During the time $p_{ss} < p_i$, an ice-salty water RIC (δ_{is}^*, Fig. 7.13e) can also be observed. When $p_s = p_i$ (Fig. 7.13f), $\delta_{is}^* = 0$ meaning that ice hits the salty drop, which eventually melts the surroundings as the freezing temperature of the melt is lowered.

Figure 7.16 describes the typical evolution of supercooled water droplets and ice crystals condensing around a salty drop initially at salt saturation on a substrate kept at $T_s = -12$ °C, as reported by Guadarrama-Cetina et al. (2013b). A water-salty water RIC initially forms (Fig. 7.15c), which grows and then shrinks after 6 s, following pathway II in Fig. 7.2. This indicates a flux balance process. In addition, a frost front appears from the upper left side of the sample after 3.75 s and propagates, forming an ice-salty water RIC on the upper side of the photo (Fig. 7.15d). This ice-salty water RIC shrinks with time. In addition, an ice-water RIC forms between ice and supercooled water (Fig. 7.15d). Ice crystal hits the

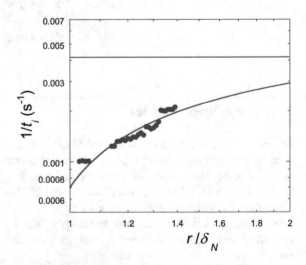

Fig. 7.13 Water concentration profile (in units of inverse rescaling time t_i) with respect to r/δ_N corresponding to data of Fig. 7.9c. The curve is a fit to Eq. (7.26) with p_0 and E as adjustable parameters (From Guadarrama-Cetina et al. 2014)

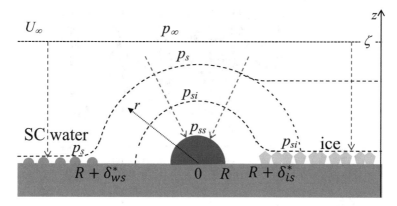

Fig. 7.14 Schematic of the condensation process (notations: see text)

supercooled droplets later, at 13.5 s, leaving only ice crystals and salty drops on the substrate (similar to Fig. 7.15e). Then, at 20 s, the time needed for salty water vapor pressure to reach ice vapor pressure, a dendrite issued from an ice crystal hits the salty drop (Fig. 7.15e). It results in the melting of ice in its vicinity due to the lowering of freezing temperature in the salty melt. In the above experiment, the evolution of the salty drop radius obeys Eq. (7.19) $R^5 = R_0^5 + Bt$ with $R_0 = 103\ \mu m$ and $B = 1.3 \times 10^9\ \mu m^5\ s^{-1}$.

7.5.2 Ice-Salt and Water-Salt RICs Evolution

According to the observations, the variation of the water-salty water and ice-salty water RICs can be evaluated by using the nucleation assumption since water drops nucleate around the preexisting salty drop assumption. As the salty drop radius remains smaller than typically 200 μm, a value less than the boundary layer thickness, in order of a few mm (Chap. 2), Eq. (7.21) can be used for the evaluation of the RIC water-salty water, δ^*_{ws}. Concerning ice-salty water, δ^*_{iw}, Eq. (7.21) must be modified to replace the water vapor pressure p_s by the ice vapor pressure p_i. It thus follows:

Water-salty water RIC:

$$\delta^*_{ws} \sim \zeta \left(\frac{p_s - p_{ss}}{p_\infty - p_s} \right) \approx \zeta \left(\frac{p_s - p_{s0}}{p_\infty - p_s} \right) \left(\frac{R_0}{R} \right)^3 \tag{7.27}$$

Ice-salty water RIC:

$$\delta^*_{ws} \sim \zeta \left(\frac{p_i - p_{ss}}{p_\infty - p_s} \right) \approx \zeta \left[\left(\frac{p_s - p_{s0}}{p_\infty - p_s} \right) \left(\frac{R_0}{R} \right)^3 - \left(\frac{p_s - p_i}{p_\infty - p_s} \right) \right] \tag{7.28}$$

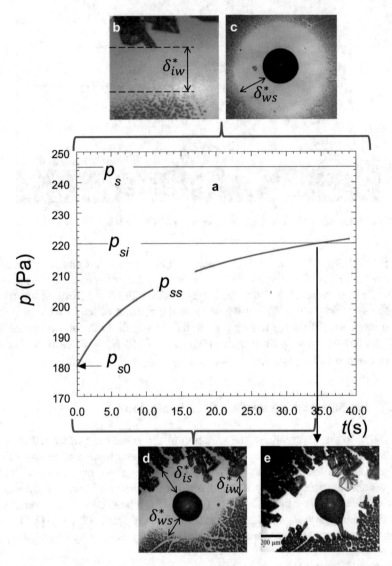

7.15 a Evolution of the salty drop water pressure at $T_s = -12$ °C using the date of Table 7.1 temperature-2 in Eqs. (7.15) and (7.19) with $R_0 = 103$ µm and B = 1.3×10^9 µm^5 s^{-1} (data from Guadarrama-Cetina et al. 2013b). **b** As $p_i < p_s$, ice crystals can evaporate, forming an ice-water RIC δ_{iw}^*. **c** Since $p_{ss} < p_s$, supercooled droplets does not condense in the water-salty water RIC δ_{ws}^* around the salty drop. **d** When $p_{ss} < p_i$, ice crystals and supercooled droplet can coexist around the salty drop, forming three RICs: ice-water δ_{iw}^*, ice-salty water δ_{is}^*, water-salty water δ_{ws}^*. **e** Later, when $p_{ss} < p_i$, ice can reach the salty drop, preventing the further stage $p_i < p_{ss} < p$ where ice—salty water RIC cannot be present any more (Photos from Guadarrama-Cetina et al. 2013b)

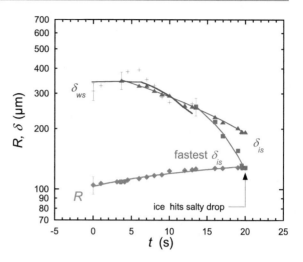

Fig. 7.16 Evolution of RIC radius δ_{ws} (SC water-salt, crosses) and δ_{is} (ice-salt, triangles and squares) correlated with salty drop radius R (circles). The lines are smoothing functions. The full line in R is a fit to Eq. (7.19) with $R_0 = 103$ μm and $B = 1.3 \times 10^9$ μm^5 s^{-1}. Ice eventually hits the salty drop at $t = 20$ s (radius $R_c = 128$ μm) (Adapted from Guadarrama-Cetina et al. 2013b)

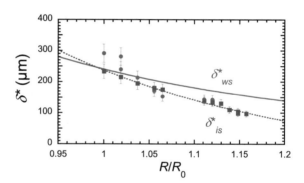

Fig. 7.17 Variation of RICs thickness for water-salty water (δ^*_{ws}, dots) and ice-salty water (δ^*_{is}, squares) with respect to reduced salty-drop radius R/R_0 (circles). The lines are fits to Eq. (7.27) (supercooled water) and Eq. (7.28) (ice) (Data are from Guadarrama-Cetina et al. 2013b)

Figure 7.17 reports the above variation with ζ as a fitting parameter. Agreement with data was good with $\zeta \approx 6$ mm for δ^*_{ws} and $\zeta \approx 10$ mm for δ^*_{is}. The ζ values are larger than a few mm, as expected. However, as already noted in Sect. 7.3.2, the equations above are scaled equations and precise comparisons with data remain always within a multiplicative coefficient of order unity but with precise values are difficult to estimate.

Border Effects

8

This chapter addresses different aspects of the growth of a droplet pattern near geometrical and thermal discontinuities. Near this frontier, the vapor concentration profile differs from what is found in the middle of the substrate, due to a different droplet arrangement.

A first experimental and numerical study of boundary effects was performed by Medici et al. (2014). Park et al. (2016) later used this effect to enhance droplet collection on bumps. Zhao et al. (2018) compared the effect of edges for condensation in humid air and pure water vapor, and Zhu et al. (2020) reported the combined thermal and diffusive effects of condensation on a cantilevered microfiber cooled at one end.

8.1 Geometric and Thermal Discontinuities

8.1.1 General

All drop growth laws studied in Chaps. 4 and 6 are concerned with (i) a uniform, infinite substrate and (ii) a substrate maintained at constant uniform temperature. The latter condition can be ensured by radiative cooling of the condensation substrate whose emissivity is close to unity, noting that water drops exhibit near unity emissivity (Beysens 2018; Trosseille 2019; Trosseille et al. 2021b) or by contact cooling with a substrate of high thermal conductivity in contact with a thermostat (conductive and radiative thermal modes are considered in Chap. 13).

Condensing substrates are of finite size and, in this aspect, two types of boundaries can be considered: (i) geometric discontinuities, e.g. the edges of the condensation substrate and (ii) thermal discontinuities, e.g. the border between cooled and non-cooled conductive support. Both geometrical and thermal discontinuities have in common a modification of the vapor concentration profile and

© Springer Nature Switzerland AG 2022
D. Beysens, *The Physics of Dew, Breath Figures and Dropwise Condensation*,
Lecture Notes in Physics 994, https://doi.org/10.1007/978-3-030-90442-5_8

a   b

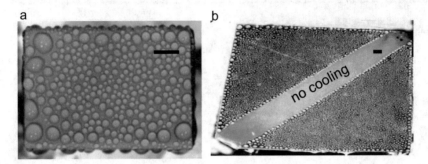

Fig. 8.1 Dropwise condensation on high conductivity substrates cooled from below. **a** PVC foil on brass parallelepiped (10×15 mm^2 area, 17.5 mm high; $t = 28$ min, air temperature $T_a = 25.9$ °C, RH $= 59\%$, substrate temperature $T_c = 7.7$ °C). **b** Polypropylene foil on duralumin diamond with non-cooled central stripe (35×35 mm^2 area, 8 mm high; $t = 26$ min 40 s, $T_a = 22.6$ °C, RH $= 44\%$, $T_c = 4.8$ °C). The bar corresponds to 2 mm (Adapted from Medici et al. 2014)

thus of the droplet evolution near the border. Thermal discontinuities and geometrical discontinuities can speed up or slow down droplet growth in their vicinity because more water vapor (or less) can be collected at the border (Fig. 8.1).

A quantification of the effect of borders on drop evolution needs to consider several factors. One first notices that drops near discontinuities always undergo coalescence with the neighboring drops that grow on the same plane. The drop advancing contact angle, being in general less than 90°, coalescences between orthogonal planes and can be neglected.

8.1.2 Linear Discontinuity

Border drops coalesce with the other drops sitting in a half-plane. The border drops can thus be considered as particular drops of the same plane but with a vapor concentration profile that is now two-dimensional (2D), in contrast to the linear profile observed in Sect. 6.1.2. Observations (Fig. 8.1) show that the corresponding drop pattern exhibits the features of scaling during growth: uniform radius and drop inter-distance on the order of the radius. Such scaling can be verified by the constancy, during the drop pattern evolution, of the border drop surface coverage:

$$\varepsilon_E = \frac{\pi \sum_E R^2}{S_E} \tag{8.1}$$

The summation is made on the border condensation surface, S_E. The latter is not constant during evolution and its determination can be performed by making use of a method based on Voronoi polygons or Dirichlet tessellation (Okabe et al. 1992). Such a polygon is the smallest convex polygon surrounding a drop whose sides are the bisectors of the lines between the drops and its neighbors. The border

Fig. 8.2 Voronoi polygon construction for corner, border and central region of the Fig. 8.1a substrate (Adapted from Medici et al. 2014)

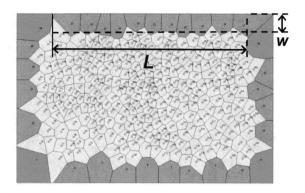

surface can thus be considered as the total surface of the Voronoi polygons of the border drops (Fig. 8.2).

Constant surface coverage is found for the border zone ($\varepsilon_E = 0.56 \pm 0.02$), see insets in Fig. 8.3 corresponding to the drop patterns in the Fig. 8.1 patterns. The mean radius can also be written as (line or area weight, Appendix G5, Eq. G5.3)

$$\langle R \rangle = \frac{\sum_E R^2}{\sum_E R} = \frac{\sum_E R^3}{\sum_E R^2} \tag{8.2}$$

8.1.3   Approximation of the Surface Coverage

A good approximation of the surface coverage can be obtained from the evaluation of the line surface coverage of the edge drops (Eq. 6.74):

$$\epsilon_1 = \frac{2 \sum_E R}{L} \tag{8.3}$$

The summation is made along the border of length L removing the two corner drops (Fig. 8.2). The border surface can then be considered as a band of length L and constant width w, the latter corresponding to the mean extent of the Voronoi polygons perpendicular to the border. From the Voronoi construction, $w = \alpha \langle R \rangle$, with α a proportionality constant on order two. One thus obtains the total surface $S_T \sim L \langle R \rangle$ and

$$\epsilon_E = \frac{\pi \sum_E R^2}{S_T} = \frac{\pi \sum_E R^2}{\alpha L \langle R \rangle} \tag{8.4}$$

Making use of Eq. (8.2), one readily obtains

$$\epsilon_E = \left(\frac{\pi}{2\alpha} \right) \varepsilon_1 \tag{8.5}$$

Fig. 8.3 Left ordinate: evolution of the mean drop radius <*R*> on corners, edge, and central area. The lines are fits to Eq. (8.7). Right ordinate: evolution of the drop surface coverage on corners, edges, and central area, fitted to a constant (solid lines: surface coverage; interrupted line: edge linear coverage). **a** Fig. 8.1a substrate ($T_a = 22.6$ °C, $T_c = 4.8$ °C, RH = 44%). **b** Fig. 8.1b substrate (same conditions) (Adapted from Medici et al. 2014)

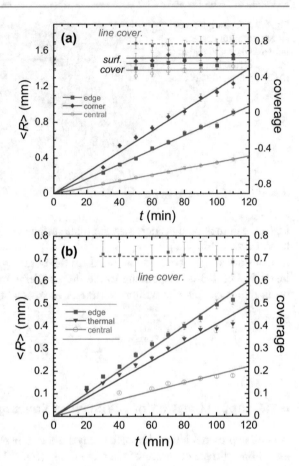

Insets in Fig. 8.3 show constant line and surface coverage with $\epsilon_E \approx 0.56$ for and $\varepsilon_1 \approx 0.75$, giving $\alpha \approx 2.1$, a value close to the expected value 2.

The vapor concentration profiles of the border drops overlap in the border direction. Then the assumption of equivalent film holds and, with $V_{T,E}$ the condensed volume near the edge

$$\sum_E R^3 \sim V_{T,E} \sim t \tag{8.6}$$

One thus eventually gets the growth law from a relation similar to Eqs. (6.40) and (6.44)

$$V_{T,E} \sim \langle R \rangle = k_E t \tag{8.7}$$

This law is similar to Eq. (6.44), however with a different prefactor $k_E > k_P$. The prefactor is estimated below by Eq. (8.13), corresponding to a different mean vapor profile on the drops. In an equivalent film approximation (Sect. 6.1.2), one

Table 8.1 Ratios of droplet radius growth rates between geometrical edges, corners, grooves (Fig. 8.4), thermal edge (Fig. 8.1a), k_i, and substrate central region, k_P. Comparison is also given with the simulated ratios (Adapted from Medici et al. 2014)

	Ratio	Expt	Simu
Corner	k_c/k_p	4.9	
Edge	k_E/k_p	2.8	1.90
Edge above groove	k_{G+}/k_p	1.6	1.66
Central groove	k_{GM}/k_p	0.86	0.43
Edge in groove	k_G/k_p	0.51	0.072
Thermal edge	k_T/k_p	2.5	2.64

has two perpendicular linear profiles, one directed perpendicular to the condensing plane and the other perpendicular to it. A rough estimation thus gives

$$k_F \sim 2k_p \qquad (8.8)$$

This factor 2 is indeed found approximately in experiments and simulations (Table 8.1).

Note that growth near a linear border ($<R> \sim t$, see Eq. 8.7) is different from one-dimensional (1D) growth (as drops growing on a very thin thread). In this case, according to Eqs. (6.61) and (6.73), one gets $<R> \sim t^{1/2}$.

8.1.4 Corner Discontinuity

This case is not very much different from the linear case, although there is only one drop growing at the corner. The corner drop can be considered as a particular drop of a linear border with a vapor concentration profile that is now three-dimensional (3D). This drop undergoes coalescence with other drops from two different borders and drops from the plane (see Figs. 8.1a and 8.2), leading to constant surface coverage (≈ 0.62 for Figs. 8.1a and 8.2 patterns). Then the film approximation still holds and growth follows Eq. (8.7), however with a coefficient $k_C > k_E > k_P$. As the corner is at the junction of 3 linear edges, one expects

$$kc \sim 4kp \qquad (8.9)$$

Experiments (Table 8.1) give a coefficient close to this value, $k_C \sim 5k_P$.

8.1.5 Droplet Growth Near a Discontinuity

In order to determine the droplet growth law near a boundary, one must estimate the local vapor gradient. The latter depends on the particular boundary geometry

Fig. 8.4 **a** Substrate geometry and **b** condensation pattern ($t = 120$ min; $T_a = 25$ °C, RH $= 78\%$, $T_c = 15$ °C). (c) Evolution of the mean drop radius <R> on (**a–b**) plate corners, edges, upper and lower groove edges and plate and groove central. The lines are fits to Eq. (8.7) (Adapted from Medici et al. 2014)

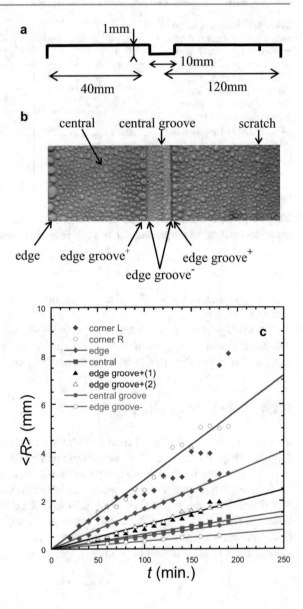

and must solve the Laplace equation, Eq. (4.17). One defines a mean gradient of concentration, $\langle \nabla c \rangle$, to put in evidence the role of the concentration gradient:

$$\langle \nabla c \rangle = \frac{1}{S} \int_S \left(\vec{\nabla} c \right)_R \cdot \vec{n} \, dS \tag{8.10}$$

The droplet volume growth rate, following Eq. (4.7), can thus be written as

$$\frac{dV}{dt} = \frac{D}{\rho_w} \int_S \left(\vec{\nabla} c\right)_R \cdot \vec{n} \, dS = \frac{DS}{\rho_w} \langle \nabla c \rangle \tag{8.11}$$

For a drop in a pattern with linear concentration profile (Sect. 6.1.2), ∇c thus becomes

$$\langle \nabla c \rangle_i = \frac{c_\infty - c_s}{\zeta_i} \tag{8.12}$$

The subscript $i = P$ (plane surface), E (edge), C (corner), etc. This formulation defines a new mean boundary layer thickness, ζ_i. The mean drop radius grows proportionally with time. Equation (6.44) still holds, however, with the new factor k_i giving $\langle R \rangle = k_i t$ (Eq. 8.7 for edges), such as, from Eq. (6.45)

$$k_i = \frac{1}{\varepsilon_{2i} G(\theta)} \frac{D}{\rho_w} \langle \nabla c \rangle_i \propto \langle \nabla c \rangle_i \tag{8.13}$$

Here, ε_{2i} is the surface coverage.

Figures 8.3 and 8.4c show the mean drop radius evolution for different types of convex or concave discontinuities, near geometric or thermal borders and linear or corner edges. The growth rate is different according to the type of discontinuities, increasing on convex forms and decreasing in concave geometries. Table 8.1 provides ratios of growth rates k_i for the different borders to the central region value k_P. Simulated ratios (see Sect. 8.2) are also given, except for corners, due to the 2D nature of the simulation. Growth enhancement with respect to the central region can reach more near 500%.

It is worthy to note that thick, low conductivity substrates cooled by contact might not exhibit edge effects. The reason is the cooling flux that now limits the condensation process everywhere on the substrate and prevents edge effects to take place (Medici et al. 2014; see also Chap. 13). A combination of both effects for cantilevered microfiber is discussed in Sect. 8.3.

8.2 Simulations

In a simulation, the diffusion equation Eq. (4.15) is solved in a fixed droplet configuration where the size of the droplets are fixed and droplets do not grow. Droplets are placed in various configurations with different concentration boundary layer geometries.

The transport of water molecules by convection is assumed to be negligible. The validity of this assumption is based on a scaling analysis of the free convection above a horizontal cooled plate. The diffuse boundary layer is evaluated in Sect. 2.2.2, Eq. (2.16) by $\zeta \equiv \delta_{FC} \approx \delta_T \approx \frac{D^{1/2}}{2} \left(\frac{\delta_F}{\beta g \Delta T}\right)^{1/4}$, where

$\delta_F = L\mathrm{Gr}^{-1/5} = L\left(\frac{g\beta\Delta T L^3}{\nu^2}\right)^{-1/5}$ (Eqs. 2.6 and 2.7) is the hydrodynamic boundary layer. Here, L ($\approx$20 mm) is the plate characteristic length, $\Delta T = T_a - T_c$($\approx$10 K) and $g = 9.81$ m s^{-2} is the earth acceleration constant. Making use of Table 1.2 data for the air volumetric thermal expansion coefficient β and air kinematic viscosity, one obtains Gr $\approx$ 13,000, $\delta_F \approx 3$ mm and $\zeta \approx 1$ mm. When the drop radius is below $\approx$1 mm, the convective mass transport remains below the diffusive transport and can be neglected.

This situation thus solves only the diffusion equation in a stationary state. As a result of the calculation, the drops receive a mass flux that characterizes their rate of growth in this fixed configuration.

8.2.1 Model Geometry and Boundary Conditions

The condensing plate is supposed to be either perfectly conductive (isothermal) to study the geometrical borders or partly conductive/adiabatic to study the thermal borders. The steady-state diffusion equation Eq. (4.16) $\Delta c = 0$ is solved in 2D for water vapor concentration c by using a finite element method implemented in MATLAB (pdetool). Note that a 2D simplification is consistent with the fact that the droplets close to edges are aligned (see Fig. 8.1). This is not the case for droplets at corners and is a simplification for the central part.

The geometry of the domains is depicted in Fig. 8.5. The boundaries of the domain are (i) the solid surfaces of the substrate where drops are present and (ii) the surface boundaries where water vapor concentration is kept constant. These latter surfaces correspond to the position of the diffusive boundary layer and are typical of the boundary layer thickness (1–2 mm).

8.2.1.1 Geometrical Discontinuities

Two substrate geometries are considered, edge (Fig. 8.5a) and edge plus groove (Fig. 8.5b), to compare with Figs. 8.1a and 8.4a configurations. On the substrate, hemispherical drops of diameter $2R = 200$ μm are distributed with a distance 50 μm between the drop perimeters (periodicity $d = 250$ μm). This configuration corresponds to a drop surface coverage $\varepsilon_2 = 0.50$, as observed during the coalescence-driven, self-similar coalescence stage (see Sect. 6.3). The drops are situated within a distance of 50 μm from the boundary.

For the edge case (Fig. 8.5ab), the thickness of the substrate is 2 mm. Zero transversal flux, i.e. zero concentration gradient, is imposed on the substrate between droplets. By symmetry, the transversal flux is zero on the boundaries $X = 3$ mm and $Y = 0$ mm. On the boundaries $Y = 3$ mm and $X = 0$, a uniform concentration layer with concentration $c_\infty = 2$ (in arbitrary units) is imposed. The layer is situated at a distance of 1 mm from the substrate, corresponding to the case $\zeta \gg d$. On a drop surface, a saturated concentration $c_s = 1$ is assumed, yielding a ratio $c_\infty/c_s = 1$ typical of the ratio p_∞/p_s used in the experiments.

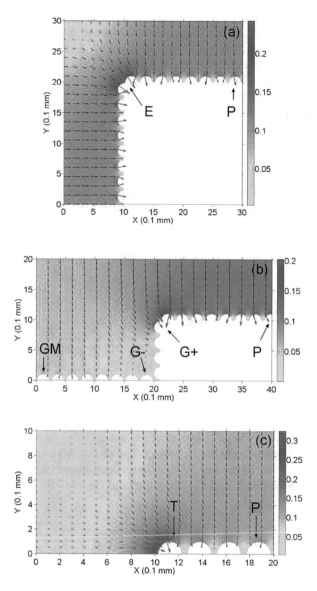

Fig. 8.5 Diffusion flux (arrows) and concentration gradient intensity (colors, arbitrary units). **a** Edge configuration. **b** Groove configuration. **c** Thermal configuration where the substrate $X <$ 1 mm is adiabatic Note the average vertical gradient in accord with the film model (Sect. 5.1.7) (Adapted from Medici et al. 2004)

The groove (Fig. 8.4a) is 1 mm in depth and 4 mm in width. For symmetry reasons, only half of the groove is simulated. The water vapor concentration conditions are the same as with the edge above, $c_\infty = 2$ at $Y = 1$ mm from the substrate $c_s = 1$ on the drops surface. In agreement with the edge case, the concentration gradient amplitude is maximized on the drop (G+) at the edge of the groove. In contrast, the gradient is smaller for the drops inside the groove, with decreasing values from the inner groove edge G− towards the inner groove middle GM. This simply reflects a larger distance towards the boundary layer border.

A thermal discontinuity is shown in Fig. 8.5c. The boundary ($Y = 0$, $X < 1$ mm) corresponds to an adiabatic substrate where drops cannot grow. As for edge and groove above, the concentration $c_\infty = 2$ at $Y = 1$ mm and $c_s = 1$ on the drops surface. The transversal concentration gradient is set to zero on the symmetry axis $X = 0$ on the solid surfaces of the substrate, and on the boundary ($Y = 0$, $X < 1$ mm). In Fig. 8.5c, similar to the edge and groove preceding cases, the concentration gradient is larger on drops situated at the edge of the (thermal) boundary (T). It decreases towards the middle (P) of the drop pattern.

8.2.2   Growth Rates

In order to determine a quantity proportional to the mean drop volume growth rate dV/dt and thus to $d<R>/dt = k_i$ from Eq. (8.11), the mean gradient $\langle \nabla c \rangle$ is calculated following Eq. (8.10), $\langle \nabla c \rangle = \frac{1}{S} \int_S \left(\vec{\nabla} c \right)_R . \vec{n} \, dS$. The integral of the concentration gradient is computed on drops marked in Fig. 8.5, P (central part of the substrate), E (edge above groove), G+ (outer edge of the groove) G− (inner edge of the groove), GM (central part of the groove) and T (thermal discontinuity). The ratio of each integral $\langle \nabla c \rangle$ to the integral $\langle \nabla c \rangle_M$ corresponding to the center of the substrate is obviously independent of the value of the vapor diffusion coefficient D and of the supersaturation and corresponds to the values k_i/k_P, with i indicating the position of the drop. According to Eqs. (8.12) and (8.13), one obtains

$$\frac{\langle \nabla c \rangle_i}{\langle \nabla c \rangle_P} = \frac{k_i}{k_P} \tag{8.14}$$

The results are listed in Table 8.1 where the calculated ratios can be compared to the corresponding experimental values. The agreement is relatively good taking into account the fact that the simulation is 2D. In particular, the edge calculation $k_E/k_P = 1.9$ corresponds to the estimated value ≈ 2 of Eq. (8.8).

There is one exception concerning k_{G-}/k_M (groove inner edge) where the calculated value is 7 times lower than the measured value. This discrepancy is due to the assumption in the simulation of a uniform boundary layer above the groove. In reality, the layer should curve towards the groove, thus enhancing the growth rate.

8.3 Competition Edge Effect—Thermal Conduction

Zhu et al. (2020) investigated a cantilevered metal microfiber cooled at its base and the effect of competition between thermal conduction resistance within the fiber and vapor diffusion at the cantilever surface. The typical geometry was (length) $L = 1 - 5.7$ mm, (diameter) $d = 127$ μm. Figure 8.6 presents three typical configurations corresponding to different cantilever length with same material. One notes that near the base, droplets are smaller or even condensation does not take place although it is the coldest part of the fiber. This effect is due to a decrease of water molecule diffusion due to the wall edge effect as discussed in Sect. 8.1.5. When going further from the base, condensation increases because (i) the wall effect diminishes and (ii) the concentration gradient increases as the tip comes closer to the boundary layer limit (see Chap. 2 and Sect. 8.2.2). Then condensation decreases due to the reduction of the cooling heat flux, to eventually vanish if the fiber is long enough. For an intermediate length where heat and diffusive flux balance, condensation appears nearly uniform, except close to the base (wall effect) and the tip, where diffusing flux is enhanced by an edge effect similar to what is found at corners (see Sects. 8.1 and 8.2).

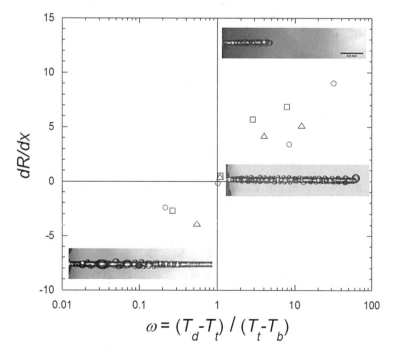

Fig. 8.6 Mean slope of the top-side condensate representing the drop radius variation from the base dR/dx with respect to $\omega = (T_d - T_t)/(T_t - T_b)$. Notations: see text. Photos: 316L stainless steel with same diameter (0.127 mm) but different lengths (1.0, 2.9, 5.7 mm) (Adapted from Zhu et al. 2020, with permission)

The condensation rate near the tip dm/dt, with m the mass of condensate, can be limited by the conductive heat flux inside the fiber. With λ the fiber thermal conductivity, T_t the tip temperature and T_b the base temperature, one has

$$\left(\frac{dm}{dt}\right)_c \sim \lambda \frac{\pi d^2}{4L}(T_t - T_b) \sim T_t - T_b \tag{8.15}$$

The condensation rate can be also limited by the diffusive flux of water monomers around the fiber. From Eq. (6.13), with c_∞, the water vapor concentration (in mass per volume) in air, c_s the saturation concentration, T_d, the humid air dew point temperature and ζ the boundary layer thickness

$$\left(\frac{dm}{dt}\right)_d = \frac{D(c_\infty - c_s)}{\rho_w(\zeta - L)} \sim T_d - T_t \tag{8.16}$$

The value of ζ (see Chap. 2) depends on the substrate geometry and whether convection is forced or free. It can be related to a convective heat exchange coefficient $a \sim (\zeta - L)^{-1}$ (see Appendix B, Eqs. B.4 and B.12). Measurement of heat transfer coefficient for free convection around cylindrical wires of varying diameter d give a dependence of $a \sim d^{-0.25}$ (Park et al. 2016).

Following Zhu et al. (2020), one can define thermal resistances, either conductive in the fiber as

$$R_{cd} = \frac{4L}{\pi \lambda d^2} \tag{8.17}$$

or convective at the fiber surface

$$R_{cv} = \frac{1}{\pi a d L} \tag{8.18}$$

Given the same heat flux density at the fiber tip, the ratio of convective and conductive thermal resistances can be expresses as

$$\omega = \frac{R_{cv}}{R_{cd}} = \frac{T_d - T_t}{T_t - T_b} \tag{8.19}$$

The relative amplitude of R_{cd} and R_{cv} can be tuned, giving rise to three condensate morphologies (Fig. 8.6). As a function of the distance from the base, x, and considering the drop radius variation dR/dx, condensation decreases ($\frac{dR}{dx} < 0, \omega < 1$), is nearly uniform ($\frac{dR}{dx} = 0, \omega = 1$) or increases ($\frac{dR}{dx} > 0, \omega > 1$). The numerical simulations by Zhu et al. (2020) confirm the experimental measurements.

Spatio-Temporal Dynamics

9

Drops growing on a substrate are not static and the fluctuations in time and space of the droplet configuration can exhibit very particular behavior. As a matter of fact, coalescence events, in addition to decreasing the drop-substrate surface of contact, makes the droplets move.

This chapter addresses the different spatio-temporal aspects occurring in a growing pattern due to coalescence: Droplet random displacement leading to a typical drop "diffusion", persistence problem in visited surface and temporal fluctuations in surface coverage. Note that the problems analyzed here are closely related to practical situations, e.g. the random motion of drops as induced by their coalescence can be a key process for drop collection (see Sect. 14.3.2) and the dynamics of the visited surface is of interest for drug spreading on vegetal leaves or surface decontamination.

9.1 Drop Self-Diffusion

Let us mark (e.g. by a dye) a droplet that has nucleated and follow its fate (Fig. 9.1a), studying how the dye spreads over the substrate by droplet growth and coalescence.

9.1.1 Number of Coalescence Events

As shown in Chap. 6, in the self-similar, coalescence-dominated regime, the mean radius of the droplets, $< R(t) >$, grows on average as a power function of time t with exponent $\gamma = 1$, while individual drop radius grows as t^β between coalescences, with $\beta = 1/3$. For a given t, the droplet size distribution scales with $< R(t) >$, i.e. all droplets are of similar radius (Sect. 6.3.2). Furthermore, the distance between droplets is also in the order $<R(t)>$ (see Eq. 6.41). Let us now

© Springer Nature Switzerland AG 2022

D. Beysens, *The Physics of Dew, Breath Figures and Dropwise Condensation*, Lecture Notes in Physics 994, https://doi.org/10.1007/978-3-030-90442-5_9

Fig. 9.1 **a** Drop motion over
a substrate thanks to the
growth and coalescence of
droplets. n: number of
coalescences, Δ :
displacement of the droplet
center); S: substrate area not
touched by a droplet. **b**
Evolution (semi-log plot) of
the number of coalescences n
that undergo 5 different
droplets (different symbols)
during their growth. (Adapted
from Marcos-Martin et al.
1995)

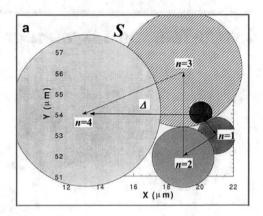

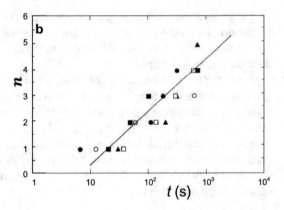

consider a particular droplet of size $\sim<R(t_i)>$, where t_i corresponds to the time
at which the ith coalescence of this droplet occurs. Its next coalescence with a
neighboring droplet will occur when its radius increases by a given (random) factor r_i, corresponding to the fact that the distance to the nearest droplet is also in
the order of $<R(t_i)>$. Since the growth of a droplet between two coalescences is
still governed by the $t^{1/3}$ law, the $(i + 1)$th coalescence will take place at time t_{i+1}
such as

$$t_{i+1} = r_i^{1/\beta} t_i \tag{9.1}$$

The time needed for n successive coalescences is thus given by

$$t_n = t_1 \prod_{i=2}^{n} r_{i-1}^{1/\beta} \tag{9.2}$$

Here, t_1 is the first coalescence time and the ratios $r_i > 1$ are random. Therefore,
t_n is log-normally distributed, with a most probable value given by

$$t_n = G^{n-1} t_1 \tag{9.3}$$

where

$$G = e^{\frac{\langle ln(r_i) \rangle}{\beta}} \tag{9.4}$$

Said differently, from Eq. 9.3, the number of coalescences experienced by a given droplet increases on average proportionally to $ln(t_n)$

$$n = 1 + \left(\frac{1}{ln(G)} \right) ln(t_n/t_1) \tag{9.5}$$

Below in Sect. 9.2.3, simplified models are presented that give additional evidence for this lnt behavior. Figure 9.1b shows that, experimentally, this law is obeyed satisfactorily. One infers the value from the slope $1/ln(G) \approx\sim 0.86$, corresponding to $G \approx 3.2$. This means that each droplet increases its size on average between two coalescences by a factor $G^{\beta} \approx\sim 1.47$. This number must be compared with the case of two droplets of the same size that coalesce where the factor is $2^{1/3} \simeq 1.3$. The agreement looks satisfactory considering the relatively small number of droplets analyzed and the fact that the droplet exhibits a radius polydispersity $g \approx 0.2 < R >$ (see Sect. 6.3.2).

9.1.2 Mean Drop Displacement

The mean displacement D of the drop center at time t (Fig. 9.1a) can be evaluated in a rather simple way given that the displacement is ruled by the last coalescence where the drop radius is the largest. The latter imposes a motion of the centers over a distance of the order R (Fig. 9.2), with $\sqrt{\langle \Delta \rangle^2} \sim R(t)$ in a random direction (neglecting, however, the correlation between positions and sizes of the different droplets, see Sect. 6.3.2). Hence, the displacement $\overrightarrow{\Delta_n}$ of the droplet after

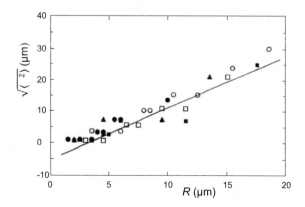

Fig. 9.2 Root mean square displacement $\sqrt{\langle \Delta \rangle^2}$ of a marked droplet with respect to its radius R, showing proportionality. (Adapted from Marcos-Martin et al. 1995)

n coalescences is given by

$$\vec{\Delta_n} = \sum_{i=1}^{n} R(t_i)\,\vec{e_i} \tag{9.6}$$

where $\vec{e_i}$ is a vector in the direction of the jump with a norm in the order of unity. Neglecting the correlations between $\vec{e_i}$, s, ($\langle \vec{e_i}.\vec{e_j}\rangle = 0$ if $i \neq j$) and the fluctuations in $R(t)$, one finds

$$\langle \Delta_n^2 \rangle = a^2 \langle e^2 \rangle \sum_{i=1}^{n} \left(\frac{t_i}{t_1}\right)^{2\gamma} \tag{9.7}$$

where we use $\langle R(t) \rangle = a(t/t_1)^\gamma$ with $\gamma = 1$ for 3D droplets on a 2D substrate (Sect. 6.3.2 and Eq. 6.46). One can estimate $\langle \Delta_n^2 \rangle$ as

$$\langle \Delta_n^2 \rangle \cong a^2 \langle e^2 \rangle \sum_{i=1}^{n} \left(G^{2\gamma}\right)^{i-1} \cong a^2 \langle e^2 \rangle \frac{G^{2\gamma}}{G^{2\gamma}-1} \left(\frac{t_n}{t_1}\right)^{2\gamma} \quad (n\,\text{large}) \tag{9.8}$$

where we use $t_n = G^{n-1}t_1$, Eq. 9.3. Equation 9.8, noting that $\langle R(t_n) \rangle = a(t_n/t_1)^\gamma$, can thus be rewritten as

$$\sqrt{\langle \Delta_n^2 \rangle} = A_\gamma \langle R(t_n) \rangle \tag{9.9}$$

With

$$A_\gamma \approx \sqrt{\langle e^2 \rangle} \frac{G^\gamma}{\sqrt{G^{2\gamma}-1}} \tag{9.10}$$

Equation 9.9 corresponds to Fig. 9.2 data and can readily be interpreted. Within a numerical prefactor, the total displacement after time t is given by the typical size of the droplets at this particular time, $R(t)$. In other words, only the last and largest jump determines the total distance traveled, which is reminiscent of the situation encountered in Levy flights (see e.g. Bouchaud and Georges 1990 and Refs. therein). In the present case, however, the size of the jumps grows continuously with time, whereas in Levy flights this is only true in a statistical sense.

In Fig. 9.2, the prefactor is found to be $A \simeq 1.66$, to be compared to Eq. 9.9, giving $\frac{\approx G^\gamma}{\sqrt{G^{2\gamma}-1}} = 1.05$ using $\gamma = 1$ and the value of $G = 3.2$ as determined above in Sect. 9.1.1). The agreement is satisfactory owing to the roughness of the arguments used. Similar results are obtained with the simplified model detailed in Sect. 9.2.3, with better quantitative agreement.

Note that the deformation of coalescing drops during coalescence can lead to further coalescence with another drop. It can therefore somewhat modify this diffusion-like behavior, as noted by Garimella et al. (2017).

9.2 Visited Surface

9.2.1 Dry Surface Power Law Decay

Another interesting question concerns the study of collective properties of the drop pattern, and especially how the ensemble of droplets wets the substrate. The question which then arises is to know which fraction f—called the "dry" fraction of the substrate—has not yet been touched by the droplets at any previous time. This fraction should not be confused with the dry fraction at a given instant of time, $1 - \varepsilon_2$ (≈ 0.55 for hemispherical drops, see Sect. 6.3.1).

This quantity has been determined by Marcos-Martin et al. (1995) by image analysis. At regular time intervals, the droplet configuration was recorded. Then all images corresponding to times prior to a given t are superimposed. The wet regions are black while regions never reached by the droplets are white. Figure 9.3a shows a typical contour of the dry region after a certain time t, well in the self-similar regime.

Fig. 9.3 a Contour separating dry substrate area (never touched by a drop) and wet substrate area (touched by a drop). The islands are the dry regions. Their shape is reminiscent of the circular shape of the drop (see Fig. 9.1a). **b** Evolution of the dry area f (log–log plot). The early time regime is not shown. (Adapted from Marcos-Martin et al. 1995)

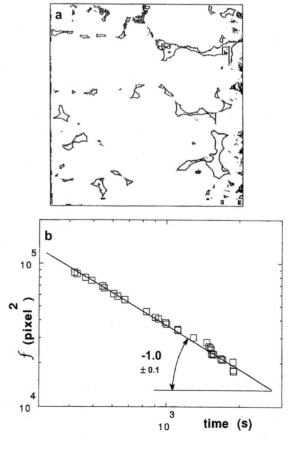

The dry fraction f is plotted in Fig. 9.3b as a function of time, clearly showing a power law decay

$$f \sim t^{-\theta} \tag{9.11}$$

with the surface dry fraction exponent $\theta = 1.0 \pm 0.1$. This power law decay is also found in the models described in Sect. 9.2.3, some of which being amenable to an analytical treatment allowing the exact determination of θ. The origin of this power law behavior can be understood as follows. Since the center of each droplet moves in a random direction when coalescence occurs, one can reasonably assume that after a few coalescences per droplet the whole spatial configuration is randomized. This configuration is seen as Poissonian with surface coverage $\varepsilon_2 = 0.55$. One can state that, after n_0 coalescences per droplet, the configuration will thus be completely random. The probability that a given point systematically escapes 'wetting' for a time T can thus be estimated as

$$f \approx (1 - \varepsilon_2)^{n/n_0} \tag{9.12}$$

Using the result $n \approx \ln t / \ln G$ $n \approx \mathrm{ln}t/\mathrm{ln}G$, Eq. 9.5, f is seen to indeed decay as in Eq. 9.11 with the exponent

$$\theta = -\frac{\ln(1 - \varepsilon_2)}{n_0 \ln G} \approx \frac{0.68}{n_0} \tag{9.13}$$

Here, one uses the values found above in Sect. 9.1, $G \approx 3.2$ and $\varepsilon_2 = 0.55$. This power law decay is thus the consequence of the multiplicative character of the probability of independent events, coupled with the logarithmic growth of this number of events with time. As such, it is expected to be quite general in coarsening problems. It appears to belong to the so-called "persistence" problem where the fraction of space persisting in its initial ($t = 0$) state is up to some later time t. It is a classic problem, which has been extensively studied over the past two decades (see e.g. Bray et al. 1994; Jain and Yamano 2019 and Refs. therein). Let us note that the motion of droplets upon coalescence is biased by, e.g. the presence of gravity effects or a gradient of surface tension, thus one should expect a much faster (exponential) completion rate.

9.2.2 Dry Surface Fractal Contour

The contour of the dry regions (Fig. 9.3a) reveals islands of various sizes and irregular aspects, resembling fractal geometry. This contour was analyzed by Marcos-Martin et al. (1995) by determining the masses in boxes of different sizes. The results are reproduced in Fig. 9.4 for various experimental times. In a regime extending from a few pixels to $\xi(t) \sim 50 - 100$ pixels, a cross-over length that is an increasing function of time, masses give a fractal dimension of the contour

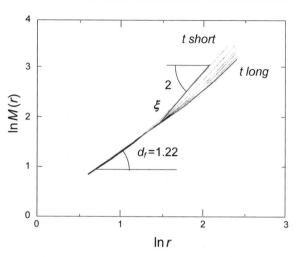

Fig. 9.4 Mass $M(r)$ contained in a box of size r as a function of r for various times t. The long time regime, characterized by a fractal dimension of 2, is reached beyond a cross-over length $\xi(t)$ which increases with t. (Adapted from Marcos-Martin et al. 1995)

$d_f \approx\sim 1.22$ and crossing over to $d_f = 2$ for $r > \xi(t)$. It is indeed expected that the fractal region is of finite extent when time is finite. The fractal nature of the contour was found in the model studied by Bray et al. (1994) but not in the simplified model of Sect. 9.2.3 which uses monodisperse drops.

9.2.3 Models and Simulations

9.2.3.1 One-Dimensional Model

In this simplified model, proposed by Derrida et al. (1990, 1991), N points are randomly distributed on a line, with periodic boundary conditions. They represent the nucleation sites, i.e. the centers of the droplets. Let hi be the distance between the points and take a uniform distribution of hi as the initial condition. Droplets grow with time and one takes snapshots of the system whenever one coalescence occurs. The distance between the centers of the coalescing droplets is equal to their common diameter, i.e. to the shortest interval h_{min}. These two coalescing droplets are replaced by one single droplet, of *same size*, centered half-way between them. The total number of droplets is thus reduced by one. This procedure is then iterated, the minimal distance h_{min} increasing at each iteration because of the continuous droplet growth. At each step, one thus searches for the two closest points and replace these points by a single point half-way between them (Fig. 9.5).

Since the system is observed only when coalescence occurs, h_{min} plays the role of effective time. Let us call L the length of the system and $N(h_{min})$ the number of droplets at this time. One obtains

$$L = \sum_{i=1}^{N} h_i \qquad (9.14)$$

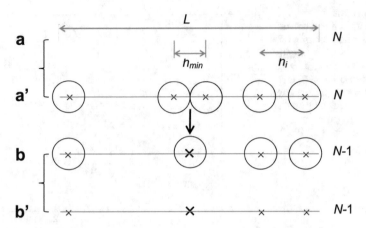

Fig. 9.5 Simplified 1-D model where only coalescence is considered. **a** N point randomly distributed. (**a'**) Check for coalescence ($h_i = h_{min}$). **b** Coalescence without growth. (**b'**) $N-1$ points rearranged by coalescence

The line coverage at "time" h_{min}, keeping in mind that h_{min} is also the droplet diameter, can be expressed as

$$\epsilon_1(h_{min}) = \frac{N(h_{min})h_{min}}{L} = \frac{h_{min}}{\langle h \rangle} \tag{9.15}$$

Here, $\langle h \rangle$ is the average inter-droplet distance at "time" h_{min}. In the scaled regime, ϵ_1 is a constant (≈ 0.75, see Sects. 6.3.1 and 6.4.2.1).

9.2.3.2 Number of Coalescences
Let us follow the evolution of a droplet in time. The minimal interval length at the nth coalescence of this particular droplet is written $h_{min}^{(n)}$. For n large enough, one may write, similar to Sect. 9.1.1, recalling that h plays the role of time:

$$h_{min}^{(n)} \cong h_{min}^{(0)} \exp(\lambda n) \tag{9.16}$$

where

$$\lambda = \left\langle \ln\left(\frac{h_{min}^{(n)}}{h_{min}^{(n-1)}} \right) \right\rangle \tag{9.17}$$

In other words, n varies logarithmically with $h_{min}^{(n)}$, similar to Eq. 9.5 in Sect. 9.1. From Eq. 9.16, it indeed comes

$$n = \frac{1}{\lambda}\ln\left(\frac{h_{min}^{(n)}}{h_{min}^{(n-1)}} \right) \tag{9.18}$$

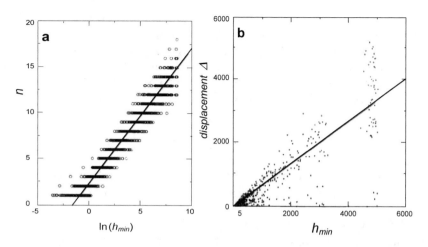

Fig. 9.6 100 histories starting with 20,000 points in the 1D model. **a** Number of coalescences n of a marked droplet with respect to logarithm of effective time h_{min}. **b** Total displacement Δ of a marked droplet versus h_{min} for 100 histories and 20,000 starting points. The average slope is 0.67. (Adapted from Marcos-Martin et al. 1995)

By analogy, with the quantity denoted by G introduced in Sect. 9.1.2, Eq. 9.3:

$$G = \exp(\lambda) \tag{9.19}$$

In order to have an estimate of G, n is plotted with respect to $\ln(h_{min})$ (Fig. 9.6a). The average slope is $1/\lambda = 1.49$, giving $G \approx 1.96$. It means that a droplet, between two coalescences, nearly doubles in size, then

$$h_{min}^{(n)} \approx 2^n h_{min}^{(0)} \tag{9.20}$$

This provides the following intuitive picture where, for each generation, every droplet experiences one coalescence, dividing the number of droplets by 2. Since in the scaling regime the line coverage is constant, Eq. 9.15 shows that between two coalescences h_{min} doubles.

9.2.3.3 Droplet Displacement

One can follow the displacement of a marked droplet. During the nth coalescence, the droplet moves over a distance $h_{min}^{(n)}/2$ to the left or to the right according to the position of the droplet with which it coalesces. One can characterize the total displacement of the droplet after n coalescences by

$$\Delta_n = \left| \sum_{i=1}^{n} \sigma \frac{h_{min}^{(n)}}{2} \right| \tag{9.21}$$

Here, $\sigma = 1$ if the droplet moves to the right and -1 to the left. Figure 9.6b shows that Δ_n varies linearly with $h_{min}^{(n)}$, with a mean slope ≈ 0.67. This value can be compared to the following estimation, following the approach of Sect. 9.1.2. One assumes that Eq. 9.16 is exact at each coalescence. Then, neglecting correlations, the total displacement is found by summing a geometrical series similarly to Eq. 9.8, leading to the following value (compare with Eq. 9.9 with $\gamma = 1$):

$$\Delta_n = \frac{G}{2\sqrt{G^2 - 1}} h_{min}^{(n)} \tag{9.22}$$

The above equation corresponds indeed to a proportionality between Δ_n and $h_{min}^{(n)}$ with a factor of 0.58, not far from the mean slope 0.67 of Fig. 9.6b.

9.2.3.4   Dry Surface Evolution

The line is progressively visited or "wet" by droplets during the evolution. Figure 9.7 gives the fraction of dry space with respect to h_{min}. in a log–log plot. It clearly shows a power law decay of the dry fraction, in the scaling regime, with exponent $\theta \approx 2.2$:

$$f \approx h_{min}^{-\theta} \tag{9.23}$$

The dynamics of the present model is in fact a deterministic version of the Glauber dynamics of the infinite-state Potts model in one dimension at zero temperature. In the Potts model, domains are defined as regions where spins have the same state. They are separated by domain walls performing random walks, hence "wetting" the space that they traverse. When two domain walls meet, they

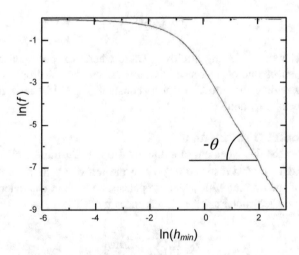

Fig. 9.7 Fraction of dry space f as a function of h_{min} for the 1D model (20,000 starting points). (Adapted from Marcos-Martin et al. 1995)

annihilate and are replaced by one single domain wall. In our model, domains are intervals and domain walls are the centers of the droplets. The centers of the droplets nearest from one another are replaced by one point half-way between them.

In the infinite-state Potts model, starting with a random initial configuration, one can measure the fraction r_t of spins that never flipped from time $t = 0$. Following the arguments developed by Derrida et al. (1994), the domain walls can be considered as particles on the line performing a Brownian motion with the reaction rule that when two particles meet, they merge into one particle: $A + A \rightarrow A$. Any given site does not flip between time 0 and time t, if and only if, it has neither been visited by the first particle on its right nor the first particle on its left. It is a well-known fact of random-walk theory that the probability p_t that a site is not visited by a random walker at its right decays as $p_t \sim t^{-1/2}$. If one wants a site to be visited neither by the particle at its right nor by the particle at its left, then

$$r_t = p_t^2 \sim t^{-1} \tag{9.24}$$

Since the mean length of domains, l, increases with time as

$$l \sim t^{1/2} \tag{9.25}$$

one readily obtains

$$r_t \sim l^{-2} \tag{9.26}$$

The quantity r_t is analogous to the dry fraction f and Eq. 9.26 is similar to Eq. 9.23, with $\theta = 2$.

One can also consider the heuristic argument presented at the end of Sect. 9.1. Suppose that one places at random "droplets" of size 1 on the line, separated by distances h_i randomly distributed between 1 and 2. These droplets represent the wet region at the first generation. The corresponding coverage is the density (2/3) times the common size (1) of the droplets. Hence, the dry fraction is 1/3. In the second generation, one distributes at random droplets of size 2, with separations between 2 and 4. Again the coverage is the density (1/3) times the size of the droplets (2), hence the dry region is 1/3. One iterates this process. The dry fraction after each step is the intersection of the uncovered regions of all previous steps. Generations are independent, thus the dry fraction after n generations is equal to $(1/3)^n = h_{min}^{-\theta}$. Since $h_{min} = 2^{n-1}$, the exponent $\theta = \ln 3 / \ln 2 \approx 1.58$.

One also can use the same reasoning with a distribution of h_i whose rescaled average is equal to $1/\varepsilon_1$ where $\varepsilon_1 = 0.75$ is the actual coverage. This leads to a dry fraction equal to $(1 - \varepsilon_1)^n$ after n generations, hence an exponent $\theta = -\ln(1 - \varepsilon_1)/\ln 2 \approx 2$, a value not very far from the numerically determined value $\theta = 2.2$.

9.2.3.5 Structure of the Dry Contour

It is possible to look at the set of points that constitute the boundary between the wet and dry regions on the whole interval. In order to determine whether this set is fractal, two (related) quantities have been calculated by Marcos-Martin et al. (1995): (i) the two-point correlation function $C(r)$, measuring the probability of finding a point of the set at a distance r from a given point of the set, and (ii) the number $N(r)$ of points of the set within a distance r from a given point of the set. $C(r)$ fluctuates around the average density of points, c. These fluctuations increase as the number of points decrease. Similarly, it was found that $N(r) \sim cr$. This implies that the set of points of the boundary of the dry region are uniformly scattered (Poisson distribution) and does not form a fractal set.

In an attempt to determine whether the absence of fractality was due to the one-dimensional (1D) character of the simulation, an extension to two-dimensional (2D) of the above 1D model was considered. Again, one takes all droplets to be of equal radii at all times with each droplet represented by its center. The dynamics of growth and coalescence is performed on these points. The growth component of this model is achieved by searching for the two nearest points. This minimal distance h_{min} is the diameter of the droplets at that stage. Coalescence of these two droplets is done by replacing the two points by a single point midway between them (their center of mass). The points are placed on a square of side 1, with periodic boundary conditions (Fig. 9.8).

The quantities $C(r)$ and $N(r)$ were measured for the boundary between the wet and dry regions. A typical example of such a contour is given in Fig. 9.9 and should be compared with Fig. 9.3a. The existence of a characteristic length scale is quite apparent, suggesting that this contour is, as in the 1D case, not fractal. In

Fig. 9.8 Contour of the dry region for the two-dimensional model with monodisperse droplets. Note the difference with Fig. 9.3a induced by the droplets monodispersity. (Adapted from Marcos-Martin et al. 1995)

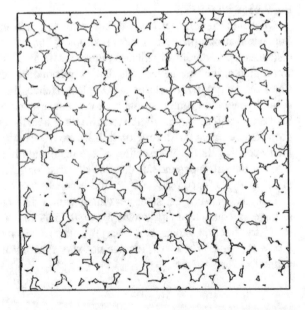

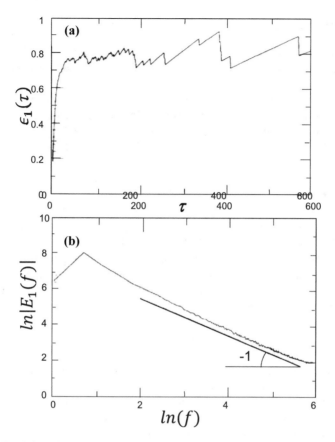

Fig. 9.9 a Evolution of the surface coverage $\varepsilon_1(\tau)$ in a typical simulation (here in Eq. 9.27 $\beta = 0.7$). R_0 is such that the asymptotic value of ε_1 is reached after about $\tau = 30$ time steps. The initial number of droplets $N_0 \approx 10^3$; a simulation lasts typically 600 time steps. **b** Fourier transform modulus $|E_1(f)|$ of the coverage $\varepsilon_1(\tau)$ shown in (**a**) with respect to frequency f (log–log plot). (Adapted from Steyer et al. 1992a)

fact, as shown in Marcos-Martin et al. (1995), $C(r)$ does not exhibit any longrange behavior typical of fractal structures and $N(r)$ grows with a slope of 2, with no fractal dimension.

Between experiments where fractality was observed and simulations, where it was not, the immediate difference is the absence of polydispersity in the drop diameters for the simplified model. It is thus very likely that the polydispersity is at the origin of the fractal nature of the dry islands in the experiments. In support of this claim, one must note that the fractal theory of dropwise condensation (Mei et al. 2009, 2015; see Sect. 6.6.3.2) is based on the multiscale size of the droplet.

9.3   Temporal Noise in Surface Coverage

Drop surface coverage is constant in the self-similar regime. However, it exhibits fluctuations due to the coalescence events. The coverage fluctuates because of two competing processes that balance each other: droplet growth, which increases surface coverage, and droplet coalescence, which decreases coverage. This phenomenon can be viewed as a hierarchical process since each pair of coalescing droplets is replaced by a new droplet. Under certain circumstances, such hierarchical processes are able to generate $1/f^a$ noise with exponent $a = 2$.

9.3.1   1D Numerical Study

Simulation by Steyer et al. (1992a) follows the same general rules as reported in Sect. 6.4.2.1 for a 1D substrate. Spherical droplets are initially placed at random on a line of length unity. Each droplet radius grows following a power law similar to Eq. 6.15, with exponent β Several values of β were used (0.1, 0.2, 0.3, 0.5, 0.7, 1.3) and gave the same results. With dimensionless time τ one gets, with R_0, the initial drop radius at time $\tau = 1$.

$$R = R_0 \tau^{\beta} \tag{9.27}$$

At each time step t of the simulation, coalescence between droplets is checked and each coalescing pair is replaced by a new droplet. Mass conservation defines the new radius, which is taken as the new prefactor of the growth law.

The relevant quantity is the fraction $\varepsilon_1(\tau)$ of the line which is occupied by the droplets (Fig. 9.9a). As discussed in Sects. 6.3.1 and 6.4.2.1, the mean line fraction $\langle \varepsilon_1(\tau) \rangle$ is asymptotically constant and reaches a value near 75%. The discrete Fourier transform of $\varepsilon_1(\tau)$ is computed according to, with p the total number of time steps and f the frequency:

$$E_1(f) = \sum_{i=1}^{p} \varepsilon_1(\tau) \exp\left[2\pi i \frac{(f-1)(t-1)}{T} \right] \tag{9.28}$$

Figure 9.9b shows that over two decades the $F(f)$ modulus follows a power law with exponent close to unity. This behavior is valid even at low frequency, the left increasing part of the spectrum being an artifact due to the two smallest discrete frequencies (those which correspond to the total duration of the simulation) and the small saturation at large frequency is due to the finite size of the time steps. In terms of power spectrum, this is a "$1/f^2$" noise since $|F(f)|^2$ is the Fourier transform of the $S(\tau)$ autocorrelation function:

$$|E_1(f)|^2 \sim \frac{1}{f^2} \tag{9.29}$$

Such $1/f^2$ noises are trivially observed in the many cases where correlations decay exponentially. In this case, the $1/f^2$ behavior is only present above some cutoff frequency. Such a type of noise is also encountered when a distribution of correlation times is assumed, see e.g. Kertesz and Kiss (1990). In contrast, here this particular $1/f^2$ noise does not exhibit any cutoff frequency.

9.3.2 Analytical Model

One now considers the Zheng et al. (1989) simplified model and the Steyer et al. (1992a) approach. The droplets are originally equidistant and monodisperse. Droplets grow according to Eq. 9.27 and thus coalesce simultaneously. It is assumed that coalescence proceeds with one of the two neighbors. This artificial rule preserves the self-similarity of the pattern. The coalescence events occur at definite times τ_i, which, for a number of i events, are given by

$$\tau_i = 2^{2i/3\beta} \qquad (9.30)$$

The τ_i are then exponentially distributed. In this model, $\varepsilon_1(\tau)$ oscillates between the values $2^{-2/3}$ and 1 (Fig. 9.10). These oscillations slow down with time because, between times τ_i and τ_{i+1}, $\varepsilon_1(\tau)$ varies as $(\tau/\tau_{i+1})^\beta$. $\varepsilon_1(\tau)$) can be written as the product of τ^β and a function $P(\tau)$ which takes the value $\tau_{i+1}^{-\beta}$ over each interval $[\tau_i, \tau_{i+1}]$. This stepwise function is the sum of Heaviside-like functions $H_i(\tau)$, which takes the value $(1 - 2^{-2/3})\tau_i^{-\beta}$ on $[0, \tau_i]$ and the value zero elsewhere. Each $H_i(\tau)$ function represents the contribution of the ith-level coalescences which occur at time τ_i.

However, in the experimental case, droplets are neither equidistant nor monodisperse, giving for $S(\tau)$ the shape in Fig. 9.9a which differs from the shape in

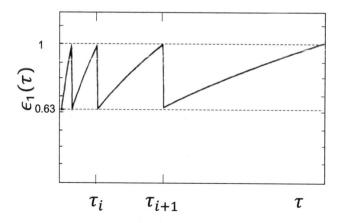

Fig. 9.10 Line coverage $\varepsilon_1(\tau)$ in the Zheng et al. (1989) simplified model

Fig. 9.10. One can take into account this difference by introducing fluctuations in the coalescence times. The ith-level coalescences are now assumed to occur at different times around τ_i. The fluctuations are supposed to remain weak enough in order to still distinguish different levels i of coalescence. As a consequence, the $h_i(\tau)$ functions should scale with the characteristic time τ_i.

A rigorous calculation can be performed if, instead of a Heaviside function for $H_i(\tau)$, one considers an exponentially decaying function with time

$$H_i(\tau) \Rightarrow \left(1 - 2^{-2/3}\right)\tau_i^{-\beta}\exp(-\tau/\tau_i) \tag{9.31}$$

The surface coverage $S(\tau)$ is the sum of all the $H_i(t)$ multiplied by τ^β

$$\varepsilon_1(\tau) = \tau^\beta\left(1 - 2^{-2/3}\right)\sum_{i=0}^{\infty} \tau_i^{-\beta}\exp(-\tau/\tau_i) \tag{9.32}$$

As $\varepsilon_1(\tau)$ converges, one can obtain its Fourier transform

$$E_1(\omega) = \left(1 - 2^{-2/3}\right)\sum_{i=0}^{\infty} \tau_i^{-\beta} \int_0^{\infty} \tau^\beta\exp(j\omega\tau - \tau/\tau_i)d\tau \tag{9.33}$$

According to the Watson theorem (Jeffreys and Jeffreys 1950), Eq. 9.33 becomes

$$E_1(\omega) = \left(1 - 2^{-2/3}\right)\Gamma(\beta + 1)\sum_{i=0}^{\infty} \frac{\tau_i}{(1 - j\omega\tau_i)^{\beta+1}} \tag{9.34}$$

Here, Γ is the gamma function (see Appendix K). The summation can be performed in two steps around the particular value $i = N = -3\beta\ln|\omega|/2\ln2$. When $i < N$, one makes the approximation $\omega\tau_i \ll 1$ and when $i > N$, $\omega\tau_i \gg 1$, which gives

$$F(\omega) = \left(1 - 2^{-2/3}\right)\Gamma(\beta + 1)\left[\sum_{i=0}^{N-1} \tau_i + \sum_{i=N}^{\infty} \frac{\tau_i^{-\beta}}{(-j\omega)^{\beta+1}}\right] \tag{9.35}$$

It is possible to compute these two sums, which form geometric series. For small ω, one obtains

$$E_1(\omega) = \frac{\Gamma(\beta + 1)}{\omega}\left[\frac{\left(1 - 2^{-2/3}\right)}{2^{2/3\beta} - 1} + \frac{1}{(-j)^{\beta+1}}\right] \tag{9.36}$$

Equation 9.36 shows that $E_1(\omega) \sim 1/\omega$, i.e. that the noise is in $1/f^2$. This noise is observed even at small frequencies, in contrast to the classical $1/f^2$ noise. A very remarkable result is that the precise value of the growth exponent β is not relevant for the existence of this $1/f^2$ noise.

The analytical model above can be extended to the case of a 2D substrate. In this case, $\varepsilon_1(\tau)$ oscillates as for the 1D substrate. The two extreme values are, however, different (see Zheng et al., 1989) but this fact does not change Eq. 9.36. As a consequence, one expects a $1/f^2$ noise to be observed also in the case of a 2D substrate and more generally in all the cases where the substrate has a lower dimensionality than that of the drops to ensure constant surface coverage and scaling (see Sect. 6.3.1). Note that this $1/f^2$ noise was observed experimentally, in a BF experiment where the drop surface coverage $S(\tau)$ can be deduced from the measurement of the light transmittancy and then Fourier analyzed (Beysens 1991).

Micro- and Nano-Patterned Substrates

10

Micro- and nanopatterned substrates were initially considered as substrates with controlled roughness. Roughness indeed amplifies the hydrophilic or hydrophobic character of the materials (see Appendix E). On such substrates, a drop can either be in a superhydrophobic, Cassie-Baxter (CB) composite air-pocket state, sitting on the head of the microstructures, or in a penetration, Wenzel (W) state, where it wets the substrate materials and fills the microstructures. In the latter state, the drop is strongly pinned, in contrast to the CB state. The equilibrium state depends on whether CB or W state corresponds to the minimum of energy. Metastable states are, however, frequent because of the presence of an energy barrier when going from one state to the other. This barrier can be overcome by, e.g. firmly pushing the drop (CB to W) or occurs thanks to its coalescence with another drop. The latter process occurs during condensation when drops grow and invade the microstructures.

Micropatterning is not only related to geometry (e.g. pillars, grooves) but can be of a chemical nature by varying the local wetting properties (e.g. hydrophilic-hydrophobic stripes). Also, both geometrical and wetting patterns can be considered as in biphilic structures (e.g. hydrophobic pillars with hydrophilic heads).

Since the first investigations in the early 2000s, condensation on such micropatterned surfaces has been the object of many studies. The general motivation to use micropatterned surfaces was to increase the water condensation yield by improving drop nucleation and collection, or enhancement of heat transfer by better condensate collection. The number of studies concerning this process is such that one can cite all. Table 10.1 gives an overview of the patterns that were developed. The geometries exhibit quite various forms such as grooves, pillars (circular or squares), cones, pyramids, arrows, etc. The designs can be made by 3D patterning or alternating hydrophobic and hydrophilic coatings, or both. Different scales can be used as, e.g. covering a micropattern with nanocarbon tubes (nanoforest) or, when using silicon, etching the pattern to produce nanograss. Figure 10.1

© Springer Nature Switzerland AG 2022
D. Beysens, *The Physics of Dew, Breath Figures and Dropwise Condensation*,
Lecture Notes in Physics 994, https://doi.org/10.1007/978-3-030-90442-5_10

Table 10.1 Examples of micro- and nano-patterned structures. (a) Narhe et al. (2004a), (b) Zhong et al. (2013), (c) Izumi et al. (2004), (cc) Bintein et al. (2019), (d) Wier et al. (2006), (e) Narhe and Beysens, (2007), (ee) Dorrer and Rühe (2007), (f) Jung and Bhushan (2008), (ff) Chen et al. (2009), (fff) Xiao et al. (2009), (g) Zhang et al. (2016), (h) Narhe (2016), (hh) Shim et al. (2017), (i) Chen et al. (2011), (ii) Mouterde et al. (2017), (j) Lau et al. (2003), (jj) Feng et al. (2004), (jjj) Zhu et al. (2006), (jjjj) Artus et al. (2006), (k) Chen et al. (2007), (l) Chen et al. (2011), (ll) Lo et al. (2014a), (m) Dai et al. (2018), (mm) Wen et al. (2018b), (n) Izumi et al. (1986), (o) Trosseille et al. (2019), (p) Chen et al. (2004), (pp) Su et al. (2009), (q) Baldacchini et al. (2006), (qq) Liu et al. (2021), (r) Anand et al. (2012), (2015), (s) Dai et al. (2018), (t) Park et al. (2016), (u) Chaudhury et al. (2014), (v) Xie et al. (2020), (w) Schaffle et al. (2003), (x) Alizadeh-Birjandi et al. (2018), (y) Varanasi et al. (2009), (yy) Su et al. (2016), (z) Hou et al. (2015).

Type	Form	Uniform coating	Refs.	Non-uniform coating hydrophobic/hydrophilic	Refs.
Smooth	–	–	–	Stripes / Dots, squares	(u) (v) / (w) (x)
1-scale	Groove		(a) (b) (c) (cc)		
	Pillar		(d) (e) (ee) (f) (ff) (fff) (g) (h) (hh)	Hybrid	(y) (yy)
	Cone, pyramide		(i) (ii)		
	Nanograss		(j) (jj) (jjj) (jjjj)		
2-scale	With nanograss		(k) (l) (ll) (m) (mm)	Hybrid with nanograss	(z)
Random	Sand-paper scratches		(n)		
	Sand-blasting craters		(o)		
Liquid-infused	Flowers / Black Silicon cones		(p) (pp) / (q) (qq)		
	Pillar		(r)		
	Groove		(s)		
	Asymmetric bumps		(t)		

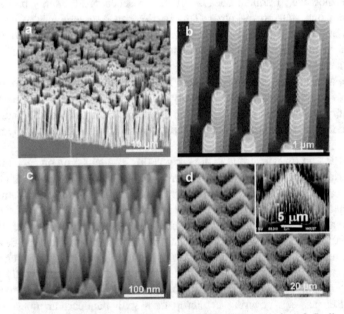

Fig. 10.1 **a** Clustered silicon nanowires (Chen et al. 2009, with permission). **b** Scalloped silicon nanowire arrays (Xiao et al. 2009, with permission). **c** Conical silicon nanowire arrays (Mouterde et al. 2017, with permission). **d** Micro-pyramids covered by silicon nanowires (Chen et al. 2011, with permission)

shows several examples. An important development was the possibility to imbibe micropatterns with liquids (see Sect. 11.2).

Below are reviewed the different modes of micropattern fabrication. Then condensation is considered on the most common patterns such as pillars, cones, grooves, and hydrophobic-hydrophilic stripes. Because drop adhesion can be made very small in some patterns, drops can jump out of the surface when coalescing with each other. Sect. 10.4 is devoted to this motion.

10.1 Fabrication of Micro-Patterns and Functional Coatings

Here one considers the most important processes of fabrication of micro- and nanostructures. Many publications since the 1990s have been devoted to this question. The field is very active with, however, a trend to report industry-related processes. The referenced works below are frequently taken from the well-documented reviews by Wen et al. (2018a), Hsu et al. (2011) and Roach et al. (2008), where more details and references can be found.

10.1.1 Functional Coatings

Microstructures are quite generally coated to give them specific hydrophobic or hydrophilic characteristics. The main problem encountered with coatings is durability.

10.1.1.1 Hydrophilic Coatings

Such hydrophilic coatings provide low or even zero values to the drop contact angle. Many methods are used such as sprays, plasma polymerization (PDMS), electrochemical deposition and solvent-mediated phase separation to produce hydrophilic coatings such as branched-polyethyleneimine and N-isopropylacrylamide films, silicon dioxide and titanium dioxide films, and graphene oxide (coating on poly(dimethylsiloxane) substrates (see e.g. Maheshwari et al. 2010; Yang et al. 2012; Kou and Gao 2011; Jeong et al. 2009; Tsuge et al. 2008; Foster et al. 2012).

10.1.1.2 Hydrophobic Coatings

Such coatings are used to increase the drop contact angle and make a rough substrate superhydrophobic. Various hydrophobic coatings can be applied (silane, long-chain fatty acids, polymer materials, rare-earth oxide ceramics and self-assembled monolayers, see, e.g. Chaudhury and Whitesides 1992; Veeramasuneni et al. 1997; Lau et al. 2003; Kulinich and Farzaneh 2004; Gnanappa et al. 2008; Paxson et al. 2014; Wen et al. 2015, 2016; Azimi et al. 2013). Well-known polytetrafluoroethylene (PTFE, Teflon) coatings have also been used (McCarthy et al. 2014). Chemical vapor deposition and plasma-enhanced chemical vapor deposition techniques can be used to grow ultra-thin polymer coatings (Paxson et al. 2014).

REO Ion Beam Sputtered thin film coatings can also be applied as hydrophobic material (Azimi et al. 2013).

Thick polymer coatings have been shown to be robust (Gnanappa et al. 2008) and ultra-thin graphene also exhibits good durability (Preston et al. 2015).

Many processes are not applicable to produce functional coatings on nanowires in view of their nanoscale spacing and high aspect ratio. However, nm monolayers have been applied on nanowires (perfluorooctyl trichlorosilane by Wen et al. (2018b) and Chen et al. (2011) and n-octadecyl mercaptan (Wen et al. 2017b, c). A 20 nm PTFE layer has also been coated on silicon nanowires using plasma reactive ion etching (Lu et al. 2017). Atomic layer-deposited coatings can protect against corrosion and aging (Yersak et al. 2016; George 2010).

10.1.2   Incompletely Ordered Roughness

The separation between random and periodic roughness is sometimes arbitrary as some processes exhibit self-organization and not fully random.

10.1.2.1   Stretching
A quite simple way to produce a rough superhydrophobic surface is to stretch a thin sheet of PTFE fibrous crystals. A large fraction of void spaces between crystals forms (Zhang et al. 2004).

10.1.2.2   Sandpaper
Paper with different roughnesses (#320, #120 and #80) were used by Izumi (1986) to scratch horizontally a smooth surface. The latter was initially a mirrorwise smooth surface that was eventually polished with a 30 nm alumina plate and then buffed.

10.1.2.3   Sandblasting
This is a non-conventional technique used by Trosseille et al. (2019) to enhance drop nucleation. It consists of impacting the surface with a flux of silica beads under pressure. The grains are directed perpendicular to the surface, explode at the impact and form a circular crater whose depth and width depend on the applied pressure (see Appendix L for details).

10.1.2.4   Phase Separation
Phase separation obtained by changing the temperature of a multi-component mixture can occur through a bicontinuous structure, the so-called "spinodal decomposition" (for fundamentals of phase transition, see, e.g. Onuki et al. 2002). The structure, initially very fine, coarsens with time. If one component solidifies during the process, the second phase can be removed to create a solid porous network. The size of the pores in the solid can be controlled by tuning the rate of phase separation with respect to the rate of solidification. This is an interesting easy and low cost production, enabling one to create various shapes by casting and coating.

An additional advantage is the possibility to be cut or abraded for refreshing the surface if contaminated.

Rough films can be produced by sol–gel techniques (Nakajima et al. 2000; Rao and Kulkarni 2002; Shirtcliffe et al. 2003). Polypropylene solutions in solvent mixtures can be also used to produce the equivalent polymer (Erbil et al. 2003). Available polymers include polyvinylchloride (Li et al. 2006), polycarbonate (Zhao et al. 2005) and polystyrene (Yuan et al. 2007) as well as some fluoropolymers (Yamanaka et al. 2006). Block copolymers have also been used to produce structured surfaces (Han et al. 2005a; Tung et al. 2007).

10.1.2.5 Crystal Growth

Various parameters can be altered during crystallization to influence the size and shape of crystals: rate of cooling, solvent evaporation or solvent addition. Rough surfaces can be observed on top of a crystallizing liquid. Hydrophobic materials will naturally lead to superhydrophobic surfaces and hydrophilic materials can be coated later on to become superhydrophobic. There are many examples of such materials (for more details, see Roach et al. 2008).

(i) *Organic materials.* The earliest studies were concerned with an alkyl ketene dimer (a kind of wax), which spontaneously forms a fractal crystalline surface (Shibuchi et al. 1996; Mohammadi et al. 2004). Fractal triglyceride surfaces have been reported (Fang et al. 2007). Random crystallization of n-hexatriacontane gives high hydrophobicity (Tavana et al. 2006).

(ii) *Inorganic materials.* Fractal aluminum oxide surfaces formed by anodic oxidation becomes superamphiphilic when coated with hydrophobising agents (Tsujii et al. 1997; Shibuichi et al. 1998). Ishizaki et al. (2007) successfully prepared a transparent superhydrophobic film in the silica-alumina system by plasma-enhanced chemical vapor deposition. Reversible superhydrophobic and superhydrophilic states have been observed by Feng et al. (2004) in packed nanorods of ZnO (a semiconductor) by alternation of ultraviolet irradiation and dark storage. Aligned SnO_2 nanorods exhibit the same properties (Zhu et al. 2006). Galvanic cell reaction is a facile method to deposit chemically Ag nanostructures on a *p*-silicon wafer at a large scale (Shi et al. 2006). When further modified with a self-assembled monolayer of *n*-dodecanethiol, a superhydrophobic surface can be obtained. Very high superhydrophobicity was achieved by Hosono et al. (2005). Inexpensive superhydrophobic surfaces on a variety of substrates can be obtained by coating at ambient temperature silicon, polymethylsilsesquioxane nanofilaments. The reaction is performed in the gas phase at room temperature and standard pressure without a carrier gas (Artus et al. 2006). Fabrication of a bionic nanoroughness metal surface was carried out by sulfur-induced morphological development (Han et al. 2005b).

(iii) *Nanostructured crystals.* Such structuration can be produced when the crystal face parallel to the surface grows faster and forms pillars on the nucleation sites or screw dislocations. This situation is often found in oxides of metals, such as zinc (Wu et al. 2005), where the structure can be tuned from vertically aligned nanosheets to flower-like microstructures (Su et al. 2009). Deposition can occur in liquid or gas phase (the range of parameters for such an organized growth is often small). Nanoflowers of SnO_2 have also been grown by a thermal oxidation of a tin organometallic precursor (Chen et al. 2004). Nanostructured flower-like crystals can also be prepared by crystallization of low-density polyethylene by adjusting the crystallization time and nucleation (Lu et al. 2004).

(iv) *Catalyzed growth.* Nanograss or nanoforest can be formed from a sputtered array of metal particles to control growth. Such fibers can be coated or chemically modified to present a hydrophobic surface. A fiber or tube starts to grow from each particle, either on top of the particle, or below it, lifting the particle as it grows. This type of growth can generate structures with a very high aspect ratio. Carbon fibers (Lau 2003), silicon and other materials have also been generated. Such growth of nanograss can be applied as a secondary coating on micro-machined posts to give two-scale roughness (see Table 10.1).

10.1.2.6   Porous Aluminum Oxide

Aluminum oxide layers can be grown on aluminum metal using anodic potentials in acid. The oxide forms hexagonal arrays of nano-pores whose sizes are determined by the potential. Its growth depends on the size mismatch between aluminum oxide and the underlying metal, together with the electric potential. The porous arrays have been used to produce nano-columns of carbon, metals, polymers (for polystyrene, see Jin et al. 2005). Note that titanium oxide forms similar rough patterns (Balaur et al. 2005).

10.1.2.7   Etching

Differences in the rates of etching of different crystal planes can produce a roughness in the order of the crystallite size or smaller. The main etching methods and their applications are listed below.

(i) *Plasma, ion etching or laser ablation* of polymers can be used at a large scale. One can cite PTFE with Busher (1992), polypropylene by Chen et al. (1999) and nanostructured polyethylene terephthalate by Teshima et al. (2003). Nanostructures of optically transparent polyethylene naphthalate and polystyrene were produced by Teshima et al. (2005). Polymethyl methacrylate nanosurfaces were made by Vourdas et al. (2007).

On a silicon wafer, a femtosecond laser was used to create micro-/nanoscale roughness (Baldacchini, 2006). Conical microstructures forming "black silicon", whose typical arrangement traps light, are formed by a high-density plasma reactor linked to multiple gas lines (SF_6 and O_2) operating with silicon at cryogenic temperatures (Nguyen et al. 2012; Saab et al. 2014). Recently, the use of PTFE-coated black silicon was extended to make a quite efficient rough substrate for water condensation (Liu et al. 2021).

(ii) *Wet etching* is used with metals, allowing a broad range of materials to be considered. Aluminum nanopits were fabricated by Narita et al. (2000), with large pits formed by electrolytic etching and smaller pits by anodic oxidation. Aluminum, zinc and copper are polycrystalline metals that have been used to form superhydrophobic surfaces. Such substrates were treated by a dislocation etchant that preferentially dissolves the dislocation sites in the grains to give a rough surface (Qian and Shen 2005; Shirtcliffe et al. 2005). Stable biomimetic superhydrophobic surfaces have been obtained on aluminum alloy wet chemical etching following modification with crosslinked silicone elastomer, perfluorononane and perfluoropolyether (Guo et al. 2006). A Titanium layer was etched using a RF plasma and using CF_4 as an etchant (Zhang et al. 2006). Steel and copper alloys were roughened by using nitric acid and hydrogen peroxide etch solution, and hydrofluoric acid/hydrogen peroxide for titanium alloys (Qu et al. 2007). Crystallization of metal alloys can also be controlled to produce needle-like narrow size distributed nanostructure crystals by selective etching of the eutectic (Hassel et al. 2007). Copper nanowires were synthetized by Chen et al. (2009).

Silicon can be etched using ethanolic hydrofluoric acid and anodic etching. Small pores form and then additional wet etching transforms the porous layer to pillars (Wang et al. 2007). Wet etching of silicon using initial deposited gold nanoclusters, followed by chemical etching and chemical removal of gold clusters, form nanocons that trap light, forming the so-called "black silicon" (Koynov et al. 2006). A silicon dual rough structure surface, which mimics the lotus leaf, was obtained using a combination of CF_4 glow-discharge etching and masking with copper nanodots (Kim et al. 2007).

10.1.2.8 Role of Inhomogeneities in Diffusion-Limited Growth

In many deposition processes that involve diffusion-limited growth (e.g. electrodeposition and gas-phase deposition), any protuberance collects more material than the surrounding area. This effect is similar to boundary effects in water condensation (see Chap. 8). Around an inhomogeneity, the concentration gradient of diffusing molecules is larger and growth is accelerated. This effect occurs on the top and side of the bump and generates a branching structure with a fractal character, which generally resembles cauliflower blossoms such as with ZnO deposition (Su et al. 2009).

10.1.2.9 Aggregation/Assembly of Particles

Colloidal particles can become close-packed on surfaces by using methods such as deposition, spin-coating, dip-coating or reverse dip-coating. The attractive Van der Waals forces between particles naturally give a tightly packed layer. The resulting roughness is in the order of the particle size. Films of spaced particles can be generated by using electrostatic repulsion.

Silica particles are often used. They form hexagonally close-packed arrays with particle sizes between a few nm and a few hundred μm (Tsai 2006). Polymer spheres are also used and naturally form ordered superhydrophobic surfaces (Shiu et al. 2004; Wang et al. 2006).

Another degree of roughness can be added to the particles, by aggregating two-sized particles together to produce raspberry-like patterns (Ming et al. 2005; Liu et al. 2007). Colloidal array of polystyrene beads can be coated with carbon nanotubes (Li et al. 2007). $CaCO_3$-loaded hydrogel spheres and silica or polystyrene beads were consecutively dip-coated on silicon wafers by Zhang et al. (2005). The former assemblies were recruited as templates for the latter self-assembly.

Layer-by-layer assembly is a method that uses the attraction between electrostatically charged species to develop multi-layer structures by sequential dipping in positive and negative polyelectrolytes (Zhai et al. 2004). The same technique can also be used to control the arrangement of particles by using alternating polymer and charged particle layers (Sun et al. 2007; Bravo et al. 2007; Zhang et al. 2007; Jaber and Schlenoff 2006). For a review of polyelectrolyte multi-layer fabrication, see Jaber and Schlenoff (2006).

10.1.3   Ordered Roughness

The need to obtain periodic and well-ordered geometrical patterns that cannot be found in self-organized patterns as those described above led to develop specific technics.

10.1.3.1   Conventional Milling

Milling and dicing is the simplest way to obtain dedicated structures. However, the minimum scale is limited to about 10 μm. Silicon can be easily milled to form pillars and grooves (Yoshimitsu et al. 2002; Narhe et al. 2004a; 2007). Metal can also be milled such as aluminum (Sharma 2019).

10.1.3.2   Spray with Masks

An easy way to form periodic biphilic structures on a flat surface or pillars can be achieved by spraying above a mask a hydrophilic or hydrophobic solution on a surface that was first hydrophobic or hydrophilic, respectively. The mask must be precisely positioned to form patterns (see e.g. Alizadeh-Birjandi 2019).

10.1.3.3   Nanoimprint Lithography

In lithography a design is transferred from a master onto a substrate. Multiple copies can thus be produced from the same master. The method involves contact between an inked stamp and the substrate. Micrometer scale is standard but it can reach a few nm thanks to nanoimprint lithography. This technique uses compression molding to create a thickness contrast pattern in a thin resist film carried on a substrate, followed by anisotropic etching to transfer the pattern through the entire resist thickness (Chou et al. 1996).

Photolithography works with a photoactive polymer layer irradiated through a mask and was followed by developing stages whereby either the exposed or unexposed polymer was removed. Different radiations can be used such as UV, X-ray, and e-beam. It is also possible to use a laser or particle beam to etch the surface directly. The patterned surface can be either used as is, or used as a mask on the substrate for deposition or etching.

Lithographic processes can produce master surfaces (Kawai and Nagata 1994; Öner and McCarthy, 2000; Zhu et al. 2006; Dorrer and Rühe 2006; Bintein et al. 2019). Electron-beam etching can be used to write on silicon (Marines et al. 2005). Silicon pillared structures are prepared using X-ray lithography (Fürstner et al. 2005). Reactive ion etching and holographic interference lithography was also used to produce structures of the order of tens of nm. Nanoimprint lithographic process has been used to groove silicon (printing a photoresist layer followed by ozone treatment and wet etching, see Pozzato 2006) and to replicate the structure of Lotus plant leave (Lee and Kwon 2007). Elastomeric pillars prepared on silicon by lithography can be deformed (du Roure et al. 2004); such bending has been used to measure forces exerted by cells on a substrate.

10.1.3.4 Use of Templates

A pattern can be replicated by templating. The principle of the method is to print or press onto the materials the cavities of a template. The template can be removed and then reused or destroyed to reveal the copy.

Bico et al. (1999) produced a master surface in the μm range. The master was then replicated using an elastomeric mold and used to cast silica features onto a silicon wafer. A template-based extrusion method was used by Feng et al. (2003) for the preparation of much thinner structures. Anodic aluminum oxide membrane (Zhang et al. 2006; Wang et al. 2012), alumina membrane (Chen et al. 2009) (Fig. 10.1a) and porous aluminum oxide (Lee and al. 2004) can be used as a template to prepare nanofibers and nanopillars or nanotubes surface.

A negative master has also been built directly from a micro-machined master (He et al. 2003). Replicas of natural superhydrophobic surfaces, as, e.g. artificial lotus leaves, have been used with success (Sun et al. 2005; Lee et al. 2006; Singh et al. 2007). Metal (nickel) mold of a lotus leaf has been fabricated (Lee and Kwon 2007).

Concerning sacrificial templates, Love et al. (2002) assembled metallic half-shells with nanometer-scale dimensions by depositing metals onto an array of spherical silica colloids. Similarly, Abdelsalam et al. (2005) used close-packed uniform sub-micrometer silica spheres as a template where gold was electrode-posited between the spheres. Removing the template led to a hexagonal array of partial spherical pores. Many fabrications used this procedure in the same way, as Li et al. (2006a, 2006b), Sun et al. (2007) and Bormashenko et al. (2006).

10.1.3.5 Silicon Micropillars and Nanograss

Due to its importance in fabrication, silicon micropillars and nanograss deserve some attention. They can be fabricated from wet and dry etching techniques (Roach et al. 2008; Wang et al. 2012). Wet etching methods give typically silicon aligned nanowires with a diameter ranging between 20 and 300 nm and with spacing of 10–200 nm (Chen et al. 2009 (Fig. 10.1a); Lu et al. 2011; Li et al. 2012; Kim et al. 2014; Alam et al. 2016). Defects can be removed by lithography (Shim, 2017). Dry etching allows finer structures to be fabricated because etching can be better controlled (Fonash 1990; Xiao 2009, Fig. 10.1b).

Conical silicon nanowires were also fabricated (Mouterde et al. 2017) (Fig. 10.1c) as well as hierarchical silicon nanowires with both micro- and nanoscale features (Li and Duan 2015; Wagner et al. 2003; Lo et al. 2014a, b) and micro-pyramids covered by silicon nanowires (Chen et al. 2011) (Fig. 10.1d).

10.1.3.6 Metal Nanowires

Metal nanowires (e.g. copper and gold) have better physical properties than silicon (high thermal conductivity, stability, and machinability), to be exploited for improving phase change heat transfer (Li et al. 2008; Chen et al. 2009 (Fig. 10.1a); Shim et al. 2017 (Fig. 10.1b); Shi et al. 2015; Kumar et al. 2017; Kim et al. 2014; Yao et al. 2011a; Anderson et al. 2012). The more convenient and adaptable process appears to be electrochemical deposition with porous templates (Li et al. 2008; Wen et al. 2017a, b, c; Chen et al. 2009; Shie et al. 2015; Wen et al. 2018b; Anderson et al. 2012; Yao et al. 2011b).

10.1.3.7 Biphilic Surfaces

Biphilic microstructures are important for condensation since, on the one hand, hydrophilic parts help nucleate water drops (see Sect. 3.2) and, on the other hand, hydrophobic region lower drop pinning and makes the drops slide more easily.

Varanasi et al. (2009) built alternating hydrophobic and hydrophilic segments on a silicon wafer (contact angles 110° and 25°, respectively). The hydrophilic regions are made up of the native oxide on a Si wafer, while the hydrophobic regions are coated with fluorosilane. Segments, 25 nm in width, were fabricated via microcontact printing using a prefabricated polydimethylsiloxane stamp. Another surface consisted of an array of hydrophobic posts with hydrophilic tops, fabricated via lithography combined with a UV-assisted surface modification approach. The hydrophilic tops are made up of deposited silicon dioxide, while the hydrophobic sidewalls and valleys are modified with fluorinated hydrocarbon molecules.

Betz (2013) manufactured biphilic surfaces on silicon wafers. They used a combination of random nanostructuring processes, microlithography and thin hydrophobic polymer coating.

Kim et al. (2018) developed a simple method to fabricate silicon-based smooth and micro-structured biphilic surfaces. They fabricated micropillar arrays on silicon wafers via standard photolithography. After etching, the surfaces were coated with a conformal layer (≈ 100 nm) of Octafluorocyclobutane C_4F_8 by chemical vapor deposition to uniformly make the surfaces hydrophobic and to ensure drop-wise condensation. In order to render the surface biphilic, they utilized the lift-off method to remove the masking photoresist on pillar tops, thereby removing all hydrophobic coatings on top of the photoresist patches. After completing the fabrication process, the surfaces were biphilic with hydrophilic pillar tops (smooth silicon) and hydrophobic pillar sides and sample base (C_4F_8 coated).

10.1.4   Liquid-Infused Microstructures

Liquid invasion of microstructures is performed (see, e.g. Anand et al. 2015) by dipping the silanized samples in a reservoir of the impregnating oil with a dip-coater. In order to prevent excess oil on the samples, the samples were withdrawn at a controlled velocity V such that the capillary number Ca $= \eta u / \sigma$ (see Eq. 5.30) was less than 10^{-4} according to Seiwert et al. (2011a, b). Another method was followed by Kajiya et al. (2016), where a drop of ionic liquid was initially deposited on the substrate. The substrate was left 1 h for the ionic liquid drop to homogeneously infiltrate the structure. The height of the lubricant was adjusted by removing part of the ionic liquid with a tissue and waiting until the thickness became homogeneous, as verified by confocal microscopy.

10.2   Pillars

Water condensation was briefly referred by Lau et al. (2003) on top of a superhydrophobic carbon nanotube forest, with condensed drops sitting in a Cassie-Baxter state. Narhe and Beysens (2007) and Dorrer and Rühe (2007) studied condensation on square micropillars. Chen (2007) considered condensation on a superhydrophobic surface with short carbon nanotubes deposited on micro-machined posts, a two-tier texture mimicking lotus leaves. Dorrer and Rühe (2008) reported condensation on Black Silicon (conical silicon nanograss) with drops in the Cassie-Baxter state. Microfabricated pillars with a high aspect ratio have shown bending induced by condensation (Narhe et al. 2016).

In the above situations, condensation proceeds through the CB state because it appears to be thermodynamically more stable than the W state. In contrast, condensation studies on spikes (Narhe and Beysens 2006) and on micro-machined rectangular pillars (Narhe and Beysens 2007) were concerned with water drops in the final W state.

Dorrer and Rühe (2007) observed surface condensing droplets in the W state whose coalescence could made them transition to the most favorable CB state. The criteria for the more stable Cassie-Baxter state than Wenzel's resumes in the equilibrium contact angle θ_c on a smooth surface (see Appendix E), whose value

$$\theta_c > \theta_c^* \tag{10.1}$$

Here (Eq. E.11):

$$\cos\theta_c^* = \frac{\phi_s - 1}{r - \phi_s} \tag{10.2}$$

r, the surface roughness, is defined as the ratio of the actual area to the projected area and ϕ_s is the area fraction of the liquid–solid contact. When $\theta > \theta_c^*$, the most stable state is CB, whereas when $\theta < \theta_c^*$, it is the W state (Lafuma and Quéré

2003). In this latter situation, if $\theta_c > 90°$, the drop can be in a metastable state, with the existence of an energy barrier to make the transition CB to W (Ishino et al. 2004; Patankar 2004; Dupuis and Yeomans 2005; Porcheron and Monson 2006). However, drop coalescence can overcome the energy barrier.

The situation is more complex during condensation where initially droplets are smaller than the structure and partially fill it. Dorrer and Rühe (2009) proposed that the size of the smallest stable drops with respect to the length scale of the roughness microstructure is a critical parameter. Rykaczewski et al. (2012b) indeed later demonstrated that the growth mechanism of individual water microdroplets is general and independent of the surface architecture.

Different nucleation and growth stages are presented below. Different regimes can be detected, dependent on the relative length scales (thickness a, spacing b, depth c) of the substrate and the drop radius, as sketched in Fig. 10.2.

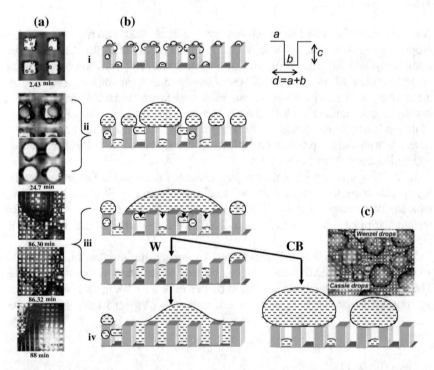

Fig. 10.2 Droplet growth on square pillars. **a** Time sequences of different growth stages of condensed water drops on a square-pillar substrate surface (square pillars with thickness $a = 32$ μm, spacing $b = 32$ μm, depth $c = 62$ μm, periodicity $d = a + b = 64$ μm). **b–c** Sketch of growth stages. Depending whether the final equilibrium stage is Wenzel (**b**) or Cassie-Baxter (**c**), two final drop configuration are observed, with coexistence of metastable states (CB for W final state and W for final CB state). (**a–b**: Adapted from Narhe and Beysens, 2007; **c**: Adapted from Dorrer and Grühe, 2007 with permission)

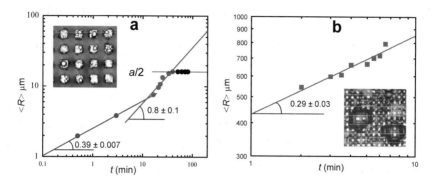

Fig. 10.3 Evolution of mean drop radius $\langle R \rangle$. **a** $\langle R \rangle < a/2$. The data obey regular power laws on a planar surface (straight lines, Eqs. 6.15, 6.44). **b** $\langle R \rangle > a/2$. Growth follows regular power law between coalescences (Eq. 6.15). (Adapted from Narhe and Beysens 2007)

10.2.1 2R < a, b, c Nucleation and Growth—Stage (i)

Depending on surface sub-cooling temperature, the value of the cap radius of the minimum droplet to grow, R^*, varies from a few nanometers to a few hundred nanometers (see Chap. 3, Eq. 3.36). The drop diameter in this regime, $2R$, is smaller than the typical size of the microstructures ($2R < a, b, c$). At nanoscale, it may happen that the base area of forming droplets can be pinned during growth, which allows droplets to grow only through contact angle increase (Rykaczewski et al. 2012). For larger, more mobile drops, the contact angle is the same as on a flat substrate (in Fig. 10.2 $q_c = 90°$). The drops on the top surface, on the sides of pillars, and on channels grow by condensation and coalescence (Fig. 10.3a) with nearly the same growth laws since the microstructure lengthscale is much lower than the diffusive boundary layer thickness (~ mm). The average drop radius follows the same regimes as detailed in Chap. 6.

10.2.2 2R ~ a, b, c—Stage (ii)

When $2R \sim a, b, c$, the drops on the top surface cover the entire surface of each pillar (Fig. 10.2, stage (ii) $t = 24.7$ min.). The top surface drops coalesce with another neighboring top surface drops. In addition, the drops on the side of the pillars coalesce with neighboring side drops and form bridges (provided that $c > b$). At the end of this stage, a few drops cover several pillars as a result of the coalescence of top surface drops with other top surface drops. These drops grow over the air present in incompletely filled channels in a CB state, stable or metastable depending on whether the condition of Eq. 10.1 is fulfilled or not.

10.2.3 2R > a, b, c Transition to W or CB State—Stages (iii–iv)

Depending on whether the final equilibrium state is W or CB, two scenarios arise when $2R \geq a, b, c$.

10.2.3.1 Wenzel State

The beginning of stage (iii) manifests itself by a remarkable drying process of the top surface of pillars, corresponding to the transition CB to W. Drops on pillar top surfaces indeed come into contact and coalesce with either neighboring drops or bridges or drops in the channel. A composite drop covering many pillars in a CB state is formed (Fig. 10.2, stage iii, $t = 86.30$ min). This drop then flows down into the channel in a very short time (< 20 ms), causing drying of the pillar's surface (Fig. 10.2, $t = 86.32$ min). This self-drying phenomenon is similar to the drying of the top of 2D grooved surfaces discussed below in Sect. 10.3.1. It is due to the coalescence of the upper drop with the drops and bridges that grow in the channels and correspond to the expected transition from CB to the most stable W state (see the discussion above at the beginning of Sect. 10.2 and Appendix E, Section E.2.2 for more details). The coalescence process with bottom droplets is able to overcome the energy barrier between these two states.

When $2R \gg a, b, c$, after the drying stage, new nucleation events take place. The water level in the channel increases up to the top surface and coalesces with drops at the top to form a large drop (Fig. 10.2, stage iv, $t = 88$ min). The drop at this stage is almost flat, having a very strong hysteresis contact angle because its edge is strongly pinned. Its growth law (Fig. 10.3b) compares well with that for a drop on a flat surface when no coalescence occurs ($R \sim t^{1/3}$, Sect. 6.1.2, Eq. 6.15).

10.2.3.2 Cassie-Baxter State

When such a configuration corresponds to the final state, drops are highly mobile and their evolution corresponds to the well-known evolution on smooth surfaces (Eqs. 6.15, 6.44). However, the aspect ratio and size of the microstructures matters. Rykaczewski et al. (2012b), by estimating the change of adhesion energy, proposed different general scenarios depending on the height and spacing of the microstructures and the typical minimum filling height between pillars, R_{no} (~500 nm), where surfaces do not retain superhydrophobic characteristics during condensation. (Fig. 10.4). It is interesting that it is only for a small aspect ratio $a/b \sim 1$ and height $c > > a \sim b$ that the Cassie-Baxter state can be observed.

Such an analysis is corroborated by the Molecular Dynamics simulations of Hiratsuka et al. (2019) and the Lattice Boltzmann simulations of a droplet dewetting transition performed by Zhang et al. (2016). The wetting transition from W to CB is indeed hampered by a large energy barrier. The kinetics of dewetting is a result of a subtle interplay of wetting and adhesion forces. In certain geometries, such as cone- or pillar-shaped textures, dewetting even appears to be spontaneous (Fig. 10.5).

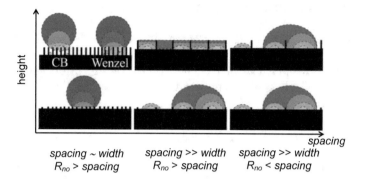

Fig. 10.4 Summary of influence of nanostructure geometry on microdroplet condensation. (Adapted from Rykaczewski et al. 2012b, with permission)

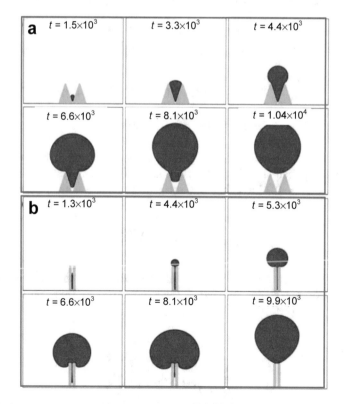

Fig. 10.5 Spontaneous dewetting transition of a microdroplet on surface with different structural topography. **a** Black silicon—like. **b** nanograss-like. (Adapted from Zhang et al. 2016, with permission)

10.3   Long-Range Coalescences

Specific structures can promote coalescence between drops that are far apart. These structures are grooves and hydrophilic stripes, which, once filled or wet, can connect distant drops and amphiphilic nanostructures with top hydrophobic and base hydrophilic. The latter structure, filled with water, can connect the drops above it.

10.3.1   Grooves

Long-range coalescence only occurs for grooves that can be continuously filled with water. This condition is fulfilled only for sufficiently large aspect ratios depth/spacing c/b (independent of thickness a), otherwise condensation is similar to what is encountered on smooth surface. The critical ratio $(c/b)_c$ separating these two regimes is directly related to the formation of groove-filling filaments, which requires that the imbibition of the grooves be thermodynamically favorable (Seemann et al. 2005; Bertier et al. 2014a, b). Indeed, when an imbibition front advances over a distance dx inside a small groove (b, $c \ll l_c$, the capillary length, see Eq. E.6) connected to a large drop with negligible Laplace pressure, the free energy of the groove and filament system varies as (for notations, see Fig. 10.6):

$$\frac{dF}{dx} = b(\sigma_{LV} + \sigma_{AW} - \sigma_{AV}) + 2c(\sigma_{BW} - \sigma_{BV})$$

$$= \sigma_{LV}[b(1 - \cos\theta_A) - 2c\cos\theta_B] \qquad (10.3)$$

Here one has used the contact angles θ_A (water/A material) and θ_B (water/B material) and applied the Young-Dupré relation (Eq. E.1). Groove-filling imbibition ($dF = dx < 0$), involving the advancing contact angles, occurs only if:

$$\frac{c}{b} > \left(\frac{c}{b}\right)_c = \frac{1 - \cos\theta_A}{2\cos\theta_B} \qquad (10.4)$$

Small deviations from Eq. 10.4 may originate from the curvature of the water surface due to contact line pinning at the groove edges (Seemann et al. 2005;

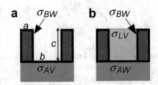

Fig. 10.6 Schematics of the groove imbibition criterion with notations. Bottom and lateral walls are made with different materials A and B. **a** Dry groove. **b** Groove filled with liquid (water)

Herminghaus et al. 2008). Filling-up of the grooves may also involve dynamical wetting motions associated with droplet coalescence as well as the groove wedges imbibition stability in the presence of wetting hysteresis. Equation 10.4, nonetheless, compares well with the experimental thresholds.

Grooved microstructures are particularly interesting because, as seen in Sect. 14.3.2, grooves can greatly improve droplet shedding by enhancing drop coalescence. Instead of many pinned tiny droplets, grooves generate a few large drops that can then slide down by gravity.

Condensation on micro-grooved substrates exhibits the same different growth stages as the pillar microstructures (see above Sect. 10.2.1), nevertheless with some specific features due to their 2D symmetry (Narhe and Beysens 2004; Zhong et al. 2013; Bintein et al. 2019; Sharma et al. 2019). With microstructures thickness a, spacing b, thickness c, periodicity $d = a + b$, one can distinguish four typical stages depending on the size of the drops with respect to microstructure dimensions (Fig. 5.16 where $a = 22$ μm, $b = 25$ μm, $c = 52$ μm, $\theta_A = \theta_B = 57°$). Equation 10.4 is fulfilled and filaments form in the channel inside the grooves.

10.3.1.1   2R < a, b, c: ucleation and Growth—Stage (i)

Nucleation of tiny water drops at the top surface of the groove (plateau) as well as in the channel initially takes place and growth laws similar to drops on planar surface are observed. In particular, since in this stage the surface coverage remains low, drops do not undergo significant coalescence events and the mean drop radius grows as Eq. 6.15, $R \sim t^{1/3}$.

10.3.1.2   2R ≥ a, b: 2D Growth—Stage (ii)

The drop size now exceeds the channels and top width, and drops grow on the plateaus along the groove length where they are pinned, with an elongated shape. At this stage one can observe a large number of elongated drops that grow in the direction perpendicular to the substrate and along the groove direction. Drop coalescences occur only along the groove direction. Such drops can thus be considered as 2D drops condensing on a 1D substrate. The growth law $R \sim t^g$ should exhibit an exponent $\gamma = \frac{1}{D_d - D_s} = 1$, with a 2D surface coverage $e_1 \approx 0.75$ (see Sect. 6.4.2.2). Experimentally (Fig. 6.14), the exponent $g = 1$ is well observed, with surface coverage $e_1 \approx 0.83$, slightly larger than the expected value. Note that drops can be more or less elongated depending on the groove characteristics.

Meanwhile the channels progressively fill in with coalescing drops. Depending on criterion in Eq. 10.4, filaments can form.

10.3.1.3   2R ≥ a: Coalescence with Channels—Drying Stage (iii)

With Eq. 10.4 is fulfilled (in Fig. 10.7 $\theta_A = \theta_B = 57°$, $c/b = 2.08 > 0.4$), at some time a channel becomes sufficiently filled with water such that it can coalesce with a drop of the plateau. The channel level increases, absorbing other plateau drops, which in turn increase the channel water level, favoring new coalescence events between plateau drops and channel in a kind of chain reaction. Coalescence events of a plateau drop with a groove is near instantaneous (<20 ms). It has been

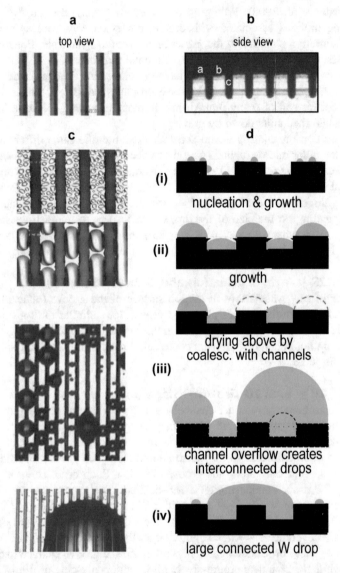

Fig. 10.7 a, b Top and side view of grooves (thickness a = 22 μm, spacing b = 25 μm, depth c = 52 μm). **a, c** Microscopic pictures and **b, d** schematic illustration. Four growth stages (see text) are found with drop contact angle 57° on patterned silicon substrate. Focus is on the top of the groves. (Adapted from Narhe and Beysens 2004)

observed by fast camera filming by Sharma et al. (2019) that the oscillations of the water in the channel helps the coalescence of plateau droplets. In symmetry, the two top surfaces adjacent to a channel dry simultaneously. Channel overflows produce a drop that sits on both plateaus aside a channel, which is filled symmetrically over a distance in order of the substrate length L. The presence of defects in the channels and on the plateaus, however, makes the useful channel length L^* < L such, as $L^* = L/p$ with $p \geq 1$ (see Sect. 14.3.2.3).

This process develops progressively on the entire grooved surface, with a few drops eventually connected to the channels. As the plateaus became dry, new drops can nucleate, grow and by coalescence-induced displacements (see Chap. 9) reach the channel and coalesce with it. One notes that the drying process is reminiscent of the coalescence-induced W-CB transition as observed with micropillars in stage (iii) in Sect. 10.2.3.1 above.

Channels connect drops, with the smallest drops (larger capillary pressure) draining out to large drops (lower pressure) in a Wenzel state over the two neighboring plateaus of the channel. Drops grow and cover successive plateaus to form a few large drops in the W state on horizontal substrates or slide down on tilted substrates when reaching the critical drop radius (see Sect. 14.3.2.3).

10.3.1.4 2R >> a, b: Large Drop Growth—Stage (iv)

The large drops develop by (i) incorporating at their surface water vapor molecules, and (ii) by collecting liquid water at their perimeter from the channels. The channels are in turn continuously fed by drops from the top surfaces. These drops are indeed in permanent random motion due to their coalescence events (see Chap. 9). The groove then acts as a well with zero drop concentration. The plateau drops eventually reach the channels and coalesce with the contained water. These grooves, in turn, feed the large drops because capillary pressure is larger in the grooves than in the drop.

Assuming for simplicity a single near-hemispherical drop that does not undergo coalescences with its neighbors, the growth law can be written as follows:

$$2\pi R^2 \frac{dR}{dt} = \left[2\pi R^2 \frac{D}{\rho_w} \left(\frac{c_\infty - c_s}{R} \right) \right] + \left[2RL \frac{dh}{dt} \right] = 2R \frac{D}{\rho_w} (c_\infty - c_s) \left(\pi + \frac{L^*}{\zeta} \right)$$
(10.5)

The left term is the volume increase. The first square bracket corresponds to the drop intrinsic growth (Eqs. 4.42). The second square bracket represents the contribution of channels, with h the condensed volume per surface area. The channels indeed collect an equivalent film on surface $2RL$ (neglecting the drop surface) with thickness h and is expressed through Eq. 6.13). Equation 10.5, once integrated, thus gives the following growth law:

$$R = \left[\frac{2D}{\pi \rho_w} \left(\pi + \frac{L^*}{\zeta} \right) (c_\infty - c_s) t \right]^{1/2} \approx \left[\frac{2DL^*}{\pi \rho_w} \left(\frac{c_\infty - c_s}{\zeta} \right) t \right]^{1/2}$$
(10.6)

Fig. 10.8 Radius evolution (log–log plot) of the large drop in stage (iv) of Fig. 10.7. The straight line is the power law Eq. 10.6. (Adapted from Narhe and Beysens 2004a)

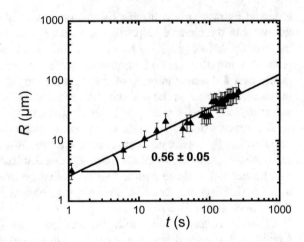

Here $\zeta \sim$ mm is the diffuse boundary layer thickness (see Sect. 2.2) Very generally $\frac{L^*}{\pi \zeta} \gg 1$, validating the approximation in Eq. 10.6.

One can also express Eq. 10.6 from the condensation rate $\dot{h} = \frac{D}{\rho_w}\left(\frac{c_\infty - c_s}{\zeta}\right)$ (Eqs. 7.13). It becomes

$$R = \left(\frac{2L^*}{\pi}\dot{h}t\right)^{1/2} \tag{10.7}$$

Figure 10.8 shows a typical drop radius evolution, with the power law exponent in agreement with Eq. 10.7.

10.3.2 Amphiphilic Nanostructures

Micropatterns can exhibit in some places hydrophobic properties and in other places hydrophilic characteristics, which makes them amphiphilic. It results in interesting properties that are detailed below with examples.

10.3.2.1 Biphilic Stripes, Squares and Dot Patterns

An alternate hydrophilic and hydrophobic stripes phenomena similar to what occurs on grooves can be observed. The hydrophilic regions play the role of the filled grooves collecting drops nucleating and growing on the hydrophobic regions (Fig. 10.9). Pinning of drops on groove edges at the origin of stage (ii) in Fig. 10.7 is obviously not observed. A droplet touching a wet hydrophilic region immediately coalesces with it. Early stages of nucleation and growth were described by Varanasi et al. (2009) on stripes of 25 μm thickness and contact angle 25° (hydrophilic) and 110° (hydrophobic). Chaudhury et al. (2014) and Peng et al. (2014) identified that, similar to what occurs on groove plateaus (Sect. 10.3.1), coalescences on hydrophobic regions could efficiently move the drops (see Chap. 9),

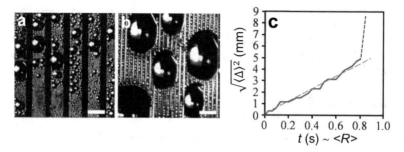

Fig. 10.9 Images of biphilic surfaces of different length scale. **a** Drops pattern corresponding to grooved substrates in Fig. 10.7 to stage (i) and **b** to stage (ii–iv). The bar is 1.2 mm. Contact angle on hydrophobic region is 112° and 0–209° on hydrophilic region. (From Garimella et al. 2017, with permission). **c** Displacement of drop center of mass on a hydrophobic stripe before it hits a hydrophilic stripe where it is sucked (interrupted line). The mean radius of the drop pattern grows linearly with time (Eq. 6.44). This evolution is similar to the evolution in Fig. 9.2. (Adapted from Chaudhury et al. 2014)

which eventually coalesce and are sucked by contact with a hydrophilic stripe (Fig. 10.9c). Garimella et al. (2017) outlined the effect of the deformation of drops during coalescence events (driven by the Ohnesorge number Oh, see Eq. 5.57), which can provoke another coalescence, and so on... leading to faster drop motion.

Wu et al. (2017) considered the collection of fog droplets by biphilic stripes (hydrophobic contact angle about 157°; hydrophilic contact angle less than 5°) on an inclined surface at 45° to horizontal. They found paradoxically that, although the minimum departure volume is smaller when stripes are parallel to gravity, higher fog harvesting efficiency occurs with stripes perpendicular to gravity, probably a result of the accumulation area that should dominate the fog harvesting efficiency rather than the minimum departure volume (see Sect. 14.3.2.4).

Historically, alternate stripes of hydrophobic/hydrophilic zones have been considered since the 1980s to enhance heat exchange coefficients with pure water vapor. Film condensation reduces the heat transfer because of its thickness (see Sect. 13.4 The earliest reports involved mm size stripes (Yamauchi et al. 1986; Kumagai et al. 1987, 1991) and considered heat exchange coefficients. More recent studies by Derby et al. (2014) considered parallel and Y-shaped stripes on copper partially coated with PTFE, showing 40° and 120° contact angles, respectively. Peng et al. (2015) showed experimentally that the performance of biphilic stripes can be greater than that of complete dropwise condensation. Yang et al. (2017) also investigated heat exchange by considering inverted V shape stripes. Xie et al. (2020) elaborated a mathematical model to evaluate the heat transfer, dominated by the hydrophilic stripe efficiency of droplet suction from the hydrophobic region.

Although the process is similar for grooves and stripes—sucking the small drops in channels—it is likely that aging of stripes will remove their hydrophilic character relatively soon in outdoor conditions. Grooves will more likely maintain their long-term properties, based uniquely on geometry.

Other biphilic structures have been considered, such as lyophilic spots by Schäfle et al. (2003) who studied diethylene glycol dropwise condensation around lyophilic patches. Details can be found in Sect. 7.1. One also notes the experimental and mathematical study by Alizadeh-Birjandi et al. (2019) of water condensation on mm size hydrophilic square spots on a hydrophobic surface.

10.3.2.2 Hydrophilic Porous Structures with Hydrophobic Heads

Instead of parallel stripes, some attempts have been made to superpose thin biphilic surfaces. When the substrate is set vertical, liquid in the hydrophilic sub-layer can thus be efficiently collected by gravity.

Oh et al. (2018) created a thin hydrophilic porous matrix of nickel inverse-opal (NIO) nanostructures (pore radius $\approx$ 250 nm and porosity (ratio void-space volume/total volume) $\approx$ 0.8 coated with a hydrophobic polyimide, where dropwise condensation can take place. Due to the nanoscale porosity of the NIO structure, the coating is made only on the NIO top surface, resulting in a superhydrophilic sub-layer and superhydrophobic top layer. Initially, nucleated drops grow on the upmost hydrophobic layer in the Cassie-Baxter state. However, due to coalescence events and contact with the lateral and bottom hydrophilic regions, they transition to the Wenzel state and are sucked by the porous hydrophilic structure (Fig. 10.10) by the effect of larger drop capillary pressure. Gravity flows inside the porous structure are presented in Sect. 10.3.2.2.

The configuration by Anderson et al. (2012) is similar, with however the use of densely packed nanowires made of hydrophilic base material with hydrophobic heads (Fig. 10.11a). In this configuration, after coalescence, initial condensed CB droplets end in a W configuration. In contrast with the previous hydrophilic support, drops are not sucked in because the rise in water height would result in the creation of many small radius interfaces with high capillary pressure at the hydrophilic-hydrophobic frontier. W-drops can thus coalesce each other while they are not in contact, producing long-range coalescence events.

The base material is made of 200 nm diameter hydrophilic gold wires closely packed with less than 25 nm separating adjacent wires and an estimated packing fraction of 85%. Wires are coated from the top with a hydrophobic fluoropolymer using plasma-enhanced chemical vapor deposition. Because the process is directional, the fluoropolymer only coats the tips of the nanowires. As the wires are not uniform, they create a secondary roughness scale that allows droplets to exhibit a high contact angle (>150°). During condensation, a liquid film nucleates in the hydrophilic gold nanowires and droplets nucleate and grow on the hydrophobic wire tips. They are in a CB state until they coalesce (Fig. 10.11b) where they transition to a W state and become connected to the wetting liquid film beneath. This film can connect two W-drops that do not touch, with the smaller drop with larger Laplace pressure filling the largest drop (Fig. 10.11c), in a process reminiscent of groove-induced long-range coalescence (Sect. 10.3.1). The fact that large drops are in a "sticky" Wenzel state is not favorable for gravity-induced shedding. However, the coalescence mediated by the sub-layer means that the largest droplets can reach the critical sliding size more easily.

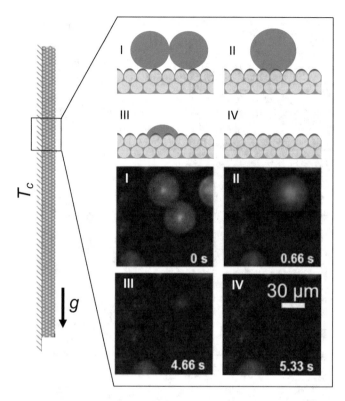

Fig. 10.10 Red color represents the hydrophobic coating. (I) Coalescence of two drops in CB state. (II) The resulting drop is W state and suction by the hydrophilic porous substrate starts. (III) Suction. (IV) The drop has completely been sucked. (Adapted from Oh et al. 2018, with permission)

10.3.2.3 Hydrophobic Structures with Hydrophilic Heads

Varanasi et al. (2009) showed that it was possible to favor the nucleation of drops at the top of biphilic structures formed by a hydrophobic microstructure of square posts (contact angle 110°) with a hydrophilic top (contact angle 25°). The rate of nucleation being strongly dependent on the contact angle (see Sect. 3.2), this configuration leads to having the hydrophobic sub-material mostly dry, with water drops only nucleating at the top (Fig. 10.12). This structuration allows nucleation near the dew point temperature (hydrophilic heads) with easy gravity removal of the drops. However, the inherent large contact hysteresis may hamper water collection by gravity (Gerasopoulos et al. 2018 and Sect. 14.4.2).

The migration of small water droplets from the hydrophobic side-wall of silicon micropillars onto its top hydrophilic surface during water condensation was observed by Orejon et al. (2016). The contact angles on the hydrophilic and the hydrophobic surfaces were 50° and 110°, respectively. Droplet migration is due to the difference in wettability between the hydrophobic sides and hydrophilic top

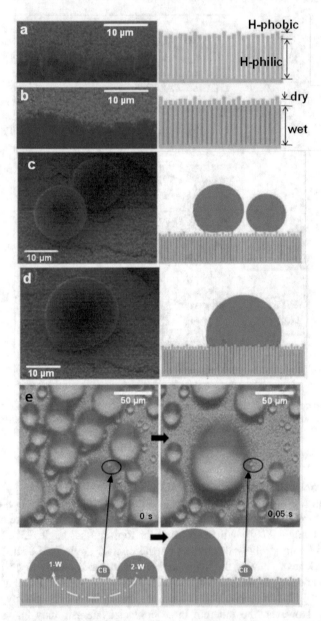

Fig. 10.11 ESEM images and schematics. **a** Dry amphiphilic surface before condensation. **b** Start of condensation in the hydrophilic base of nanowires and formation of a wetted sub-layer. **c** Formation and coalescence of CB droplets on hydrophobic nanowire tips. **d** Resulting coalesced drop is in W state, connected to wetted sub-layer. **e** Pre- and post-coalescence of W-drops apart steady CB droplets showing long-range coalescence through the water sub-layer. The smaller drop with larger capillary pressure flows into the larger drop with smaller capillary pressure. (Adapted from Anderson et al. 2012, with permission)

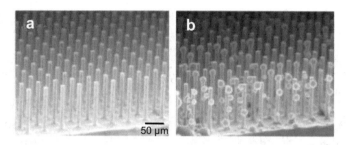

Fig. 10.12 Hydrophobic square pillars with hydrophilic top heads. **a** Dry structure. **b** Beginning of condensation. (Adapted from Varanasi et al. 2009, with permission)

surfaces of the micropillars. It corresponds to a shift in condensates from a Wenzel to a Cassie-Baxter state and promotes further droplet detachment.

The migration of condensate is due to (i) coalescence with a droplet resting on the top of the pillar, in a way similar to what is observed with droplets of a hydrophobic stripe coalescing with hydrophilic stripes (see Fig. 10.9) and (ii) wettability gradient between the side and top of each pillar, which exerts on the drop a net force towards the hydrophilic top. The origin of such wettability gradient-induced motions is detailed in Appendix M. For a near spherical cap droplet on a flat surface moving towards the more hydrophilic region, the force originating from the difference in wettability between drop edge A (hydrophobic, receding contact angle θ_{rA}) and drop edge B (hydrophilic, advancing contact angle θ_{aB}) can be written from Eq. M.8 as

$$F_d^* = \pi R^2 \sigma_{LG}\left(\frac{d\cos\theta_d}{dx}\right) - 2\sigma_{LG}R(\cos\theta_{rA} - \cos\theta_{aB}) \qquad (10.8)$$

Here θ_d is the mean dynamic contact angle defined by (Eq. M.7), $\cos\theta_d = \frac{1}{2}(\cos\theta_a + \cos\theta_r)$. Equation 10.8 becomes, writing d $\cos\theta_d/dx \approx (\cos\theta_r - \cos\theta_a)/R$:

$$F_d^* \approx \left(\frac{\pi}{2} - 2\right)R\sigma_{LG}(\cos\theta_{rA} - \cos\theta_{aB})_{LV} \qquad (10.9)$$

However, the present situation, which involves a curved side-wall and a flat upper surface, complicates the analysis. Therefore Orejon et al. (2016) limited their analysis to three characteristic dimensions, the maximum projected droplet chord length, $2R$, the length l of the droplet contact line on the hydrophilic region and the diameter a of the pillar (Fig. 10.13). The conditions for migration can indeed be considered as a combination of two factors: (1) pinning and hysteresis of the contact line (Eq. 10.9), which can be represented by the ratio l/R, and (2) the equilibrium geometry of a droplet resting on a curved edge, represented by the ratio of curvature droplet/pillar ($= 2R/d$). Figure 10.14 reports the regions where migration or no migrations are experimentally observed. As expected, the larger

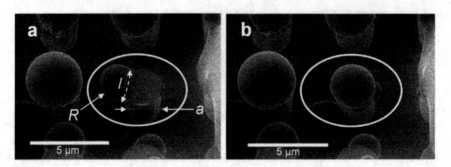

Fig. 10.13 ESEM images of droplet migration from the hydrophobic side of one micropillar onto its hydrophilic top. (Adapted from Orejon et al. 2016, with permission)

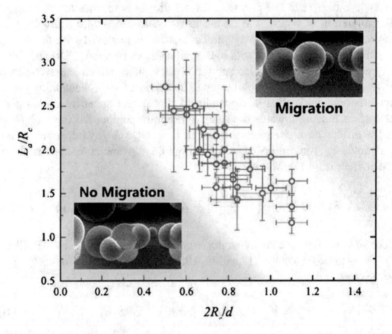

Fig. 10.14 Variation of the ratio of the contact line length on the hydrophilic top surface and droplet radius of curvature (*l/R*) with respect to the ratio of the radii of curvature of droplet and pillar (2*R/a*). (Adapted from Orejon et al. 2016, with permission)

the relative contact line contribution on the hydrophilic region and the smaller the relative radius of pillar curvature gives the best conditions for droplet migration.

10.4 Coalescence-Induced Jumping Drops

Ballistospore mushrooms are known to discharge a spore from the tip of their sterigma resulting from the coalescence of condensates at the base of the spore (Buller's drop, see Buller 1909; Turner and Webster 1991; Pringle et al. 2005). The same phenomenon can be observed during condensation on a superhydrophobic surface when the capillary forces released upon drop coalescence exceeds the pinning forces. The first observation was made by Boreyko and Chen (2009) on a superhydrophobic substrate composed of two-tier roughness with carbon nanotubes deposited on silicon micropillars and coated with hexadecanethiol. Since then, many other observations have followed on many superhydrophobic and biphilic materials (Boreyko and Chen 2013a, b: silver nanoparticles deposited on copper plate plus hexadecanethiol coating; Miljkovic et al. 2013a: silanized copper oxide surfaces; Enright et al. 2014: fluoropolymer-coated carbon nanotube; Tian et al. 2014: closely packed ZnO nanoneedles; Hou et al. 2015: biphilic micropillars covered by superhydrophobic nanograss with top covered by hydrophilic SiO_2; Kim et al. 2015: nanostructured superhydrophobic Al surface; Chen et al. 2016: silanized nanostructured truncated microcones on silicon; Mouterde et al. 2017: Silanized nanocones; Mulroe et al. 2017: thermally dewetting platinum films on silicon wafers coated with SiO_2 and later with a fluorinated silane; Gao et al. 2019: Al-based superhydrophobic; Lee et al. 2019: Nanostructured copper surfaces further functionalized by dodecanoic acid; Yan et al. 2019: Hierarchical superhydrophobic CuO surface and biphilic microhill structures and CuO nanowires; Liu et al. 2021: Black silicon coated with PTFE). A comprehensive study was recently carried out by Yan et al. (2019) who developed a microdroplet generation and visualization system capable of creating stationary liquid droplets ranging from 40 to 1000 μm radii on plain superhydrophobic, hierarchical superhydrophobic and biphilic (hydrophilic top) surfaces.

10.4.1 Jump Velocity

The origin of drop motion is the formation during coalescence of a liquid bridge between drops and the excitation of the drop oscillations modes (see Sect. 5.3.4 for drop coalescence). The expansion of the bridge has a key role in propelling the coalescent drop (Fig. 10.15). This effect therefore occurs only in low enough viscosity materials where such oscillations are present (see Sect. 5.3.4, Fig. 5.12). It results that the initial jump velocity u_0 is always directed perpendicular to the surface (Fig. 10.15). Note that when the radii of the coalescing droplets are different, this configuration has little effect on jumping direction (see Fig. 10.15b) due to the negligible lateral momentum generation by the symmetry-breaking surface.

In this process, a fraction f of surface energy E_s is transformed into kinetic energy E_c. As jumps are observed on surfaces with large contact angles, the two coalescing drops can be assumed nearly spherical, with radii ρ_1 and ρ_2 and the resulting composite drop with radius $\left(\rho_1^3 + \rho_2^3\right)^{1/3}$. The released surface energy

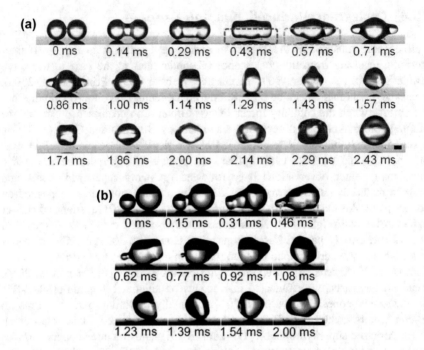

Fig. 10.15 Pictures of the coalescence behavior of two condensing droplets. **a** With same radius showing initial separation of the coalescing droplet with the surface (red rectangle) and the liquid bridge hitting the surface (blue rectangle), propelling the droplet out of plane perpendicular to the surface. Scale bar: 200 μm. **b** With unequal radius. The liquid bridge impinging on the surface is framed in blue (Scale bar: 400 μm). (Adapted from Yan et al. 2019, with permission)

corresponds to the difference in interface between the two initial drops and the composite drop, that is, with σ_{LV} the liquid–air surface tension:

$$\Delta E_s = 4\pi f \sigma_{LV} \left[\left(\rho_1^2 + \rho_2^2 \right) - \left(\rho_1^3 + \rho_2^3 \right)^{2/3} \right] \tag{10.10}$$

A more refined treatment can be found in Wang et al. (2011) who took into account the drop contact angle and drop W or CB state.

The gained kinetic energy can be written, with ρ_l the drop density and u_0 the initial drop velocity:

$$E_c = \frac{2\pi \rho_l}{3} \left(\rho_1^3 + \rho_2^3 \right) u_0^2 \tag{10.11}$$

Ignoring dissipation effects due to liquid viscosity (see below), one deduces u_0 by equaling Eqs. 10.10 and 10.11:

$$u_0 = f \left[\frac{6\sigma_{LV}}{\rho_l} \frac{\left(\rho_1^2 + \rho_2^2 \right) - \left(\rho_1^3 + \rho_2^3 \right)^{2/3}}{\rho_1^3 + \rho_2^3} \right]^{1/2} \tag{10.12}$$

When droplets are of equal radius ρ, Eq. 10.12 simplifies into:

$$u_0 = f \left[\frac{6\sigma_{LV}}{\rho_l} \frac{1 - 2^{-1/3}}{\rho} \right]^{1/2} \approx 1.11 f U \tag{10.13}$$

with U the scaling velocity:

$$U = \left(\frac{\sigma_{LV}}{\rho_l \rho} \right)^{1/2} \tag{10.14}$$

Equations 10.12–10.14– show that smaller droplets will have a larger velocity. Since droplets jump perpendicularly to the surface, the effective kinetic energy is equal to the maximum equivalent kinetic energy of the vertical velocity component. Lattice Boltzmann simulation by Peng et al. (2013) including viscous effects shows that it corresponds to about 25.2% of this velocity component. It thus corresponds to the fraction $f \approx 6\%$ of the surface energy transformed in kinetic energy, in accordance with experiments (see, e.g. Boreyko and Chen 2009; Enright et al. 2014; Yan et al. 2019).

Neglecting viscosity effect means that the Ohsegorne number (Eq. 5.3) remains much less than one. Using the numerical values for water at 5 °C (Table 1.2), one finds

$$\text{Oh} = \frac{\eta}{\sqrt{\rho_l \sigma_{LV} \rho}} \approx 1.76 \times 10^{-4} \rho^{-1/2} \tag{10.15}$$

Oh thus increases with the decrease in drop radius to reach Oh = 0.1 (the value where viscosity cannot be neglected any more), corresponding to $\rho \approx 3$ μm. For water drops where Oh is below this value, one thus expects a decrease of velocity and eventually the cessation of jumps (Fig. 10.17, insert). Several studies included the effect of viscosity in their theory and simulations of coalescence (see e.g. Wang et al. 2011; Peng et al. 2013; Enright et al. 2014; Vahabi et al. 2017). Enright et al. (2014) performed a 2D axisymmetric numerical simulations of the coalescence process in which both viscous and inertial forces are present. A fit of the velocity numerical and experimental data in the range [4–70] μm gave the following expression:

$$u_0 = f(\text{Oh})U \tag{10.16}$$

with f (Oh) showing a weak variation:

$$f = 0.2831 - 1.5285 \, \text{Oh} + 3.4026 \, \text{Oh}^2 \tag{10.17}$$

Figure 10.16 contains the expression Eq. 10.16 together with small droplets data from Enright et al. (2014) and large droplets data from Yan et al. 2019. The

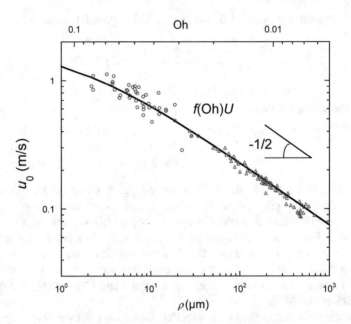

Fig. 10.16 Jump velocity u_0 as a function of coalescing droplets with radius r of equal size (lower abscissa) and Ohsegorne number (upper abscissa) representing the effect of viscous dissipation (log–log scale). The continuous curve $f(Oh)U$ is Eq. 10.16, which behaves as $\sim r^{-1/2}$ for large r or small Oh. Triangles: data from Yan et al. 2019; circles: data from Enright 2014)

latter were obtained by injecting between the drops a nanodrop, which nevertheless increases the jumping velocity by a few percent. The agreement is remarkably good. When ρ is large (or Oh small), one comes across again Eqs. 10.13, 10.14 where $u_0 \sim \rho^{1/2}$ Note that when the substrate exhibits a hierarchical biphilic structure with hydrophilic heads, there is an obvious cut-off for droplets size smaller than the spacing between heads (see Yan et al. 2019).

The effect of different coalescing drop radius should follow Eq. 10.12, however with the Oh correction Eq. 10.16. This equation resumes in

$$u_0 = f(Oh)F(x)U \tag{10.18}$$

with

$$F(x) = \left[6\frac{1 + x^2 - \left(1 + x^3\right)^{2/3}}{1 + x^3} \right]^{1/2} \tag{10.19}$$

where $x = \frac{\rho_2}{\rho_1}$, where ρ_1 is the lower radius. Its maximum corresponds to $x \approx 0.851$, with maximum value ≈ 2.780. However, Eq. 10.18 does not properly describe the experimental data (see Fig. 10.17). Data peaks at $x \approx 0.95$, instead of

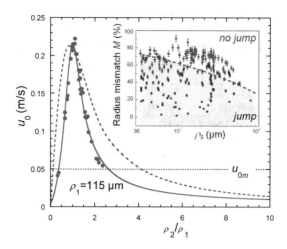

Fig. 10.17 Typical jump velocity u_0 as a function of the ratio of radius of coalescing drops (semi-log plot) for minimum radius $r_1 = 115$ μm. Interrupted curve: Eq. 10.18. Dots: data from Yan et al. (2019), corrected from a 3% overestimation due to the coalescing technics used. Data are fitted (continuous line) to Eq. 10.21 with $f = 0.276$. Horizontal interrupted line: minimum observed velocity u_{0m} defining jump– no jump. Insert (semi-log plot): Map showing when jump can occur as a function of radius mismatch with respect to the radius r_s of the smallest coalescing droplet. The interrupted line corresponds to Eq. 10.23. (From Yan et al. 2019, with permission)

≈ 0.85 as expected from Eq. 10.18. The latter equation represents the effect of the smaller fraction of excess surface energy converted to droplet kinetic energy when droplets have different size as compared to equally sized droplets. One sees that Eq. 10.18 exhibits a maximum at a lower value of x (≈ 0.85) than experimental data, although the maximum remains indeed close to the experimental value. In addition, the experimental values are more peaked than those from Eq. 10.18. This difference can be attributed to the weaker reaction force of the surface towards the droplet bridge when droplets have different sizes. Indeed, in this latter configuration (Fig. 10.15b), the liquid bridge is smaller. It can eventually become so small that it does not impinge on the surface, thus preventing jumps to occur.

Yan et al. (2019) fit the experimental data (Fig. 10.17) to Eq. 10.19 modified by an empirical modified Gaussian function $g(x)$, defined by

$$g(x) = \left(\frac{1}{1 + 33.6\left(\frac{x-1}{x+1}\right)^2} \right)^{1/2} \exp\left(-2.205\left(\frac{x-1}{x+1}\right)^2\right) \quad (10.20)$$

making $g(x = 1) = 1$. Equation 10.18 thus becomes

$$u_0 = f(Oh)F(x)g(x)U \quad (10.21)$$

It is then possible to draw a map where jump can be observed as a function of the radius mismatch between the coalescing droplets and the radius of the smallest coalescing drop. Jumps can be detected when jump velocity is above a minimum value u_{0m} (≈ 0.05 m.s^{-1} on Fig. 10.17). For simplicity, a radius mismatch can be defined as

$$M = 100\left[\frac{\max(\rho_1, \rho_2) - \min(\rho_1, \rho_2)}{\max(\rho_1, \rho_2)}\right] = 100\left(1 - \frac{1}{x}\right)(\%) \qquad (10.22)$$

The condition $u_0 > u_{0m}$ corresponds to a condition on smaller droplet radius ρ_s after making use of Eqs. 10.14:

$$\rho_s \le \frac{f^2 \sigma_{LV}}{\rho_l u_{0m}^2} F^2(x) g^2(x) \qquad (10.23)$$

Equation 10.23 is shown in the insert of Fig. 10.17 together with the experimental data corresponding to jump or no jump. A mismatch up to 80%, corresponding to a volume ratio of 125 between coalescing droplets, has been observed for the smallest droplets.

10.4.2 Jump Direction

Because the jump is due to the surface reaction of the growing bridge between the coalescing drops, the initial velocity is always directed perpendicularly to the surface (Figs. 10.15b and 10.16a), in spite of the difference in coalescing droplet radius. The only requirement is that the bridge can hit the surface to produce a reaction force to the coalescent droplet. This remains true as long as the drop radius is much larger than the local geometry of the substrate, with lengthscale d. When $\rho_s \le d$, the effect of locally inclined structures makes the jump direction different from 90° (Fig. 10.16b; see Rykaczewski et al. 2013; Yan et al. 2019) (Fig. 10.18).

10.4.3 Ballistic Motion

The ballistic motion of the jumping droplet (mass m), sent with initial velocity u_0 at angle a from a plane surface can be described by the following equations (Appendix N, Eqs. N.17, N.18), accounting for air friction:

$$x = \frac{m}{k_a} u_0 \left[1 - \exp\left(-\frac{k_a}{m}t\right)\right] \qquad (10.24)$$

$$z = -\frac{mg}{k_a}t + \frac{m}{k_a}\left(u_{0z} + \frac{mg}{k_a}\right)\left[1 - \exp\left(-\frac{k_a}{m}t\right)\right] \qquad (10.25)$$

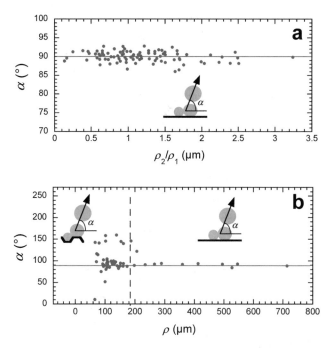

Fig. 10.18 Jump direction α with horizontal as a function of coalescing drop radius. **a** On a super-hydrophobic surface, as a function of the ratio ρ_2 / ρ_1 of the coalescing drop radius showing constant 90° angle. **b** On a hierarchical surface (50 μm height micro-hills separated by lengthscale $d \sim 50$ μm) as a function of the coalescing drop radius ρ. The jumping velocity direction deviates from 90° and becomes scattered when $\rho < d$ (left of interrupted line). (Data from Yan et al. 2019, with permission)

with the air friction coefficient (Eq. N.15):

$$k_a = 6\pi \eta_a \varrho \qquad (10.26)$$

Here η_a is air dynamic viscosity ($\approx 1.7 \times 10^{-5}$ Pa.s, see Table 1.2). Figure 10.19a reports on the motion of a water droplet of radius ≈ 122 μm ejected at 90° from a horizontal surface with initial velocity ≈ 0.18 m.s^{-1}. The maximum height of the jump is ≈ 1.5 mm. For about the same initial conditions, the lateral displacement from a vertical surface is 20 mm at 100 mm below to reach 26 mm at 400 mm from the jump location (Fig. 10.19b). These values ensure efficient removal of droplets from a vertical plate; the smaller droplet the better since smaller droplets have a larger initial velocity (see Fig. 10.16a) with lower air friction.

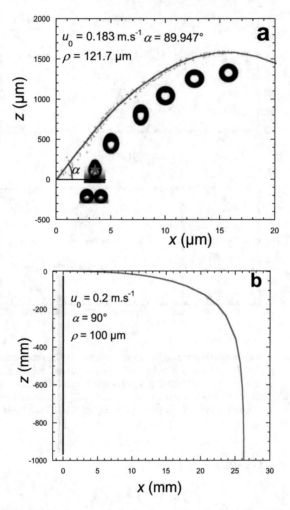

Fig. 10.19 Ballistic motion of a water drop. Continuous line are from Eqs. 10.24, 10.25 accounting for air friction. **a** Horizontal plane. (Adapted from Supplementary Informations in Yan et al. 2019, with permission. **b** Vertical plane

10.4.4 Effect of Substrate Tilt Angle

The effect of substrate tilt angle on drop size distribution and condensation rate has been systematically studied by Mukherjee et al. 2019 (Fig. 10.20). For a horizontal superhydrophobic surface, jumping droplets return to the surface. When the surface is vertical, jumping droplets can escape. In addition, jumping droplets that return to the surface can be removed thanks to gravitational shedding (see Chap. 14). These different surface orientations affect the droplet size distribution. In particular, the maximum droplet size, a key factor for heat transfer (see Chap 13) was found to be

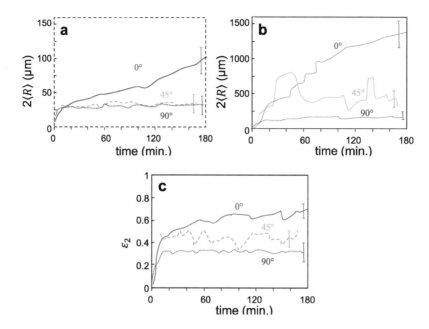

Fig. 10.20 Evolution of droplet size distributions at three typical substrate tilt angles with horizontal: horizontal 0° (full red curve), vertical 90° (blue dotted line), 45° (green interrupted line). **a** Mean droplet diameter, **b** maximum droplet diameter, **c** surface coverage. (Data from Mukherjee et al. 2019)

an order of magnitude smaller for inclined surfaces when compared to a horizontal surface (Mukherjee et al. 2019).

It should be noted that for upside-down substrates at high condensation rates, drops jumping from upside-down surfaces (tilt angle with horizontal = 180°) can return to the surface against gravity (Fig. 10.21). This counter-intuitive effect is due to vapor flow entrainment. Entrainment can be due to buoyancy effects on vapor near the surface and also related to the flow required for mass conservation of the condensing vapor (Miljkovic et al. 2013b, Boreyko et al. 2013a).

Fig. 10.21 Small droplet returning against gravity to an upside-down surface because of vapor entrainment. (From Mukherjee et al. 2019 with permission)

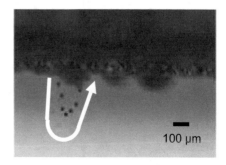

Liquid, Liquid-Infused and Soft Substrates

11

This chapter is concerned with condensation at the surface of liquid and soft substrates. The liquid substrate can be either contained in a container or infused on a patterned surface (liquid-infused surface LIS). Several differences with a solid surface are observed, as an exceptional surface smoothness, the possibility for the drop to deform the liquid and soft surfaces, and the fact that the drop contact line can move freely on liquid and LIS surfaces. These characteristics can have a strong impact on nucleation and growth processes during vapor condensation.

11.1 Liquid Substrate

The first report of water vapor condensation on a water immiscible liquid (paraffin oil) was reported by Mérigoux (1937; 1938) and Brin and Mérigoux (1954) for situations where oil did not encapsulate water drops. Condensation on liquids can somewhat differ from what is found with solid substrates, the condensation surface being capable of undergoing deformation due to the drop weight, which depends on its size. The shape of droplets on such deformable surfaces (see Lyons, 1930) is thus markedly different from those deposited on solids.

For very small drops at the interface of oil/air, gravity effects are negligible and the shape of the drop and the fact that it will sink or not will depend on the spreading coefficient of oil on water (Appendix E, Eq. E.2):

$$S_{ow/a} = \sigma_{wa} - \sigma_{wo} - \sigma_{oa} \qquad (11.1)$$

The σ_{ij} represents the interfacial tension between phases $i\,j$, with i or $j = o, w$, a denoting oil, water, air, respectively. The criterion for cloaking is given by a positive spreading coefficient (Fig. 11.1a), $S_{ow/a} > 0$, whereas $S_{ow/a} < 0$ implies the formation of a water lens (Fig. 11.1b).

© Springer Nature Switzerland AG 2022
D. Beysens, *The Physics of Dew, Breath Figures and Dropwise Condensation*,
Lecture Notes in Physics 994, https://doi.org/10.1007/978-3-030-90442-5_11

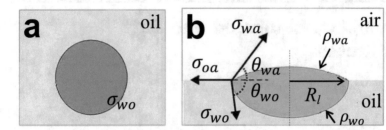

Fig. 11.1 Two water/oil drop configurations. The subscripts o, a, w, refer to the oil, water, air phases, respectively. **a** Water droplet cloaked by oil because either the spreading coefficient S_{ow} > 0 or its weight surpasses the buoyancy forces. **b** Water lens at the surface of oil when S_{ow} < 0 and its weight remains lower than buoyancy forces

11.1.1 Droplet Nucleation

Water droplet nucleation at the surface of oil is heterogeneous. The oil-air interface is in general considered smooth at the molecular level. However, random thermal fluctuations induce capillary waves, whose mean amplitude $\delta\xi$ can be evaluated (Buff et al. 1965):

$$\langle \delta\xi \rangle^2 = \frac{k_B T}{4\pi\sigma_{oa}} \ln\left(1 + \frac{1}{2}\sigma_{oa}^2 k_M^2\right) \tag{11.2}$$

Here k_B is the Boltzmann constant, T is absolute temperature and k_M is an upper cut-off wavenumber, inversely proportional to the interface width L, $k_M \approx \pi/L$. For low surface tension oil ($\sigma_{oa} \sim 30$ mN.m^{-1}) and samples of lengthscale~cm, one obtains $\langle\delta\xi\rangle \sim 2$ Å, indeed a dynamic roughness in the order of the molecular lengthscale.

The nucleation energy barrier for heterogeneous condensation (see Chap. 3) can be thought to be lowered with respect to their solid counterpart with identical solid surface chemistry because oils have higher surface energies (Eslami and Elliott, 2011; Anand et al., 2012) and thus nucleation rates are enhanced (Eqs. 3.15, 3.35 and 3.37; Xiao et al. 2013). However, the role of oil properties on nucleation is more subtle because the energy barrier depends on both the surface energy and the geometry. Anand et al. (2015) investigated the similarities and differences of nucleating on solid or oil as is explained below.

The critical cluster molecule number, n_h^*, (Sect. 3.2, Eqs. 3.12 and 3.36) is dependent on supersaturation, surface tension and drop contact angle. When comparing water nucleation rates on two different substrates i and j, the critical supersaturation, SR_i^*, needed to obtain the same nucleation rate and thus same n_h^* or critical radius ρ_h^* (see Eqs. 3.16, 3.36) depends, at given temperature, only on surface properties (surface tensions $\sigma_{wi,j}$ and contact angles, θ_{ij}).

Let us consider $\psi(\theta_c)$, the drop form factor relating the drop curvature radius ρ and volume V (see Eqs. D.5 and D.6 in Appendix D):

$$\psi(\theta_c) = \frac{3V}{\pi\rho^3} = 3F(\theta_c) = 2 - 3\cos\theta_c + \cos\theta_c^3$$

$$= (2 + \cos\theta_c)(1 - \cos\theta_c)^2 \tag{11.3}$$

From Eqs. 3.12 and 3.36 one can write the critical supersaturation of a water drop on substrate i as a function of the surface energy W_i^S as

$$\ln\left(SR_i^*\right) \propto \frac{\sigma_{wi}\psi(\theta_i)^{1/3}}{n_h^*} \propto \frac{W_i^S}{n_h^*} \tag{11.4}$$

At the same temperature and critical cluster molecule number n_h^*, the surface energies on substrates i and j, W_i^S and W_j^S, are thus related to the supersaturations and contact angles on substrates i and j as

$$\omega_{i/j} = \frac{\ln\left(SR_i^*\right)}{\ln\left(SR_j^*\right)} = \frac{W_i^S}{W_j^S} \tag{11.5}$$

The ratio $\omega_{i/j}$ determines whether it is thus more favorable to nucleate on substrate i, which is the case when $\omega_{i/j} < 1$. By evaluating in each configuration state the above ratio w_j^i of surface energies, a regime map of states can be obtained of the most favorable for nucleation. Four different configurations are considered (Figs. 11.2 and 11.3) depending on whether nucleation occurs at the interface water–air or water–oil, the substrate is solid or liquid, or if the drop shows up as a sphere, or hemisphere or a lens.

11.1.1.1   Evaluation of Surface Energies

(i) *State I* corresponds to heterogeneous nucleation on solid in air (Fig. 11.2a). This case was treated in Sect. 3.2. The surface energy, $W_{ws/a}^S$, where the subscript s means solid, is given by Eq. 3.29, where $\psi(\theta_c) = 4f(\theta_c)$:

$$W_{ws/a}^S = \pi\rho^2\sigma_{wa}\psi_I(\theta_{ws/a}) = \pi\sigma_{wa}\psi_I^{1/3}\left(\frac{3V}{\pi}\right)^{2/3} \tag{11.6}$$

(ii) *State II* relates to homogeneous cluster nucleation in an oil environment (Fig. 11.1a), corresponding to what has been treated in Sect. 3.1 with humid air. The corresponding surface energy, W_{ho}^S, is written as (Eq. 3.7):

$$W_{w/o}^S = 4\pi\sigma_{wo}\rho^2 = 4\pi\sigma_{wo}\left(\frac{3V}{4\pi}\right)^{2/3} \tag{11.7}$$

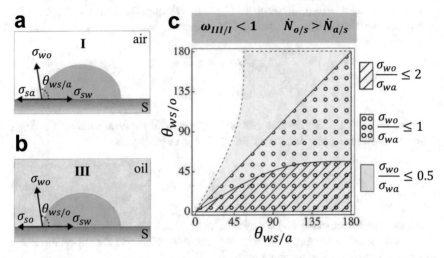

Fig. 11.2 **a** Nucleation in air on a solid. **b** Nucleation in oil on a solid. The subscripts w, a, o and s denote droplet, air, oil and solid interfaces, respectively. **c** Contact angles regime map showing conditions where nucleation rate within the oil ($\dot{N}_{oil}$) is larger than nucleation in humid air ($\dot{N}_{air}$) for different surface tension ratios σ_{wo}/σ_{wa}. (Adapted from Anand et al. 2015)

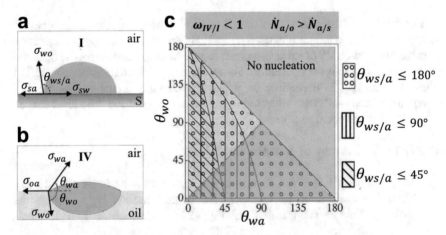

Fig. 11.3 **a** Nucleation on solid. **b** Nucleation on oil. The subscripts w, a, o and s denote droplet, air, oil and solid interfaces, respectively. **c** Contact angles regime map showing conditions where nucleation at air/oil interface ($\dot{N}_{a/o}$) is larger than nucleation at the air/solid interface ($\dot{N}_{a/s}$) for different solid surface wettabilities ($\theta_{ws/a}$). The region with a blue background corresponds to oils that satisfy the criterion $\sigma_{wo} \leq \sigma_{wa} (\theta_{wo} \leq \theta_{wa})$ while the region with a green background corresponds to oils with $\sigma_{wo} \geq \sigma_{wa} (\theta_{wo} \geq \theta_{wa})$. (Adapted from Anand et al. 2015)

(iii) *State III* deals with heterogeneous water nucleation in oil (Fig. 11.2b). For oil with a positive spreading coefficient in the presence of water $S_{os/w} = \sigma_{sw} - \sigma_{so} - \sigma_{ow} > 0$, thus heterogeneous nucleation of water on a solid surface is unlikely: any molecule reaching the surface will be replaced by the surrounding oil molecules. If oil does not spread on the solid surface in the presence of water ($S_{os/w} < 0$), this case corresponds to State I where air is replaced by oil. Equation 11.6 becomes

$$W^S_{ws/o} = \pi \rho^2 \sigma_{wo} \psi_{III}(\theta_{ws/o}) = \pi \sigma_{wo} \psi_{III}^{1/3} \left(\frac{3V}{\pi}\right)^{2/3} \quad (11.8)$$

(iv) *State IV* corresponds to nucleation at the oil/air interface. Water forms a lens if water does wet oil, that is if the spreading coefficient $S_{wo/a} = \sigma_{oa} - \sigma_{wo} - \sigma_{wa} < 0$ and oil does not cloak the condensate $S_{ow/a} < 0$ (Fig. 11.3b). Gravity effects are negligible for such small droplets, thus the oil surface is effectively planar up to the three-phase contact line. The total surface energy can be written, with superscript l for lens:

$$W^S_{wo/a} = \sigma_{wo} S^l_{wo} + \sigma_{wa} S^l_{wa} - \sigma_{oa} S^l_{oa} \quad (11.9)$$

With R_l the base radius of the water lens, ρ_{wo} the radius of curvature of the lower section of lens, and ρ_{wa} the radius of curvature of the upper section of lens (Fig. 11.1), the balance of tensions at the contact line can be written as (Fig. 11.3b):

$$\sigma_{wo}\sin\theta_{wo} = \sigma_{wa}\sin\theta_{wa}, \sigma_{oa} = \sigma_{wa}\cos\theta_{wa} + \sigma_{wo}\cos\theta_{wo},$$
$$R_l = \rho_{wo}\sin\theta_{wo} = \rho_{wa}\sin\theta_{wa}.$$

In addition, one obtains

$$\cos\theta_{wo} = \frac{\sigma_{oa}^2 + \sigma_{wo}^2 - \sigma_{wa}^2}{2\sigma_{oa}\sigma_{wo}}; \cos\theta_{wa} = \frac{\sigma_{oa}^2 + \sigma_{wa}^2 - \sigma_{ow}^2}{2\sigma_{wa}\sigma_{oa}}$$

The volume of the water lens, V^l, is the sum of two caps with contact angles θ_{wo} and θ_{wa}. Using Eq. D8 from Appendix D, it becomes $V^l = \frac{\pi}{3}\lambda R_l^3$ where

$$\lambda = \sin\theta_{wo}\frac{2 + \cos\theta_{wo}}{(1 + \cos\theta_{wo})^2} + \sin\theta_{wa}\frac{2 + \cos\theta_{wa}}{(1 + \cos\theta_{wa})^2} \quad (11.10)$$

Surface areas can be evaluated as

$$S^l_{wo} = \frac{2\pi R_l^2}{1 + \cos\theta_{wo}} = 2\pi\rho_{wo}^2(1 - \cos\theta_{wo}),$$
$$S^l_{wa} = \frac{2\pi R_l^2}{1 + \cos\theta_{wa}} = 2\pi\rho_{wa}^2(1 - \cos\theta_{wa}) = 2\pi\rho_{wo}^2(1 - \cos\theta_{wa})\frac{\sin^2\theta_{wo}}{\sin^2\theta_{wa}},$$

$$S_{oa}^l = \pi R_l^2 = \pi \rho_{wo}^2 \sin^2\theta_{wo}.$$

The total surface energy, Eq. 11.9, can thus be expressed as

$$W_{wo/a}^S = 2\sigma_{wo}\pi\rho_{wo}^2(1 - \cos\theta_{wo}) + 2\pi\rho_{wo}^2(1 - \cos\theta_{wa})\frac{\sin^2\theta_{wo}}{\sin^2\theta_{wa}}$$

$$-\sigma_{oa}\pi\rho_{wo}^2\sin^2\theta_{wo} \tag{11.11}$$

This expression can be transformed to

$$W_{wo/a}^S = \pi\left(\frac{3V^l}{\pi\lambda}\right)^{2/3}\xi\sigma_{wo} = \left(\frac{3V^l}{\pi\lambda}\right)^{2/3}\xi\frac{\sin\theta_{wa}}{\sin\theta_{wo}}\sigma_{wa} \tag{11.12}$$

where

$$\xi = \left(\frac{2}{1 + \cos\theta_{wo}} - \cos\theta_{wo}\right) + \frac{\sin\theta_{wo}}{\sin\theta_{wa}}\left(\frac{2}{1 + \cos\theta_{wa}} - \cos\theta_{wa}\right) \tag{11.13}$$

11.1.1.2 Comparison Between States III-I

For spherical caps (States I, III in Fig. 11.2), one obtains for Eq. 11.5 and using Eqs. 11.6 and 11.8:

$$\omega_{III/I} = \frac{\sigma_{wo}\psi_{III}^{1/3}(\theta_{ws/o})}{\sigma_{wa}\psi_I^{1/3}(\theta_{ws/a})} \tag{11.14}$$

Condensation in oil is enhanced if $\omega_{III/I} < 1$ because the energy barrier in III (o/s) is smaller than in I (o/a). Equation 3.37 indicates that the nucleation rate $\dot{N}_{o/s} > \dot{N}_{a/s}$. The corresponding map is reported in Fig. 11.2. It should be noted that, in contrast to what is currently assumed (see, e.g. Xiao et al. 2013), even if the surface tension water/oil is less than the surface tension water/air (i.e. $\sigma_{wo}/\sigma_{wa} < 1$), nucleation in the air environment can be preferred to nucleation within oil. At the same time, even if $\sigma_{wo}/\sigma_{wa} > 1$, nucleation within oil may be enhanced if the contact angles terms in Eq. 11.13 makes $\omega_{III/I} < 1$.

11.1.1.3 Comparison Between States IV-I

When oil does not cloak the water condensate ($S_{ow/a} = \sigma_{wa} - \sigma_{wo} - \sigma_{oa} < 0$, see Eq. 11.1) and condensate (water) does not wet oil ($\sigma_{oa} - \sigma_{wo} - \sigma_{wa} < 0$) (Eq. 11.1 where oil and water are exchanged), Eq. 11.5 becomes, using Eqs. 11.9 and 11.15:

$$\omega_{IV/I} = \frac{\xi}{\left(\psi_I(\theta_{ws/a})\lambda^2\right)^{1/3}}\frac{\sin\theta_{wa}}{\sin\theta_{w0}} \tag{11.15}$$

Nucleation at the oil-air interface (State IV) is preferable over nucleation at the solid-air interface (State I) if $\omega_{IV/I} < 1$ (Fig. 11.3c).

11.1.1.4 Comparison Between States IV-III

In the same way as above, one derives from Eqs. 11.8 and 11.9 for nucleation at the oil-air interface (State IV) as compared to nucleation at the solid-oil interface (State III):

$$\omega_{IV/III} = \frac{\left(\psi_{III}\left(\theta_{ws/o}\right)\lambda^2\right)^{1/3}}{\xi} \tag{11.16}$$

Nucleation at the oil-air interface State IV) is preferable over nucleation at the solid-oil interface (State III) if $\omega_{IV/III} < 1$. The maps satisfying $\omega_{IV/III} < 1$ as a function of the contact angles are shown in Fig. 11.4c.

One should note that a low nucleation energy barrier may not correspond to a large nucleation rate, since the latter also depends on other factors such as nucleation site density, the diffusion coefficient of molecules and the sticking probability of molecules (the so-called accommodation coefficient) in a given oil or air environment. The transport of a gaseous species through the liquid occurs by the sorption and diffusion mechanism. The maximum volume of vapor absorbed per unit volume of liquid, c_s, obeys Henry's Law:

$$c_s = H_v p_v \tag{11.17}$$

Here H_v is Henry's constant of solubility of vapor in the liquid at a given temperature T_v, and p_v is the partial pressure of vapor above the liquid, at air temperature T_a (see, e.g. Merkel et al., 2000). Let us cool the liquid to temperature $T_c < T_a$, then, as long as condensation does not occur in air region near

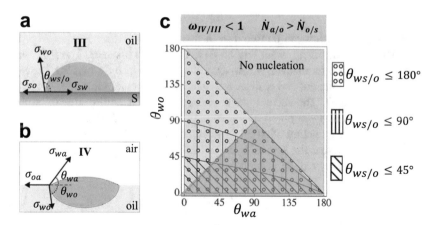

Fig. 11.4 **a** Nucleation in oil on a solid. **b** Nucleation on oil. The subscripts w, a, o and s denote droplet, air, oil and solid interfaces, respectively. **c** Contact angles regime map showing conditions where nucleation at the air/oil interface ($\dot{N}_{a/o}$) is larger than nucleation at the oil/solid interface ($\dot{n}_{o/s}$) for different solid surface wettabilities ($\theta_{ws/o}$). The region with a blue background corresponds to oils that satisfy the criterion $\sigma_{wo} \leq \sigma_{wa}(\theta_{wo} \leq \theta_{wa})$ while the region with a green background corresponds to oils with $\sigma_{wo} \geq \sigma_{wa}(\theta_{wa} \geq \theta_{wo})$. (Adapted from Anand et al. 2015)

the subcooled liquid, the partial pressure p_v near the liquid–air interface remains unaffected. The maximum amount of vapor that can be absorbed is $c_{sc} = H_c p_v$, where $H_c (> H_v)$ is Henry's constant at temperature T_c. Cooling the liquid makes it therefore under-saturated and droplet formation by condensation cannot occur. The experiments by Sushant et al. (2015) confirm that it is unlikely that nucleation can occur within bulk oil. The formation of droplets in oil is thus directly related with the saturation dynamics in the air. In the case of subcooled oil exposed to air, the region of supersaturation lies in the air beyond the oil-air interface and hence nucleation of a droplet occurs at the oil-air interface, regardless the nature of oil.

After nucleation at the oil-air interface, the droplet can grow at the expense of surrounding humid air. Two different cases can occur depending on whether oil cloaks (or not) the water condensate.

11.1.2 Droplet Growth for Cloaking Oil

For droplets growing at the oil-air interface on water-spreading oils, a layer of oil may extend between the vapor and the drop if the spreading velocity of oil on the condensing droplet (u_s) is greater than the droplet growth velocity, $u_d = d\rho/dt$, with ρ the radius of curvature of the upper segment of the lens. In the opposite case, the drop is expected to behave similar to condensation on a solid surface. The growth of the lens radius occurs by vapor condensation in the same process as for a drop on a solid surface. From Eqs. 4.45 and 4.46, with c_∞ the water concentration in air and c_s the water saturation concentration at oil temperature is

$$u_d = \frac{d\rho}{dt} \propto (c_\infty - c_s)t^{-1/2} \tag{11.18}$$

A detailed expression can be found in Sushant et al. (2015).

The spreading of oil on water can be described by two stages. The first stage is concerned with a monolayer whose spreading is driven by a balance between a surface tension gradient and a shear stress at the oil-droplet interface (Bergeron and Langevin 1996). The second stage corresponds to nanofilm growth, whose thickness is up to a few hundred nanometers (Carlson et al. 2013).

Concerning the spreading of an oil monolayer on water, Bergeron and Langevin (1996) have shown that the spreading front location follows Joos law from which the spreading velocity can be deduced as

$$u_{s,m} = \sqrt{\frac{3S_{ow/a}}{4\sqrt{\eta_o \rho_o}}} t^{-1/4} \tag{11.19}$$

Here η_o is oil dynamic viscosity, ρ_o is oil density and $S_{ow/a}$ is the spreading coefficient of oil on water as measured in air (Eq. 11.1).

The spreading rate of the nanofilm is driven by the capillary forces opposed either by inertial or viscous forces as determined by the oil Ohnesorge number

(Oh $= \frac{\eta_o}{\sqrt{\rho_o L \sigma_{oa} \varrho}}$, Eq. 5.3) with drop characteristic length, ρ, and surface tension oil/air, σ_{oa}. For small water droplets ($\rho < 100$ nm) on oil of viscosity $\eta_o > 10$ cSt, Oh > 1, implying that the spreading of oil on droplets during growth occurs in the viscous regime. In such a viscous regime, Carlson et al. (2013) have shown that the spreading velocity of a nanofilm is given by

$$u_{s,n} \propto \left(\frac{\sigma_{wo}}{\eta_o}\right)^{0.3} t^{-0.35} \qquad (11.20)$$

The prefactor is evaluated in Sushant et al. (2015).

Comparing the droplet growth rate Eq. 11.17 with the spreading rates of monolayer and nanofilms, Eqs. 11.18 and 11.19, one obtains

$$\frac{u_{s,m}}{u_d} \propto \sqrt{\frac{3 S_{ow/a}}{4\sqrt{\eta_o \rho_o}}} t^{0.25}; \ \frac{u_{s,n}}{u_d} \propto \left(\frac{\sigma_{wo}}{\eta_o}\right)^{0.3} t^{0.15} \qquad (11.21)$$

These velocity ratios show that the spreading of the nanofilm will eventually overcome droplet growth. Typical values for the time to spread over a 100 nm droplet are found in the range 10^{-16}–10^{-3} s (Sushant et al. 2015), showing that droplets that are nucleated at the oil-air interface are very soon cloaked by oil. As a result, the droplet will be submerged within the oil in order to minimize its surface energy and condensed droplets will eventually be located in the oil in a fully submerged state. When submerged, the drops can arrange as a dense network or coalesce to lead to only a few large drops.

To summarize (Fig. 11.5), nucleation and submergence of water drops on cloaking occur in the following steps. (a) droplet nucleation at the oil-air interface, (b) cloaking of the droplet by oil, (c) submergence of the droplet within the oil due to capillary forces, thereby creating a fresh oil-air interface, (d) the cycle (a)-(c) is repeated with a new generation of droplets forming at the oil interface and submerging. The interaction between the old and the new generation of droplets may lead to re-organization of submerged droplets.

Although the oil layer acts as a barrier for water molecules to be accommodated in the drop, one might think that molecules could permeate and diffuse through the oil film in the presence of a concentration gradient across the film. Droplet growth within the oil requires a gradient of concentration of water molecules around the drop, that is, a water concentration (the solute) larger than the saturation concentration at oil temperature, c_s. In practice, due to the drop curvature (Kelvin equation), the concentration at the drop surface is even larger than c_s. As discussed at the end of the preceding Sect. 11.1.1, the vapor saturation in oil is limited by the sorption mechanism. The above explanations make it unlikely for vapor to achieve supersaturation in the film. Drops, once cloaked by oil, cannot therefore grow from diffusion of water molecules. However, drops can coalesce when they are at contact, provided that the oil film between their surface is evacuated by drainage, a process whose dynamics depends on oil viscosity.

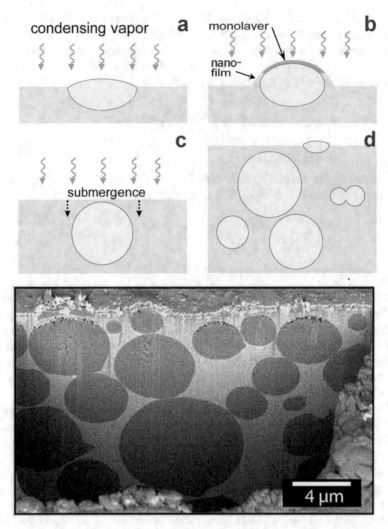

Fig. 11.5 Nucleation and growth of a condensate (water) on a cloaking liquid (oil). **a** Droplet nucleation, **b** Cloaking, **c** Submergence **d** The cycle **a–c** is repeated. **e** Cross-sectional view by Cryo-FIB-SEM technique[1] in 1000 cSt oil corresponding to **d**. (Adapted from Anand et al. 2015)

11.1.3 Droplet Growth for Non-cloaking Oil

Non-cloaked nucleated drops grow at the expense of the surrounding vapor, similarly to what is found on a solid surface (see Chaps. 4 and 6). One notices,

---

[1] Cryogenic—Focused Ion Beam—scanning electron microscope (Cryo-FIB-SEM) imaging uses focused ion beam (FIB) milling for serial block face imaging in the scanning electron microscope (SEM). It is an efficient and fast method to generate volume data for 3D analysis.

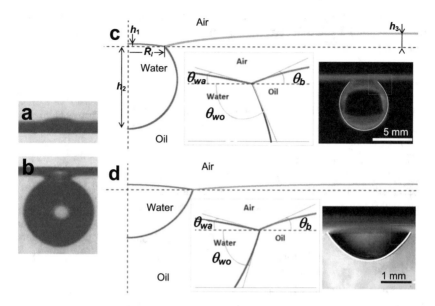

Fig. 11.6 Water droplets on an oil surface. (**a-b**) With obtuse angle θ_{wo}. **a** From above, **b** from below ($\theta_{wo}\theta_{wo}$ adapted from Knobler and Beysens, 1988). **c–d** The three contact angles are arranged so that the sum of the three tensile forces acting on the contact line is zero (rectangle). The curves on the photos are the result of the numerical model (see text). **c** Drop with obtuse angle. **d** Drop with acute angle. (Adapted from Phan et al. 2012 and Phan 2014, with permission)

however, a number of differences. The first concerns buoyancy. The density of oil is generally lower than the density of water, and thus the lens will be affected by disturbances and eventually sinks in oil. In addition, the shape of the oil–water interface is bent near the contact line (Fig. 11.6). This bending creates short-ranged attractive elastic interactions between the drops. Another difference with drops on solids is the fact that the surface of exchange with vapor is only the upper part of the lens. During growth, this surface of exchange increases due to the effect of growth but also decreases by the effect of gravity and buoyancy forces evolution (Fig. 11.11). The coalescence process is also affected. The contact line is not pinned and can move freely. When the contact angle of the lens below the oil surface, θ_{wo}, is above 90°, coalescence proceeds in oil and a delay is necessary to drain oil between the water drop surfaces before coalescence can take place.

11.1.3.1 Buoyancy and Drop Stability Against Disturbances

When oil density is smaller than water density, which is the most general case, the floating liquid droplet necessitates a subtle balance among three deformable interfaces (Fig. 11.6). Following Phan et al. (2012), the vertical force on a water lens, F, is written as the resulting of gravity and buoyancy forces, since the balance of the three tension forces along the contact line is zero:

$$F = g\left[V_0\rho_w + \left(\pi R_l^2 h_3 - V_1\right)\rho_a - \left(\pi R_l^2 h_3 + V_2\right)\rho_o\right] \qquad (11.22)$$

Here ρ_w, ρ_o, ρ_a are water, oil and air densities, respectively. V_0 is the water droplet volume, V_1 and V_2 are the volumes of air/water and water/oil fractions, respectively, with

$$V_0 = V_1 + V_2 \tag{11.23}$$

The parameter h_3 is the amplitude of bending of the oil-air interface (see Fig. 11.6). In Eq. 11.21, the first two terms are due to gravity and the last is due to buoyancy.

The geometrical factors V_1, V_2, R_l, h_3 can be determined from a model involving the angles $\theta_{wa}, \theta_{wo}, \theta_b$, as defined in Fig. 11.6. Because the balance of the three tensions should be zero at the contact line, one obtains the following two relations by projecting them on vertical and horizontal axes:

$$\sigma_{wa}\sin\theta_{wa} - \sigma_{wo}\sin\theta_{wo} + \sigma_{oa}\sin\theta_b = 0$$
$$\sigma_{wa}\cos\theta_{wa} - \sigma_{wo}\cos\theta_{wo} - \sigma_{oa}\cos\theta_b = 0 \tag{11.24}$$

A numerical solution for F as a function of θ_{wo} can be found (Phan et al. 2012). The model integrates the Young–Laplace equation along the three interfaces, from the initial points to the contact line, and matches three contact angles. It resumes in a system of ordinary differential equations. Once the three interfaces air/oil, air/water, oil/water are matched, V_1, V_2, R, and h_3 are substituted in Eq. 11.21 and F can be evaluated.

The model can thus find the equilibrium state, that is the angles $\theta^*_{wa,wo,b}$ where $F = 0$. However, the equilibrium configuration is not always experimentally obtainable. In contrast with the opposite system of oil droplets at a water surface, which is in stable equilibrium (the droplet returns to equilibrium configuration if disturbed), for a water droplet in less denser oil the droplet might sink if disturbed. Therefore, the equilibrium state would never be observed, the water droplet becoming less and less stable with increasing volume. Although the model predicts an equilibrium configuration for very large droplets (Fig. 11.7a), additional volume decreases the equilibrium angle θ^*_{wa} and makes the system less stable, the force becoming very sensitive to angle disturbances as shown in Fig. 11.7b. If the system is disturbed, the geometry is so much changed that going back to the equilibrium configuration is not possible, making the drop sink. Phan et al. (2012) found that for a water drop in vegetal oil with parameters $\rho_o = 915.8$ kg.m^{-3}, $\sigma_{wa} = 44.9$ mN.m^{-1}, $\sigma_{oa} = 21.7$ mN.m^{-1}, $\sigma_{wo} = 31.8$ mN.m^{-1}, the minimum angle for a stable droplet is $\theta^*_{wa} \sim 4°$ (Fig. 11.7a). This angle corresponds to a 3.4 mm drop radius. Note that the configuration characterized by an acute contact angle, $\theta_{wo} < \pi/2$ (Fig. 11.7b) corresponds to having a droplet that can sustain stronger disturbances and is therefore kinetically more stable than the previous configuration (Phan 2014).

Practically speaking, the above analysis shows that, during condensation, it will not be possible to sustain at the surface of oil drops whose size exceeds a few mm.

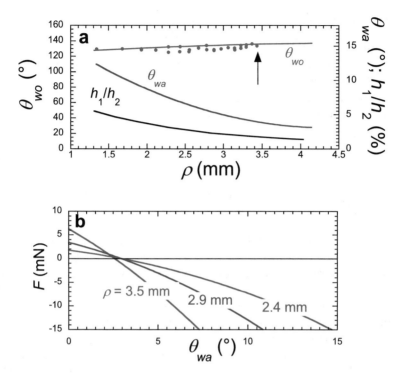

Fig. 11.7 **a** Left ordinate: Angle θ_{wo} (the most accurate determined experimentally). Right ordinate: Ratio h_1/h_2 of a water lens upper and lower heights and angle θ_{wa}, with respect to the equivalent spherical drop radius $\rho = (3V/4\pi)^{1/3}$. Dots: Experimental data. Continuous curves: From the calculation. The arrow indicates the maximum radius before a drop sinks. **b** Vertical force with respect to angle θ_{wa} for different drop radii. (Data from Phan et al. 2012)

The drop can take two different shapes depending on the value of the θ_{aw} angle. If $\theta_{wa} < 90°$ (acute angle), the drop resembles a lens on each side of the oil-air interface (Fig. 11.6d). Coalescence will occur at the contact line without delay. If $\theta_{wa} > 90°$ (obtuse angle), the main portion of the drop is in oil and coalescence will proceed within oil, with a delay to drain oil between the water drop interfaces.

11.1.3.2 Attractive Interactions ("Cheerios" Effect)

The fact that drops, once nucleated, are suspended, bends the oil-air interface, and elastic energy is stored. The curvature of the substrate causes long-range elastic interactions between the drops that are able to aggregate and/or coalesce. Aggregation (a "Cheerios effect", reminiscent of breakfast cereal circle shapes floating and sticking to each other) is the same phenomenon as observed for small solid spheres or cylinders at the surface of water (Allain and Cloitre 1988; Kralchevsky and Nagayama 2000; Karpitschka et al. 2016).

A complete calculation of the dependence of the force on the interdroplet distance is extremely difficult. However, some approximations are valid, such as the

Nicolson method (Nicolson 1949; Chan et al. 1981; Kralchevsky and Nagayama 2000). This method applies when the bond number $Bo = (\rho/l_c)^2$ (ρ: drop equivalent radius; l_c: capillary length at the interface oil/air (see Appendix E, Section E.1.4) is small enough. For paraffinic oil where $\sigma_{oa} = 22$ mN.m^{-1} and density $\rho_o = 870$ kg.m^{-3}, this condition is $\rho < l_c = 1.6$ mm.

According to the Nicolson method, the attractive force F between two droplets separated by distance d is, with ρ_a the air density:

$$F_a = \frac{4\pi\rho^6(\rho_o g)^2}{3d\sigma_{oa}}\left[\frac{\rho_a}{\rho_o} + 0.25\left(1 - p^2\right)^{1.5} - 0.75\left(1 - p^2\right)^{0.5}\right]^2 \tag{11.25}$$

The quantity p is a key parameter and is defined as $p = R_l/\rho_{wo}$, that is the ratio of R, the base radius of the water lens and ρ_{wo}, the radius of curvature of the lower section of lens (Fig. 11.6c). From simple geometry, $p = \sin\theta_{wo}$ (Figs. 11.1, 11.6c). This parameter p is related to the contact angles of the drop with the surface and is therefore very sensitive to the wetting properties of the oil–water interface. By varying p the force can be magnified by two orders of magnitude. Since the interaction force can be directly deduced from Eq. 11.24, the parameter p can be used as a quantitative measurement of the interaction between the drops. Note that any temperature gradient around a drop would be of radial symmetry and thus would not affect the force, Eq. 11.24.

(i) When p in Eq. 11.24 is *large* (~0.4 – 0.8) and $\theta_{wo} > 90°$ (obtuse angle), droplets strongly attract each other but, due to the presence of an oil film oil at the contact point, they do not coalesce immediately. Condensation experiments on paraffin oil were carried out by Steyer et al. (1993) under large supersaturation corresponding to a temperature difference of 18 K. Drops condense with $\theta_{wo} = 127°$ and $p = 0.74$ (Knobler and Beysens, 1988). Since the nucleation rate is important, many droplets are present at the surface of the oil and have time to form islands before coalescence takes place. As the link between the drops is not rigid, a droplet can easily slide around another droplet and organize in a compact hexagonal structure (the more stable configuration in 2D, see Fig. 11.8a). The islands, in turn, attract each other in a much stronger way than the single droplets. This is because the attractive force between the islands is proportional to the 6th power of their size as seen in Eq. 11.24. After having collided, the islands can rotate more and less freely around the point of contact in order to minimize the elastic energy. This is made possible because the friction between the droplets is negligible. As a consequence, the orientation of the hexagonal structure is not very different from one island to the other (Fig. 11.8b) and the drops arrange themselves to form a hexatic phase. A hexatic phase is a 2D phase that is between solid and isotropic liquid phases (Halperin and Nelson 1978; Nelson et al. 1982). It does not have an analogue in 3D and is characterized by two order parameters: a short-range positional order and a quasi-long-range orientational (six-fold) order.

The spatial correlation function of the order parameter is

$$G_6(r) = \langle\exp[6i\theta(r)]\rangle \tag{11.26}$$

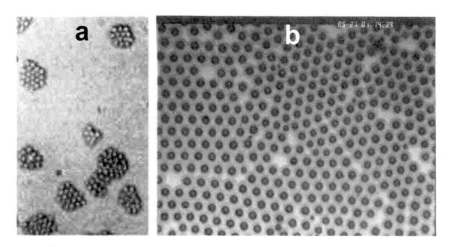

Fig. 11.8 **a** Formation of islands at the beginning of condensation (scale is given by the smallest dimension of the picture $= 650$ μm). **b** Merged islands (scale is given by the smallest dimension of the picture $= 350$ μm). (From Steyer et al. 1993)

where $\theta(r)$ is the angle with respect to some fixed direction of the segment between two neighboring droplets whose centers are separated by r. In a hexatic phase, $G_6(r)$ decays as a power law with exponent η:

$$G_6(r) \sim r^{-\eta} \tag{11.27}$$

The value of η can be interpreted in terms of structural defects (Nelson et al. 1982). With c the concentration of defects in the structure (fraction of polygons that do not have six sides):

$$\eta = \frac{9}{\pi}c \tag{11.28}$$

Steyer et al. (1993) experimentally determined the G_6 function (Fig. 11.9). In order to determine which drops are neighbors, they used Voronoi polygons (see Sect. 6.1.2). The range over which the hexatic order extends is in the order of 10 drops, which is close to the diameter of the islands (Fig. 11.8a). This shows that the orientational order persists only from one island to the nearest neighbor island. From the measured defect concentration, $c \approx 0.35$, giving from Eq. 11.27 $\eta \approx 1$, in good agreement with the value (0.91) deduced from the G_6 decay (Fig. 11.9). The large value of η is the consequence of the large number of defects in the pattern.

The formation of holes is unavoidable because the islands are too densely packed to rotate freely. At a larger scale, where the droplets are not resolved (Fig. 11.10a), these holes have a fractal-like geometry. Let us first define the radius

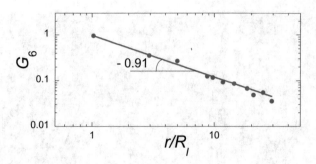

Fig. 11.9 60° angle orientation correlation function (G_6) as a function of the reduced distance r/R_l in a log–log plot. R is the lens radius (10 μm here). (From Steyer et al. 1993)

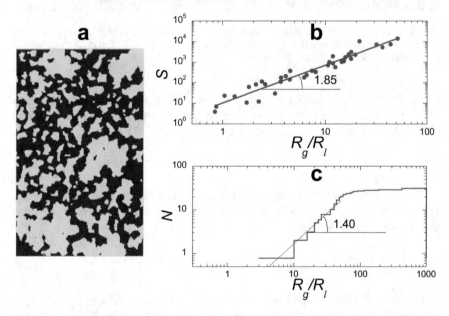

Fig. 11.10 Fat fractal arrangement of holes. **a** Fractal structure highlighted by blackening the zones with droplets (scale is given by the smallest dimension of the picture = 650 μm). **b** Surface S of holes with respect to the hole gyration radius R_g in units of lens radius R (log–log plot). **c** Number N of holes whose size is smaller than R_g in units of lens radius R (log–log plot). (From Steyer et al. 1993)

of gyration of a hole, R_g, such as

$$R_g = \sum_i \left[\left(x_i - x_g \right)^2 + \left(y_i - y_g \right)^2 \right]^{1/2} \tag{11.29}$$

where x_i and y_i are the coordinates of an element i of the hole and x_g and y_g are the Cartesian coordinates of the hole center of mass. Let us now consider the total surface $S_N(R_g)$ of these N holes, which have a radius of gyration smaller than a given value. A characteristic of fractals is the dependence of the apparent size on the resolution scale. It defines the exponent d_f such as, in the limit $R_g \rightarrow 0$:

$$d_f = \left. \frac{\ln\left[S_N(R_g)\right]}{\ln(R_g)} \right|_{R_g \rightarrow 0} \tag{11.30}$$

Or:

$$S_N(R_g)\big|_{R_g \rightarrow 0} = R_g^{d_f} \tag{11.31}$$

Steyer et al. (1993) used a two-step approach to determine d_f, by considering that

$$S_N(R_g) \sim S(R_g)N(R_g) \tag{11.32}$$

Here $S(R_g)$ is the variation of holes surfaces S with respect to R_g and $N(R_g)$ is the number of holes whose radius of gyration is less than R_g. The function $S(R_g)$ obeys a power law (Fig. 11.10b) with exponent δ:

$$S(R_g) \sim R_g^{\delta} \tag{11.33}$$

The value found experimentally was $\delta = 1.85 \pm 0.12$ (uncertainty: one standard deviation). The fact that δ is close to two means that the holes are compact.

The size distribution function $N(R_g)$ also obeys a power law (for values not too close to the picture size, see Fig. 11.10c) with exponent $\alpha = 1.40 \pm 0.05$ (uncertainty: one standard deviation):

$$N(R_g) \sim R_g^{\alpha} \tag{11.34}$$

From Eqs. 11.30–11.33, one eventually obtains $d_f = \delta + \alpha = 3.25 \pm 0.17$ (a value not to be confused with the space dimensionality).

11.1.3.3 Drop Growth Laws

Although buoyancy and gravity effects are expected to be weak for small drop sizes ($<<$ mm), these effects lead, however, for larger drops ($>$ mm) to a significant variation of drop lens radius R_l with the drop equivalent radius $\sim V^{1/3}$ (Fig. 11.11). The increase of R_l as induced by growth is indeed counterbalanced by increasing buoyancy effects (as discussed in Sect. 11.1.3.1), which makes R_l to decrease after its initial increase. Then the ratio between the drop surface on which water molecules are accommodated ($\sim R_l^2$) and the total drop surface ($\sim V^{2/3}$) decreases with time in a nearly linear manner (Fig. 11.11).

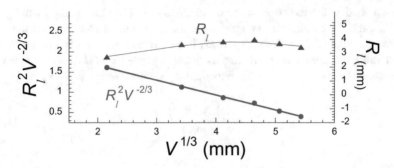

Fig. 11.11 Variation with respect to mean drop radius $\sim V^{1/3}$ (V is drop volume) of the lens radius R_l (right ordinate) and the ratio of lens and drop surface areas $\sim R_l^2/V^{2/3}$. The line is a linear fit and the curve is a smoothing function. (Data taken from pictures in Phan et al. 2012)

(ii) For small drops and low nucleation rates as in the Knobler and Beysens (1988) experiments (supersaturation corresponded to temperature difference of 8 K), islands only rarely form and the pattern looks similar to what is found on smooth solid substrates solids (see Sect. 6.3.2, Figs. 6.7 and 11.12a). The mean radius $\langle R_l \rangle$ grows as $\langle R_l \rangle \sim t^x$ with expected values $x = 1/3$ for small surface coverage and $x = 1$ for large coverage (Eqs. 6.15 and 6.46, 6.62).

The apparent drop surface coverage on oil, ϵ_2^*, as measured from the top view of the drop pattern, relates to ϵ_2, the surface coverage inside oil, the coverage which is concerned with coalescence, through the ratio $p = R_l/\rho_{wo} = \sin\theta_{wo}$ as used in Eq. 11.24:

$$\epsilon_2^* = p^2\epsilon_2 \tag{11.35}$$

The limiting value of the surface coverage in the coalescence-driven regime is about $\epsilon_2^* \approx 0.28$ (Fig. 11.9b). This value corresponds to $\epsilon_2 \approx 0.50$, close to the expected value ≈ 0.55 for spherical droplets on solids (contact angle > 90°, see Fig. 6.19). Note that the bump in surface coverage, which correlates with the beginning of coalescence-driven growth, is not seen on solids (see Fig. 6.6). This is due to the delay of removing the oil film when drops are at contact, preventing immediate coalescence and thus allowing a more compact arrangement to be obtained.

11.2 Liquid-Imbibed Substrate (LIS)

A liquid surface is an ideal smooth surface to prevent contact line pinning during condensation, but it obviously flows when the reservoir is inclined with respect to horizontal, unless the liquid is trapped in a microstructure. Liquid invasion of microstructures was initially considered by Quéré (2005) in order to create non-sticking conditions for rolling drops. The hydrodynamics of micropatterns invaded

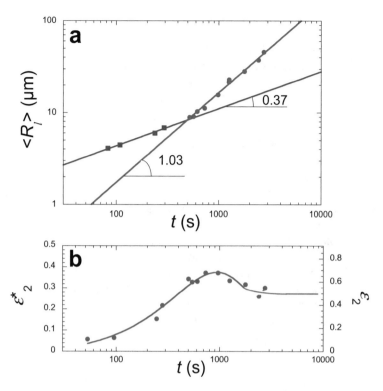

Fig. 11.12 **a** Evolution of the average apparent radius $\langle R \rangle$ with fits to the power law $\langle R \rangle \sim t^x$ with expected values $x = 1/3$ for small surface coverage and $x = 1$ for large coverage (Eqs. 6.15, 6.46 and 6.62). **b** Evolution of surface coverage from above air-oil interface (ϵ_2^*, left ordinate) and from below the oil surface (ϵ_2, right ordinate). The line is simply a guide for the eye. At time t_0 the interactions between droplets become important. (From Knobler and Beysens 1988)

by a liquid was then considered by Seiwert et al. (2011a, 2011b). Bio-inspired slippery surfaces were fabricated by Wong et al. (2011) by invading micro-nano porous structures with non-volatile oil. Condensation on such surfaces was first investigated by Anand et al. (2012, 2015), Smith et al. (2013), and followed by many studies mainly motivated by possible heat transfer enhancement (see e.g. Kajiya et al. 2016; Seo et al. 2020).

The main interest of non-wetting surfaces containing micro/nanotextures impregnated with lubricating liquids is therefore to make condensed water droplets, which are immiscible with the lubricant, easily move along these surfaces. However, when dealing with condensation, the nucleation and growth process is more complex than on solid and liquid surfaces as one must consider both surfaces and their possible interactions with, in addition, the low depth of oil in the microstructure as compared to the drop size. Resistance to coalescence can also be present, as a film of oil must be evacuated before drop contact.

11.2.1 Water Drop Nucleation

For oil not cloaking water, nucleation at a solid or oil interface is preferred depending on the contact angles solid/water/air, solid/water/oil, water/air, and water/oil (Sect. 11.1.1). For cloaking oil, due to limits of vapor sorption within a liquid, nucleation is most favored at the liquid–air interface (Sect. 11.1.2).

11.2.2 Oil–Water-Solid Interactions

In the following, we analyze the oil–water-solid interactions when the drop is larger than the LIS spacing. When it is not the case, drops either nucleate and grow on the oil interface or on the solid substrate if the latter emerges from oil.

Depending whether oil has a low or moderate surface energy, oil can engulf the water droplet or form an annular ring around it (Fig. 11.13). In addition, oil and water can also partially or completely wet the substrate. Following the Smith et al. (2013) approach, it results in 12 possible different situations as depicted in Fig. 11.13, depending on the spreading coefficient and LIS roughness. For the interface outside of the droplet (oil-solid-air), three configurations are possible depending on whether the solid is dry (A1), partially (A2) or completely

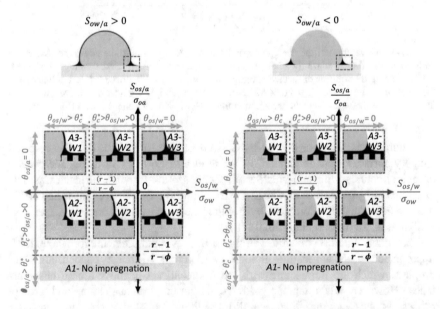

Fig. 11.13 Map of possible thermodynamic states of a water droplet sitting on a lubricant-impregnated surface. The top two drops show whether the droplet is cloaked by oil. For each cloaking/non-cloaking case, six states are possible depending on wetting properties of oil on texture in the presence of air (vertical axis) and water (horizontal axis). The corresponding interfacial energies are estimated from Table 11.1. (Adapted from Smith et al. 2013, with permission)

Table 11.1 Schematics of wetting configurations outside and underneath the drop (column 2) with total interface energies per unit area for each configuration (column 3). Equivalent stability criteria are in the last three columns. (Adapted from Smith et al., 2013, with permission)

Interface	Configuration	Interfacial energy per unit area	Equivalent stability criteria		
Oil/Solid/Air	Dry	$E_{A1} = r\sigma_{sa}$	$E_{A1} < E_{A2},$ E_{A3}	$S_{os/a}$ $< -\sigma_{oa}\dfrac{r-1}{r-\phi}$	$\theta_{os/a} > \theta_c^*$
	Impregnated, emerged	$E_{A2} = (r-\phi)\sigma_{os} + \phi\sigma_{sa}$ $+ (1-\phi)\sigma_{oa}$	$E_{A2} < E_{A1},$ E_{A3}	$-\sigma_{oa}\dfrac{r-1}{r-\phi}$ $< S_{os/a} < 0$	$\theta_c^* > \theta_{os/a} > 0$
	Encapsulated	$E_{A3} = \sigma_{oa} + r\sigma_{os}$	$E_{A3} < E_{A1},$ E_{A2}	$S_{os/a} \geq 0$	$\theta_{os/a} = 0$
Oil/Solid/Water	Impaled	$E_{W1} = r\sigma_{sw}$	$E_{W1} < E_{W2},$ E_{W3}	$S_{os/w}$ $< -\sigma_{ow}\dfrac{r-1}{r-\phi}$	$\theta_{os/w} > \theta_c^*$
	Impregnated, emerged	$E_{W2} = (r-\phi)\sigma_{os} + \phi\sigma_{sw}$ $+ (1-\phi)\sigma_{ow}$	$E_{W2} < E_{W1},$ E_{W3}	$-\sigma_{ow}\dfrac{r-1}{r-\phi}$ $< S_{os/w} < 0$	$\theta_c^* > \theta_{os/w} > 0$
	Encapsulated	$E_{W3} = \sigma_{ow} + r\sigma_{os}$	$E_{W3} < E_{W1},$ E_{W2}	$S_{os/w} \geq 0$	$\theta_{os/w} = 0$

(A3) wet by oil. Underneath the droplet, the possible configurations are impaled (W1), impregnated with emergent features (W2), and encapsulated (W3). The stable configurations are those that exhibit the lowest total interface energy. Below, one calculates those energies, outside and underneath the drop.

11.2.2.1 Outside the Drop

The energy E_{A1} of state A1 (dry; Fig. 11.13; Table 11.1) is

$$E_{A1} = r\sigma_{sa} \tag{11.36}$$

with the parameter r being the ratio of the total surface area to the projected area of the solid (see Appendix E, Eq. E.9). Considering square posts with width a, height b, and edge-to-edge spacing c (Fig. 10.2), it becomes

$$r = 1 + \frac{4ab}{(a+c)^2} \tag{11.37}$$

With ϕ the area fraction of the liquid–solid contact (see Appendix E, Eq. E.10 with $\phi_s \equiv \phi$):

$$\phi = \frac{a^2}{(a+c)^2} \tag{11.38}$$

The energy E_{A2} of state A2 (impregnated, emerged) is

$$E_{A2} = (r - \phi)\sigma_{os} + \phi\sigma_{sa} + (1 - \phi)\sigma_{oa} \tag{11.39}$$

The energy E_{A3} of state A3 (encapsulated) reads as

$$E_{A3} = \sigma_{oa} + r\sigma_{os} \tag{11.40}$$

(i) *State A2* (oil impregnated, emergent post tops) is the most stable. A2 state is preferred if its energy is lower than the other states. State A2 thus corresponds to $E_{A2} < E_{A1}, E_{A3}$. This inequality resumes, from Eqs. 11.35, 11.38 and 11.39, to.

$$E_{A2} < E_{A1} \text{ when } \frac{\sigma_{sa} - \sigma_{os}}{\sigma_{oa}} > \frac{1 - \phi}{r - \phi} \tag{11.41}$$

$$E_{A2} < E_{A3} \text{ when } \sigma_{sa} - \sigma_{os} - \sigma_{oa} < 0 \tag{11.42}$$

Equations 11.40 and 11.41 can be interpreted in different terms by making use of the Young-Dupré relation (Appendix E, Eq. E.1):

$$\sigma_{sa} - \sigma_{s0} = \sigma_{oa}\cos\theta_{os/a} \tag{11.43}$$

Let us consider the definition of the critical angle θ_c^* where Cassie-Baxter state is preferred to Wenzel's (Appendix E, Eq. E.11):

$$\cos\theta_c^* = \frac{\phi - 1}{r - \phi} \tag{11.44}$$

and the spreading coefficient Eq. 11.1 of oil on solid (see Appendix E, Eq. E.2):

$$S_{os/a} = \sigma_{sa} - \sigma_{so} - \sigma_{oa} \tag{11.45}$$

Thus Eqs. 11.40, 11.41 reduces to the conditions:

$$E_{A2} < E_{A1}, E_{A3} \text{ when } \theta_c^* > \theta_{os/a} > 0 \text{ or } -\sigma_{oa}\frac{r - 1}{r - \phi} < S_{os/a} < 0 \tag{11.46}$$

(ii) *State A3* (oil encapsulates the structure) is the most stable. One can follow the same type of calculation to find that A3 is the most stable when:

$$E_{A3} < E_{A2} \text{ when } \theta_{os/a} = 0 \text{ or } S_{os/a} \geq 0 \tag{11.47}$$

And:

$$E_{A3} < E_{A1} \text{ when } \theta_{os/a} < \cos^{-1}\left(\frac{1}{r}\right) \text{ or } S_{os/a} > -\sigma_{oa}\left(1 - \frac{1}{r}\right) \tag{11.48}$$

The latter condition is also fulfilled by Eq. 11.47. The only condition that remains for having state A3 (encapsulated) the most stable is

$$E_{A3} < E_{A1}, E_{A2} \text{ when } \theta_{os/a} = 0 \text{ or } S_{os/a} \geq 0 \qquad (11.49)$$

(iii) *State A1* (dry) the most stable. The same procedure as above leads to

$$E_{A1} < E_{A2}, E_{A3} \text{ when } \theta_{os/a} > \theta_c^* \text{ or } S_{os/a} < -\sigma_{oa}\frac{r-1}{r-\phi} \qquad (11.50)$$

11.2.2.2 Beneath the Drop

The corresponding possible states (Table 11.1) are W1 (impaled), W2 (impregnated, emerged) and W3 (encapsulated). The corresponding energies take the same form as above, replacing the subscript *a* (air) by *w* (water):

(i) *State W1*

$$E_{W1} = r\sigma_{sw} \qquad (11.51)$$

(ii) *State W2*

$$E_{W2} = (r - \phi)\sigma_{os} + \phi\sigma_{sw} + (1 - \phi)\sigma_{ow} \qquad (11.52)$$

(iii) *State W3*

$$E_{W3} = \sigma_{ow} + r\sigma_{os} \qquad (11.53)$$

The criteria for determining the most stable states thus become.

(i) *State W2* (impregnated, emerged post tops) is the most stable:

$$E_{W2} < E_{W1}, E_{W3} \text{ when } \theta_c^* > \theta_{os/w} > 0 \text{ or } -\sigma_{ow}\frac{r-1}{r-\phi} < S_{os/w} < 0 \qquad (11.54)$$

(ii) *State W3* (oil-encapsulated) is the most stable:

$$E_{W3} < E_{W1}, E_{W2} \text{ when } \theta_{os/w} = 0 \text{ or } S_{os/w} \geq 0 \qquad (11.55)$$

(iii) *State W1* (water displaced by oil, impaled droplet) is the most stable:

$$E_{W1} < E_{W2}, E_{W3} \text{ when } \theta_{os/w} > \theta_c^* \text{ or } S_{os/w} < -\sigma_{ow}\frac{r-1}{r-\phi} \qquad (11.56)$$

One notes that θ_c^* does not depend on the environment as it is only a function of r and ϕ (see Eq. 11.43).

One can draw (Fig. 11.13) the above various possible states on a map, accounting for the criterions above and adding the cloaking criterion $S_{ow/a} = \sigma_{wa} - \sigma_{wo} - \sigma_{oa} > 0$ (Eq. 11.1). For both drop cloaking and non-cloaking states, one distinguishes six possible states depending on the oil-solid interactions in the presence of air (vertical axis: normalized spreading coefficient $S_{os/a}/\sigma_{oa}$) and water (horizontal axis: normalized spreading coefficient $S_{os/w}/\sigma_{ow}$). The value of these coefficients with respect to $-\frac{r-1}{r-\phi}$ gives the criterions for impregnation, encapsulation or impaling.

11.2.3 Droplet Growth

Nucleation and growth is modified when a water droplet is cloaked by oil, as analyzed in Sect. 11.1.2. Due to limits in vapor sorption within a liquid, nucleation is most favored at the liquid–air interface. On spreading liquids, droplet submergence within the liquid occurs thereafter. Droplet growth occurs through the vapor transport in the liquid films and although the viscosity of the liquid does not affect droplet nucleation, it plays an important role in droplet growth.

For condensing droplets that are not cloaked by oil and whose cap radius R is still lower than the typical pattern spacing b and depth c: $R < b, c$, the situation initially corresponds to what is encountered with small water droplets at the air/solid interface when the latter emerges from oil (State A2, see Sect. 11.2.2; for growth see Chaps. 4 and 6). Between the posts and when oil wets the solid surface (state A3), droplets nucleate and grow on the oil interface (State A3; for growth see above Sect. 11.1). It was observed by Anand et al. (2015) and Kajiya et al. (2016) that, upon condensation, droplets having nucleated on the post-tops subsequently disappeared within the oil spacing due to capillary forces originating from the Laplace pressure of the oil cloak around the droplet. As a result of the submergence of such droplets and of droplets nucleating on oil itself, the water droplets displace oil, leading to oil draining out of the LIS and flooding the surface. However, if later $R > b$, droplets transition to the post-tops and oil flows back within the texture. For low viscosity oil cases (10 cSt and 100 cSt LIS) the oil displacement appears less severe as compared to the 1000 cSt LIS. This is because in the former cases, the oil can drain quickly between the submerged droplets allowing them to coalesce more rapidly and grow at a faster rate. On 1000 cSt LIS, significant suppression of condensation growth is observed due to the increased drainage time.

Non-cloaked condensing droplets when $R > b$ exhibit in States W1 and W2 the same contact line pinning. The only difference is the time delay to evacuate the oil film between them when the droplets come into contact. Growth laws are therefore the same as on smooth solids in Chaps. 4 and 6. Figure 11.14 presents an example of growth laws and surface coverage evolution for condensation in State

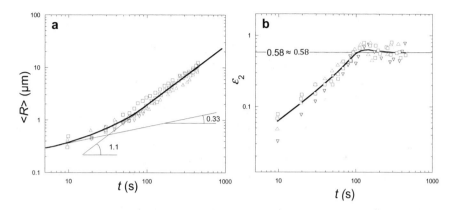

Fig. 11.14 Evolution of the mean drop cap radius <R> and drop surface coverage ε_2. The pillars have a rectangular or circular cross-section and identical spacing. Rectangular Section: blue square, 20 μm; red square, 10 μm. Circular cross-Section: Inverted black triangles, 20 μm, green triangles, 10 μm. (From Kajiya et al. 2016)

A2-W2 where initially $<R> \sim t^{1/3}$ for low surface coverage and $<R> \sim t$ for surface coverage reaching its upper limit $\epsilon_{2,\infty}$. The bump in surface coverage, which is correlated with the beginning of coalescence-driven growth, is hardly seen on solids (Fig. 6.6). Here drops are pinned and cannot rearrange as on liquids during the time that the oil film is removed when drops are in contact (see Fig. 11.12). The limiting value is about 0.58, close to the 0.55 random packing limit, and corresponds to the values obtained on solid substrates (0.57 experimentally in Fig. 6.6 and 6.57 in the Fig. 6.7 simulation).

The same non-cloaked condensing droplets when $R > b$ in State W3 do not show line pinning because of the oil film between the water and substrate. This case resembles condensation on the interface air/oil as studied in Sect. 11.1.3.3 when gravity effects are not present. Removing those gravity effects prevents drops from sinking in oil and also removes the elastic attraction between them. The same growth laws as seen on solid substrates are thus expected.

The fact that the contact line is pinned only on the posts in State W1, on the post-tops in State W2 and not pinned in State W3 allows for easy motion of the droplets. As a result (see Sect. 14.6), drop collection by gravity becomes easier when the substrate is tilted from horizontal.

When droplets are cloaked by oil in State W3, weak anisotropy in capillary forces acting on condensed droplets at the vicinity of a large pre-existing drop can also cause such droplets to move towards the large drop. This process was observed and studied by Sun and Weisensee (2019).

11.3 Soft or Deformable Substrates

Droplet condensation on soft or deformable substrates, that is on a substrate whose shear modulus G is below typically 500 kPa, shows some aspects of condensation on liquid surfaces in the sense that capillary forces and disjoining pressures can deform the substrate around them (Fig. 11.15). Yu and Zhao (2009) provided a detailed description of the drop/substrate geometry. Henkel et al. (2021) proposed a gradient-dynamics model for liquid drops on elastic substrates that captures the main qualitative features of statics and dynamics of soft wetting and which can be applied to ensembles of droplets. The model has the form of a gradient dynamics on an underlying free energy that reflects capillarity, wettability and compressional elasticity.

Sokuler et al. (2010b) and Phadnis et al. (2017) experimentally characterized the main features resulting from condensation on such soft substrates. On the same polymeric material (e.g. PDMS), G can be modified by varying the cross-linker ratio.

The first result is concerned with the increased nucleation density when compared with the same substrate with a larger G (Fig. 11.16). This increase is due to the decrease of the air–water interface area as compared to a rigid film, which compensates for the energy increase as related to the elastic deformation and the increase of the substrate-water interface area (see Sect. 3.2). Molecular dynamic simulations by Che et al. (2019) confirmed this approach.

Another interesting feature is the time delay of coalescence, similar to what is seen at the surface of liquids (see Sect. 11.1.3) where it can even be suppressed (Karpitschka et al. 2016) as a result of a ridge or dimple between neighboring droplets (Fig. 11.15). After coalescence, the return to equilibrium is delayed

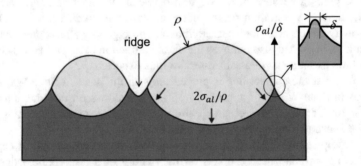

Fig. 11.15 Schematic on a deformable surface of condensed droplets and the wetting ridges between them that slows down coalescence or even prevents it. The air/liquid surface tension σ_{al} pulls the materials surface upward along the periphery of the drop while the Laplace pressure compresses the materials underneath the drop. δ is the contact line width

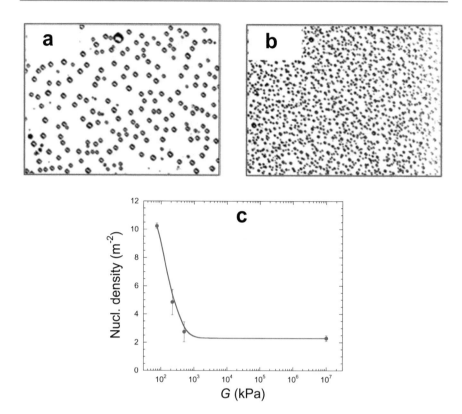

Fig. 11.16 **a** Nucleation density 45 s after cooling has started with $G = 75$ kPa, **b** Same with $G = 50$ GPa. **c** Variation of nucleation sites density with respect to substrate shear modulus G. The curve is a smoothing function. (Adapted from Phadnis et al. 2017, with permission)

because the relaxation not only involves the contact line motion as on solids (see Sect. 5.3.5) but also the displacement of the wetting ridges (Sokuler, 2010b), which adds friction in the contact line motion. It results in the formation of highly irregularly shaped drops during coalescence on the softer substrates (Phadnis et al. 2017).

Phase Change Materials

12

Materials that undergo a liquid–solid phase change in the usual water condensation range, such as cyclohexane or benzene, are not very common. They are examples of Phase Change Materials (PCM) and are also called Phase Switching Materials (Chatterjee et al. 2019). Such materials lead to a very particular drop stick–slip or crawling behavior, as originally reported by Steyer et al. (1992b). The release of the latent heat of condensation during a certain time is indeed able to melt the substrate at the contact line and beneath the drop, which causes drop depinning. Thanks to Marangoni thermocapillary forces, the drop moves rapidly on the substrate until the melted layer freezes, stopping the drop motion by pinning again the contact line. After some time, a condensation-induced melting layer forms once more and the droplet motion is reinitiated.

This process has implications on the drop pattern evolution and, unexpectedly, on ice nucleation, when the substrate temperature is set below 0 °C. Very long icing delays (about four days) have indeed been observed on such materials due to their specific properties (Chatterjee et al. 2019).

12.1 Observations

In-plane water droplet motion during condensation on PCMs (Fig. 12.1) was initially reported by Steyer et al. (1992b) during condensation of water on solid cyclohexane kept at a temperature T_c below its melting temperature $T_m = 6.52$ °C (Lide, 1998). The drop motion is an outcome of melting and refreezing of the cyclohexane surface beneath the droplets. This mechanism is similar to an inverted Leidenfrost effect on a liquid (Hall et al. 1969; Song et al. 2010; Vakarelski et al. 2012) or a solid (Anand and Varanasi, 2013; Antonini et al. 2013) where the stored heat in an object or droplet can cause vapor generation from the surrounding medium. (The Leidenfrost effect is the levitation on an insulating vapor film of a liquid drop close to a surface which is significantly hotter than the liquid boiling

© Springer Nature Switzerland AG 2022

D. Beysens, *The Physics of Dew, Breath Figures and Dropwise Condensation*,
Lecture Notes in Physics 994, https://doi.org/10.1007/978-3-030-90442-5_12

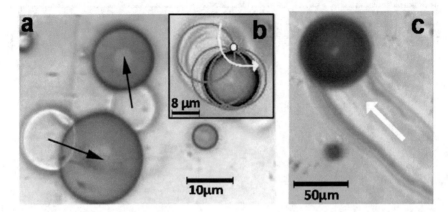

Fig. 12.1 Optical observation of water condensation on PCMs and types of motion of condensing droplets. The arrows indicate the direction of motion. **a** In-plane jumping motion on benzene (T_m= 5.5 °C). Experimental conditions: $T_a = 23$ °C, RH = 100%, flux 0.2 L/min, $T_c = 3$ °C. (Adapted from Narhe et al. 2009). **b** In-plane rotation around a pinning defect on cyclooctane (T_m= 14.5 °C). Experimental conditions: $T_a = 25$ °C, RH = 80%, $T_c = 5$ °C, natural convection. Circles correspond to a 33 ms time interval. (Chatterjee et al. 2019, from video supplementary information). **c** Crawling motion on cyclohexane. Experimental conditions: $T_a = 21$ °C, RH = 30%, T_c = 2 °C, natural convection. (Adapted from Narhe et al. 2015)

temperature; the inverted Leidenfrost effect corresponds to a hot object as e.g. a droplet levitating above a cold liquid where heat from the droplet causes the cold liquid to evaporate, inducing the object levitation).

Heat released during condensation-assisted droplet growth results in a phase change from solid to liquid of the substrate as opposed to liquid to vapor in the conventional Leidenfrost effect, or solid to vapor in some cases of the inverted Leidenfrost effect (Anand and Varanasi, 2013; Antonini et al. 2013). In the classical Leidenfrost problem, the droplet is in a non-wetting state with no distinct contact line (Biance 2003; Quéré 2013). On the other hand, condensing droplets on a PCM such as benzene and cyclohexane (Fig. 12.1) are in contact with the underlying substrate (which alternates between the liquid and solid state). Despite this difference, the condensing droplets on PCMs move freely on the underlying liquid layer as if there were no contact line pinning, similar to the free motion of Leidenfrost droplets on the vapor layer.

One may note that the motion of condensed droplets on PCMs near their melting point is representative of an *isothermal* micro-steam engine. Usual steam engines are adiabatic, however this condition cannot be maintained on the micro-scale where typical heat diffusion times become too small.

12.2 Slip-Stick Process

12.2.1 General Behavior

During the slip-stick process, condensation proceeds on the drop surface and latent heat is continuously released to the substrate. Latent heat is dissipated at the drop-substrate contact surface where PCM can locally melt and depin the drop, which can freely move on a thin liquid layer. The corresponding drop slip-stick behavior (stick: drop is pinned; slip: drop moves on the liquid layer) can be divided into two different stages (Fig. 12.2), characterized by (I) fast slip motion and (II) slow stick relaxation. Stage I starts (time $t = 0$) from the time where PCM melts and a film forms beneath the drop, with a non-equilibrium water/liquid/solid cyclohexane contact angleθ'_S. This is indeed a non-equilibrium situation where liquid and solid cyclohexane phases are in contact under a temperature gradient. The drop moves very rapidly and stops (time t_j) when the film freezes. The distance traveled by the drop, s_j, is in the order of the drop contact radius R (Figs. 12.1a, 12.8a, 12.9b, and Sect. 12.4.3). Stage II follows where the contact line recedes until PCM melts another time below the drop. At the beginning of stage I, the drop contact angle with the solid substrate is the advancing water/solid contact angle θ_{Sa} as the drop grows with the contact line pinned on the substrate. After having moved in stage II, the drop is again pinned and the contact angle with the solid substrate is between the advancing and receding contact angles θ_{Sa} and θ_{Sr}, respectively.

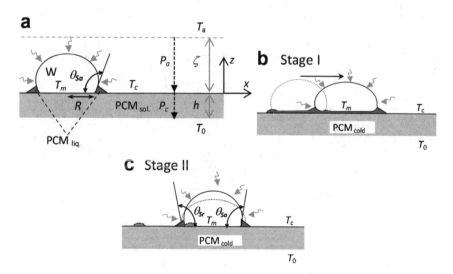

Fig. 12.2 Schematic of the in-plane jumping process. For notations, see text. **a** Continuous substrate heating process. **b** Stage I, melting, moving, and freezing-induced stopping. Melting of solid PCM occurs at the water-substrate contact area, producing a liquid film beneath the water drop that completely wets solid PCM. **c** Stage II, relaxation of the contact line. (Adapted from Narhe et al. 2015)

Thanks to condensation, this angle progressively relaxes to θ_{Sa} (Fig. 12.10a). The drop contact radius therefore immediately increases when the substrate melts and then follows an evolution corresponding to a balance between the condensation-induced growth that increases R and the effect of contact angle relaxation, which makes R decrease (Fig. 12.4).

The presence of a ridge is shown by using optical microscopy and cryogenic—focused ion beam—scanning electron microscope imaging (Cryo-FIB-SEM, see Fig. 11.5 and footnote). Figure 12.3a shows that a wetting ridge is clearly visible at the drop perimeter in the side view (Fig. 12.3a), even when the drop is immobile. The top view of water droplets (Fig. 12.1) shows that a part of this wetting ridge adheres to the substrate when the drop moves. Such a formation left on the substrate cannot be attributed to plastic deformation of solid cyclohexane at the drop perimeter, as initially hypothesized by Steyer et al. (1992b), since in the cryo-FIB-SEM images a flat interface is clearly visible without any visible distortion beneath the droplet (Fig.11b–c). In addition, such stick–slip motions can be observed very far (tens of degrees) below the melting temperature (Chatterjee et al.

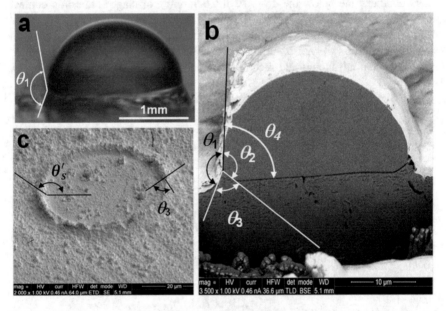

Fig. 12.3 Optical and cryo-FIB-SEM observations of water condensation on cyclohexane and contact angle estimations. **a** Side view of a 2 μL water droplet ($T_a = 21$ °C, RH = 27%, $T_c = 1$ °C). A wetting ridge of liquid cyclohexane is visible. This ridge corresponds in **b** and **c** to the cryofreezing of the layer at the drop perimeter. **b** Cryo-FIB-SEM images of a droplet (section) showing the droplet contact line morphology and the estimation of contact angles (mean over several drops give $\theta_1 \approx 149°$, $\theta_2 \approx 148°$, $\theta_3 \approx 63°$, $\theta_4 \approx 90°$) at the triple line of contact (water/vapor/liquid cyclohexane). A distinct wetting ridge is visible. **c** Cryo-FIB-SEM image of the cyclohexane surface after droplet detachment, which corresponds to the non-equilibrium contact angle $\theta'_s \approx 144°$. (Adapted from Narhe et al. 2015)

2019) where plastic deformation is negligible. It is therefore clear that the wetting ridge is formed as a result of the melting of cyclohexane due to the release of the latent heat of condensation. The ridge contact angles as measured in Fig. 12.3 ($\theta_1 \approx 149°$, $\theta_2 \approx 148°$, $\theta_3 \approx 63°$) corresponds to what is expected from the measurement of surface tensions water/vapor/liquid cyclohexane $\theta_1 = 152°$, $\theta_2 = 165°$, $\theta_3 \approx 43°$ (Narhe et al. 2015). The non-equilibrium water/liquid/solid cyclohexane contact angle (Fig. 12.3) is $\theta'_s \approx 144°$, not far from θ_2. The apparent contact angle $\theta_4 \approx 90°$.

12.2.2 Drop Evolution

For a single, isolated drop, the water vapor profile around the drop surface is radially symmetric and the drop radius evolution follows the $R \sim t^{1/2}$ growth law (Sect. 4.2.4, Eq. 4.45). An exponent 1/3 is, however, usually found for drops in a pattern (Sect. 6.1.2, Eq. 6.15). From the definition of the fraction of drop surface coverage (Eq. 6.29), the mean drop distance (no jump) is, for sessile drops on a square lattice, $\langle d \rangle = \sqrt{\pi/\varepsilon_2} R$. For drops sliding on distance $s_j = f R$, with $f \approx 1.5$ (solid benzene, Narhe et al. 2009) or $f \approx 1.2$ (solid cyclohexane, Fig. 12.8, data from Narhe et al. 2015), the surface coverage corresponds to drops of mean radius $< R + s_j \gtrsim (1+f) < R >$ or $\langle d \rangle = \sqrt{\pi/\varepsilon_2}(1+f)R$. Then the ratio of surface coverage jump/no jump is in the ratio $1/(1+f)^2$ (≈ 0.16 for benzene, 0.21 for cyclohexane). The vapor profiles near the droplets are altered, making the drop radius grow with the 1/2 exponent. In Fig. 12.4 the droplet evolution at the end of the relaxation period, stage II, is indeed found to fit

$$R = R_0 + k(t - t_0)^{1/2} \tag{12.1}$$

Here R_0 is the drop radius at time t_0.

12.3 Substrate Heating and Melting

The experimental configuration is assumed to be a thermostat at temperature T_0 (e.g. a regulated Peltier element) on which is deposited a PCM layer of thickness h (Fig. 12.2a). The substrate temperature, T_c, is different in the places where there is no drop (bare substrate) and in the locations where condensation takes place (presence of a drop). Below, these two different situations are analyzed.

12.3.1 Bare Substrate

At places where condensation does not take place, the T_c value results from (i) the balance of two opposite heat fluxes with power (i) P_a, corresponding to convective air heating, and (ii) P_c, due to the cooling flux from the thermostat below the cyclohexane layer.

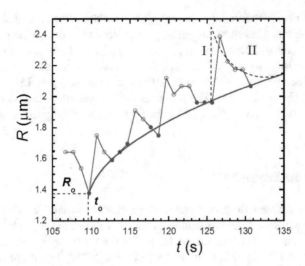

Fig. 12.4 Drop radius evolution (open circle). Full circles correspond to data at the end of the relaxation period stage II. Full line is a fit to Eq. 12.1 with $R_0 = 1.29$ μm, $t_0 = 109.4$ s and k $= (0.17 \pm 0.01)$ μm.s$^{-1/2}$. The interrupted line corresponds to motion and relaxation (schematics) like Fig. 12.9. Experimental conditions: $T_c = 3.4$ °C, $T_a = 24.2$°C, RH $= 100$ %, flux 0.17 L/min. Drop sliding and spreading (stage I) and relaxing (stage II) are clearly seen, although the data were taken every 1 (Adapted from Narhe et al. 2015)

The heat flux P_a occurs within a thermal boundary layer δ_T whose value is nearly the same as the mass diffusion boundary layer thickness ζ (Eqs. 2.13 and 2.16). The heat flux P_a can thus be written as (see Sect. 13.1.1, Eq. 13.6):

$$P_a/S_c = \lambda_a \frac{T_a - T_c}{\zeta} \tag{12.2}$$

Here S_c is the considered surface area and λ_a is the air thermal conductivity. In the case of natural convection, ζ can be evaluated by performing a scaling analysis of convection above the horizontal cooled plate (see Sect. 2.2.2). Its value is only weakly temperature-dependent and common values are $\zeta \sim$ mm, leading, for typical $T_a - T_c \sim 20$ K, to $P_a/S_c \sim 200$ W.m^{-2} (air conductivity value: see Table 12.1; detailed values: see Narhe et al. 2015).

Heat removal occurs through the solid PCM layer, of thickness h, whose lower boundary is at constant temperature T_0. The cooling heat flux P_c can thus be written as

$$P_c/S_c = \lambda_{sp} \frac{T_c - T_0}{h} \tag{12.3}$$

Here λ_{sp} is the solid PCM thermal conductivity.

Assuming near-stationary conditions (no temperature changes are observed in the Narhe et al. 2015 experiments), heating and cooling loads are equals $P_a =$

Table 12.1 Relevant physical properties of Cyclohexane, Water and air (Lide 1998)

	Latent heat (kJ.kg^{-1})	Liquid–air surface tension (mN.m^{-1})	Kinematic viscosity (10^{-6}m^2.s^{-1})	Thermal conductivity (W.m^{-1}.K^{-1})	Specific heat (kJ.kg^{-1}.K^{-1})	Density (kg.m^{-3})
Cyclohexane (liquid l, 6.5° C)	31.8	26.9	1.57	0.1	1.92	778
Cyclohexane (solid x, 3 °C)	31.8	–	–	0.2	1.23	856
Water (liquid l, 6 °C)	2.5×10^3	72.2	1.15	–	4.20	1000
Air	–	–	14	0.024	–	1.2

P_c. Equations 12.2 and 12.3 gives a realistic value $T_c - T_0 \sim 1$ K when PCM is cyclohexane, whose physical properties are contained in Table 12.1.

12.3.2 Substrate Under Drops

Let us now consider a drop that has just moved to its new position. Its temperature at contact with the substrate, whose initial temperature is T_c, results from the balance between P_a, P_c, condensation corresponding to the water latent heat production P_{Lw}, and, when PCM melting occurs, the PCM melting latent heat P_{Lp}. Latent heat flux can be written as

$$P_{Li} = \rho_i L_i \frac{dV_i}{dt} \tag{12.4}$$

The subscript i stands for water (w) or PCM (p). The other parameters are the density of liquid water, ρ_w, the density of liquid PCM, ρ_{lp}, the latent heat of water condensation, L_w, and the latent heat of PCM melting, L_p.

The condensed volume of water can be written, with $G(\theta_c) = \frac{2-3cos\theta_c+cos\theta_c^3}{3sin\theta_c^3}$ the trigonometric function Eq. D.8 in Appendix D:

$$V_w = \pi G(\theta_c) R^3 \tag{12.5}$$

The volume evolution of a water drop where condensation proceeds on its surface and the constant contact angle θ_c is the time derivative:

$$\frac{dV_w}{dt} = 3\pi G(\theta_c) R^2 \frac{dR}{dt} \tag{12.6}$$

As the contact line of the droplet is pinned on the substrate, the contact angle varies during growth, from the receding to the advancing value, causing the water vapor profile to somewhat vary. For the sake of simplification, one considers $\theta_c \approx 90°$ where this variation is weak, which enables this variation to be neglected below.

Using Eq. 12.1 with $R_0 = t_0 = 0$ in Eq. 12.6, one eventually finds

$$\frac{dV_w}{dt} = \frac{3}{2}\pi G(\theta_c)k^2 R \tag{12.7}$$

12.3.3 Melting Time

In the classical Leidenfrost problem, the vapor layer prevents the droplet from boiling and the droplet is maintained at nearly the boiling temperature. Similarly, in the case of a condensed droplet on the PCM surface, with the process of stick/slip being in a steady state, the water drop temperature remains at the PCM melting temperature T_m. In the melting time t_m, condensation heats the substrate beneath the drop from T_c to T_m, corresponding to an energy Q_p, and then melts it, corresponding to an energy $P_{L_p}t_m$. As a result, the time t_m required to heat and melt a volume v_p of solid PCM corresponds to an energy balance between heating loads (air: $P_a t_m$; water condensation: $P_{L_w}t_m$) and cooling loads (thermostat: $P_c t_m$; PCM solidification: $P_{L_p}t_m$; sensitive heat Q_p of solid PCM from T_c to T_m). One thus obtains the following equation, where heating and cooling energies are equalized:

$$\left(P_a + P_{L_w}\right)t_m = Q_p + \left(P_c + P_{L_p}\right)t_m \tag{12.8}$$

In Eq. 12.8, Q_p, the sensitive heat of the PCM layer, can be written as

$$Q_p = \rho_{ps}v_p C_{ps}(T_m - T_c) \tag{12.9}$$

where C_{ps} is the specific heat of solid PCM. In the stationary state $P_a = P_c$. Other terms in Eq. 12.8 are, with ρ_{ps} the density of solid PCM:

$$P_{L_p}t_m = v_p \rho_{ps} L_p \tag{12.10}$$

and:

$$P_{L_w} = \rho_w L_w \left(\frac{dV_w}{dt}\right) \tag{12.11}$$

Using Eq. 12.7, the equality Eq. 12.8 eventually gives the following relation:

$$t_m = \frac{2}{3\pi}\frac{\rho_{ps}}{\rho_w}\frac{L_p + C_{ps}(T_m - T_c)}{L_w}\frac{1}{k^2 G(\theta_c)}\frac{v_p}{R} \tag{12.12}$$

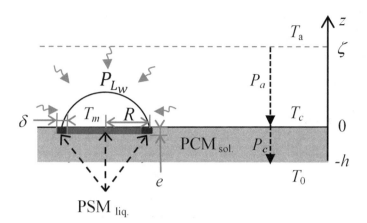

Fig. 12.5 Heat transfer (notations: see text) and melted volume as either a uniform layer of thickness e and radius R, or annulus of width δ, thickness e and radius R

Two configurations can be envisaged for the melted volume (Fig. 12.5): (i) a uniform layer of thickness e below the drop where heat exchange occurs corresponding to $v_p = \pi e R^2$ and (ii) an annular volume of thickness e and width δ around the drop contact line, $v_p = 2\pi e \delta R$. These two evaluations are not in contradiction since the temperature rise is the highest near the contact line (see Chap. 13). For high supersaturation with a large temperature difference between the air-substrate and ($T_c \ll 0$ °C, see Chatterjee et al. 2019), melting occurs only at the drop perimeter and the liquid layer spreads very rapidly below the drop, forming a non-uniform layer. For low water vapor supersaturation as in Steyer et al. (1992b) and Narhe et al. (2009, 2015), the melted layer is near uniform. These two configurations are analyzed below.

12.3.3.1 Uniform Layer (Low Supersaturation)
In this case, Eq. 12.12 becomes

$$t_m = \alpha R \tag{12.13}$$

with

$$\alpha = \frac{2}{3} \frac{\rho_{ps}}{\rho_w} \frac{L_p + C_{ps}(T_m - T_c)}{L_w} \frac{e}{k^2 G(\theta_c)} \tag{12.14}$$

Proportionality is expected between t_m and R assuming that the thickness e remains independent of R. Figure 12.6 data (cyclohexane substrate at low vapor supersaturation) show that a linear relationship is indeed observed, validating the assumption of constant layer thickness.

From the prefactor in Eq. 12.13, the melted layer thickness e can be obtained for conditions of Fig. 12.4 and 12.6. Figure 12.6 data gives $\alpha = 6 \times 10^4$ s.m^{-1}

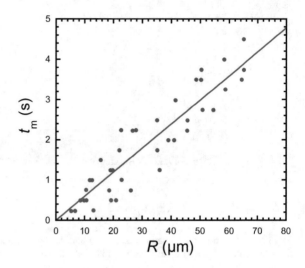

Fig. 12.6 Time t_m between jumps vs drop radius R (cyclohexane; $T_c = 3.4\ °C$, $T_a = 24.2\ °C$, RH = 100%, flux $F = 0.17$ L. min.$^{-1}$). The line is a fit to Eq. 12.13 with slope $\alpha = 0.06 \pm 0.02$ s.μm^{-1}). (Data from Steyer et al. 1992b)

and Fig. 12.4 gives $k^2 = 2.9 \times 10^{-14}$ m^2.s^{-1}. Using data from Table 2.2 and Table 12.1, one obtains, with $\theta_c \sim 115°$ (Narhe et al. 2015), $e \approx 80$ nm.

12.3.3.2 Annular Layer (High Supersaturation)

One obtains the following estimation from Eq. 12.12:

$$t_m = \left[\frac{4}{3} \frac{\rho_{ps}}{\rho_w} \frac{L_p + C_{ps}(T_m - T_c)}{L_w} \frac{1}{k^2 G(\theta_c)} \right] e\delta \qquad (12.15)$$

It is likely that the annulus section area $e\delta$ varies with R. However, quantitative information would need dedicated experiments and/or heat transfer simulations. Both are still lacking.

12.3.4 Surface Temperature Increase

One can estimate a maximum temperature rise due to the deposition of latent heat on solid PCM. This value will give the upper substrate temperature where melting can occur. Depending on the assumptions above regarding the volume geometry, two estimations can be performed (see Chatterjee et al. 2019).

12.3.4.1 Uniform Layer (Low Supersaturation)

Condensation heat release is assumed to be uniformly distributed below the droplet surface in area $S_c = \pi R^2$. The drop is also supposed to be in contact with a semi-infinite PCM solid surface at location $z = 0$ whose temperature, T_c, is T_0 at $z \to -\infty$ is and T_0 on its surface ($z = 0$) (Fig. 12.5). Heat transfer from droplet to solid surface (power P_{L_w}) is supposed to be governed purely by conduction, which is a good approximation (see Sect. 13.5.4). The governing equations along with the boundary conditions are given by the following:

$z = 0$ (droplet location):

$$T_c(z = 0, t = 0) = T_0 \tag{12.16}$$

$z \to -\infty$:

$$T_c(z \to -\infty) = T_0 \tag{12.17}$$

Under the drop, heat diffusion obeys the equation:

$$\frac{\partial T_c}{\partial t} = D_{ps} \frac{\partial^2 T_c}{\partial z^2} \tag{12.18}$$

where t is time and D_{ps} the thermal diffusivity of the PCM solid substrate. Heat flux below the drop per unit of surface area, q, is written as

$$q = -\lambda_{sp} \left(\frac{\partial T_c}{\partial z} \right)_{z=0} = \frac{P L_w}{\pi R^2} = \frac{\rho_w L_w}{\pi R^2} \left(\frac{dV_w}{dt} \right) \tag{12.19}$$

One uses Eq. 12.11 to express $P L_w$ as a function of the drop growth rate dV_w/dt. Using the time dependence of V_w, Eq. 12.7, one obtains

$$q = F_{01} t^{-1/2} \tag{12.20}$$

where

$$F_{01} = \frac{3}{2} G(\theta_c) k \rho_w L_w \tag{12.21}$$

The condensation heat release therefore causes an imposed flux on the solid at $z = 0$. The solution of the above equations for applied flux per unit area of form:

$$q = F_0 t^{n/2} \tag{12.22}$$

at $z = 0$ and $t > 0$ (where n may be -1, 0 or a positive integer) is well known. It is given by Carslaw and Jaeger (1959). With $\Delta T(z,t) = T(z,t) - T_0$, one obtains, Γ being the gamma function (see Appendix K):

$$\Delta T(z, t) = \frac{F_0 D_{ps}^{1/2} \Gamma(1 + n/2)}{\lambda_p} (4t)^{(n+1)/2} i^{n+1} \text{erfc} \left(\frac{z}{2\sqrt{D_{ps} t}} \right) \tag{12.23}$$

Here i^nercf represents the n-iterated integral of the erfc function (see Appendix K. The surface temperature ($z = 0$) then becomes

$$\Delta T(0, t) = \frac{F_0 D_{ps}^{1/2}\Gamma(1+n/2)}{\lambda_p \Gamma(3/2+n/2)} t^{(n+1)/2} \qquad (12.24)$$

Comparing it with Eqs. 12.20 and 12.22, it comes $n = -1$ and $F_0 = F_{01}$. Temperature rise on the surface is then not time dependent. With $\Gamma(1/2) = \sqrt{\pi}$ and $\Gamma(1) = 1$ (see Appendix K), one obtains

$$\Delta T(0) = \frac{F_{01}}{\lambda_p}\sqrt{\pi D_{ps}} \qquad (12.25)$$

The corresponding value is ~2.7 °C for condensation on solid cyclohexane under the conditions of Fig. 12.4 where $F_{01} \sim 1.33 \times 10^3$ W.m^{-2}s$^{1/2}$, using the data from Table 12.1 and $\theta_c \sim 115°$ (Narhe et al. 2015). This temperature rise is consistent with what is observed in such low supersaturation experiments where $T_m \sim 5 - 6$ °C and T_c remains a few degrees above 0 °C.

12.3.4.2   Annular Layer (High Supersaturation)

Condensation heat release occurs along the dropletcontact line in a region of size δ. In this case, the governingheat transfer equation remains the same as above, but Eq. 12.19 is modified as

$$q = \frac{P L_w}{2\pi R\delta} = \frac{\rho_w L_w}{2\pi R\delta}\left(\frac{dV_w}{dt}\right) \qquad (12.26)$$

giving rise to

$$q = F_0 = \frac{k}{2\delta}F_{01} \qquad (12.27)$$

with

$$F_0 = \frac{3k^2}{4\delta}G(\theta_c)\rho_w L_w \qquad (12.28)$$

This case corresponds in Eq. 12.22 to $\alpha = 0$. The surface temperature can now be written as

$$\Delta T(0, t) = \frac{F_0 D_{ps}^{1/2}\Gamma(1)}{\lambda_p \Gamma(3/2)} t^{1/2} \qquad (12.29)$$

Using $\Gamma(1) = 1$ and $\Gamma(3/2) = \sqrt{\pi}/2$ (see Appendix K), one obtains

$$\Delta T(0, t) = \left(\frac{3k^2}{2\delta}\frac{G(\theta_c)\rho_w L_w \sqrt{\pi D_{ps}}}{\lambda_p}\right) t^{1/2} \qquad (12.30)$$

or using the growth law Eq. 12.1 with $R_0 = 0$ and $t_0 = 0$:

$$\Delta T(0, R) = \left(\frac{3k}{2} \frac{G(\theta_c) \rho_w L_w \sqrt{\pi D_{ps}}}{\lambda_p} \right) \frac{R}{\delta} \quad (12.31)$$

Using the same numerical values as above for condensation on solid cyclohexane, $\Delta T(0, R) \sim 0.9 R/\delta$. The extent of temperature change thus depends upon the extent of annulus width δ. There is up to now no determination of this width. One expects such a region to be less than 1 μm. If one considers a droplet of 100 μm diameter, the temperature change can thus be $\Delta T \sim 90\ °C$. Before reaching such a huge temperature increase, PCM should melt. Stick–slip behavior of supercooled water droplets can indeed be observed even at temperatures as low as $-15\ °C$ (Chatterjee et al. 2019).

12.4 Stage I: Fast Droplet Motion

In this stage I, the establishment of a film of molten PCM depins the drop contact line. The drop then moves rapidly on a layer of molten PCM. This period is very fast (typically a few ms) and ends when the PCM film refreezes. At this point, viscous dissipation increases dramatically, leading to termination of drop motion. Figure 12.7 shows the drop displacement on cyclohexane. One notes that the drop velocity is constant during the motion and suddenly goes to zero.

12.4.1 Relaxation of Capillary Forces

The motion cannot be due to the relaxation of capillary forces
$F \sim 2\pi \sigma_{wv} R$ when the contact angle changes at depinning (σ_{wv} refers to the surface tension of water in air). This force can only give an initial velocity U_0 to the drop, whose velocity should decrease with time because of the friction in the film. Motion will stop after a distance s_c, which can be estimated by equating the drop kinetic energy:

$$E_c = \frac{1}{2} \rho_w V_w U_0^2 \quad (12.32)$$

and film viscous dissipation:

$$E_v = \rho_w \nu U_0 R^2 s_j / e \quad (12.33)$$

Here ν is the kinematic viscosity of water ($= 1.5 \times 10^{-6}\ m^2 \cdot s^{-1}$; see Table 1.2). With the data of Fig. 12.7 ($R \approx 80$ μm, $U_0 = U = 3 \times 10^{-2}\ m \cdot s^{-1}$) one gets $s_j = \pi U_0 eR/3\nu \approx 0.1$ μm for a drop radius of $R \sim 80$ μm with typical velocity $U_0 \equiv U = 3 \times 10^{-2}\ m.s^{-1}$ (Fig. 12.7d). This displacement is due to capillary forces

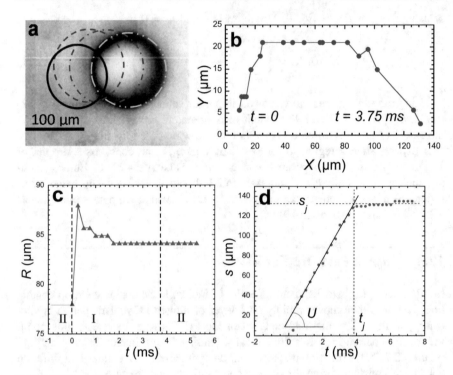

Fig. 12.7 Example of drop motion on solid cyclohexane ($T_a = 23$ °C, RH $= 100\%$, $T_c = 3$ °C, flux 0.2 L/min.). **a** Drop depinning (78 μm diameter; black circle) and moving (interrupted circles). **b** Trajectory of the drop center of mass. Each point corresponds to 0.25 ms time difference. Note the different scales in ordinates. **c** Evolution at short times of the drop-projected radius related to a change in contact angle. **d** Evolution of the distance s traveled by the drop in (a, b) showing constant velocity before brusquely stopping at $s = s_j$ for $t = t_j$. (Adapted from Narhe et al. 2015, with permission)

and is thus negligible when one compares the value (0.1 μm) with the value found in Fig. 12.7b (~140 μm).

12.4.2   Thermocapillary Motion

The fact that the capillary force contributes only very little to the motion, where velocity remains constant, shows that the drop movement is rather due to a thermocapillary effect. Temperature gradients are indeed present at the surface of the drop, at the liquid PCM ridge around the drop, and at the interface drop-liquid PCM beneath. Due to the change of position of the drop after PCM melting, nonsymmetric temperature gradients appear that move the drop until the film freezes towards the cold end, that is, outside the drop.

Let us assume the simplest case where thermocapillary motion is dominated by the interfaces air–water and air–liquid PCM. The typical fluid velocity at the

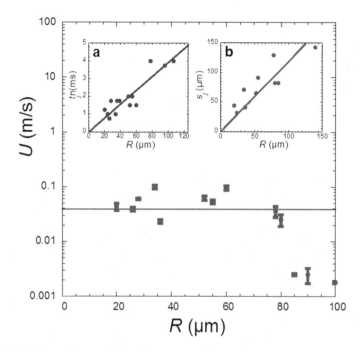

Fig. 12.8 Variation of drop velocity with respect to drop radius R. The line corresponds to $U = 0.039$ m.s^{-1}. Inset (a): Variation of jump time t_j with respect to R. The line is a fit with slope 39 ± 2 s.m^{-1}. Inset (b) Variation of jump distance s_j with respect to R. The line is a fit with slope $f = 1.19 \pm 0.1$. Experimental conditions: condensation on cyclohexane with $T_a = 23$ °C, RH = 100%, $T_c = 3$ °C, flux 0.2 L/min. (Data from Narhe et al. 2015)

interface of the liquid layer (thickness e, shear viscosity η, surface tension derivative $d\sigma/dT$) submitted to a transverse thermal gradient (dT/dx) can be expressed as (Fedosov, 1956):

$$U = \frac{e}{4\eta}\left(\frac{d\sigma}{dT}\right)\left(\frac{dT}{dx}\right) \tag{12.34}$$

Assuming from a scaling assumption that the characteristic lengths for the film and ridge are the same as the typical length scale in the temperature gradient and that the temperature difference is on order T_m-T_c, it follows:

$$U \sim \frac{1}{4\eta_{PCM}}\left(\frac{d\sigma}{dT}\right)(T_m - T_c) \tag{12.35}$$

In this formulation, the drop velocity is indeed independent of its radius (Fig. 12.8) for not too large drops. For large drops, U dramatically decreases due to early pinning on substrate inhomogeneities.

Considering water and cyclohexane properties of Table 12.1, with T_m - T_c ~ 1 K, $(d\sigma/dT)_{air-water} = -1.5 \times 10^{-4}$ N.m^{-1}, $(d\sigma/dT)_{air-cycloh.} = -5.3 \times 10^{-4}$ N.m^{-1}

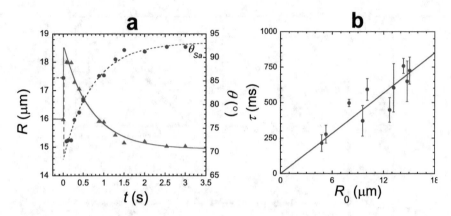

Fig. 12.9 Long time drop radius relaxation in stage II. Condensation on cyclohexane. **a** The curve is a fit to Eq. 12.41 with $\varepsilon = 0$ and $\tau = 0.7$ s. ($R_0 = 16$ μm). The evolution of the apparent contact angle follows. **b** Variation of the drop size relaxation time τ with respect to the drop radius R_0. The line is a fit to Eq. 12.40, which shows proportionality between τ and R_0 with slope 4.7×10^3 s.m^{-1}. (Adapted from Narhe et al. 2015)

and $\eta_{cycloh.} = 1.2 \times 10^{-3}$ Pa.s (Lide 1998), one obtains $U \sim 0.06$ m.s^{-1}. This value compares relatively well with the data in Fig. 12.9 where $U \approx 0.04$ m.s^{-1}.

12.4.3 Freezing Time and Jump Length

The fact that the drop stops after having traveled on length $s_j \sim R$ can be explained as the distance where the drop surface moves on a fresh, non-heated substrate area, which makes the melted layer freeze nearly instantaneously. The freezing time, $\delta t,$ is indeed quasi-instantaneous as the following estimation shows. When the drop moves, then the volume fraction $v_f \sim s_j^2 e$ of the PCM liquid layer below the drop, which is in contact with the fresh substrate, freezes and mixes with the still molten layer. Complete freezing occurs when $v_f \sim R^2 e$ or:

$$s_j = fR \tag{12.36}$$

corresponding to time:

$$t_j \sim \left(\frac{1}{U}\right) R \tag{12.37}$$

Proportionality is actually seen between drop jump s_j and R (Fig. 12.8, inset b). Concerning condensation on cyclohexane, $f \sim 1.2$ (Fig. 12.9b) and for condensation on benzene, $f \sim 1.5$ (Narhe et al. 2009). Proportionality is also observed between drop jump time t_j and R (Fig. 12.8, inset a) concerning condensation on cyclohexane: $t_j/R \sim 40$ s.m^{-1}, a value on order of $1/U \sim 26$ s.m^{-1}).

The time δt required to freeze this layer can be estimated as follows. While the drop is moving on the substrate, cooling of the drop and layer beneath is performed by the substrate at temperature T_c. The lower part of the layer is then at T_c and its upper part at T_m. Both the layer and drop are close to the same temperature T_m. This means that only the PCM latent heat is removed, the freezing process being at constant temperature. It also means that the surface temperature of solid PCM has to adapt to T_m by a thermal boundary layer of extent $\delta_T = \sqrt{D_{ps}t}$ (D_{ps} is the solid PCM thermal diffusivity). Freezing of the liquid layer will last the time δt, corresponding to the thermal boundary layer $\delta_T = \sqrt{D_{ps}\delta t}$ when the sensitive heat gained by the diffuse boundary layer will be equal to the latent heat needed for freezing. With S the surface area of contact of the liquid layer, the energy W needed to freeze a liquid layer of thickness e is

$$\frac{W}{S} = \rho_{ps} C_{ps}(T_m - T_c)\delta_T = \rho_{pl} L_p e \qquad (12.38)$$

which gives the freezing time:

$$\delta t = \left[\frac{\rho_{pl}}{\rho_{ps}} \frac{L_p}{C_{ps}(T_m - T_c)} \right]^2 \frac{e^2}{D_{ps}} \qquad (12.39)$$

Using the values (Table 1.2) concerning cyclohexane and with T_m - $T_c \sim 1$ K, a time $\delta t \approx 20$ μs is found for a uniform layer with thickness $e \approx 80$ nm as found in Sect. 12.3.3.1. This short time scale guarantees near instantaneous local freezing of the melted PCM after the drop's departure. It also explains the presence of ridge-like features (Fig. 12.1) that are left behind the droplet when it moves to a new location.

When freezing of the liquid PCM does not occur, a steady motion triggered by thermocapillary forces occurs. This is the so-called crawling motion as shown in Fig. 1c.

12.5 Stage II: Contact Line Slow Relaxation

After the drop has moved, the liquid imprint of its previous position fades gradually. The imprint corresponds to the residues of the wetting ridge around the drop. This ridge is initially liquid and freezes after some time, and disappears by evaporation (when liquid) and sublimation (when solid). The drop radius slowly diminishes to ideally recover its initial value on solid PCM, with a small increase due to condensation (see Fig. 12.4).

This process corresponds to the general situation of the motion of the contact line of a drop on a solid (see Sect. 5.3.3.1). The relaxation of the shape of the composite drop formed after the coalescence of two drops is presented in Sect. 5.3.5.2. The relaxation process of this non-equilibrium shape is very slow due to the enhanced dissipation at the contact line location. It follows an exponential decay (Eq. 5.61) with typical time (Eqs. 5.62 and 5.63):

$$\tau = K\left(\frac{\eta}{\sigma}\right)R \qquad (12.40)$$

The overall evolution of the drop radius thus follows, approximating the growth during the relaxation time to a linear evolution with (small) slope ε:

$$R \approx R_0 + \varepsilon t + R_1 \exp(-t/\tau) \qquad (12.41)$$

Here R_0 is the drop radius at the onset of motion ($t = 0$). The second term represents the drop slow growth due to condensation (Eq. 12.1) and the third term is due to contact line relaxation. The parameter K is in the range of 10^6–10^7 (see Sect. 5.3.5). Figure 12.9a shows the exponential drop radius relaxation for condensation on cyclohexane. Note that a difference exists between the initial radius (before jump) and final radius (relaxed value). This difference indeed depends on the time between two successive jumps (which varies for every jump), the relaxation being more or less completed. As expected from Eq. 12.40, τ is proportional to R (Fig. 12.9b). The data fit well to Eq. 12.41 with $K \approx 3 \times 10^6$, a value in agreement with those reported in Sects. 5.3.3.1 and 5.3.5.2.

Because the evolution of the drop volume (V) due to condensation (corresponding to the factor ε in Eq. 12.41) is in general quite small, one can relate the drop radius to the apparent contact angle θ_4 by using the expression of the drop volume with contact angle, Eq. D.7 in Appendix D. This equation allows the drop radius and the contact angle to be related when the drop volume is kept constant. For near-hemispherical drop $\theta_4 \approx 90°$ (water on cyclohexane, (Fig. 12.3b), one obtains the correlated evolution of R and θ_4, $\Delta\theta \approx 2\Delta R/R$ (Fig. 12.9a).

Thermal Effects

13

Condensation generally proceeds from a lowering of the surface temperature, either by contact with a cooler surface, the more frequent or outdoor because of a radiation deficit between the surface and the surrounding environment (see Appendix O). The thermal processes involved in both are the deposit of the latent heat and the different thermal resistances concerning condensed water, the substrate and air.

When a film forms through contact cooling underneath, which is the most general case encountered in industrial installations, Nusselt reduced the complexity of the real process to a simple model where the only resistance for the removal of the heat released during condensation occurs in the condensate film (see e.g. Incropera and DeWitt 2002). When a drop pattern forms, the substrate remains bare between drops and the yield can be increased greatly.

Radiative cooling, however, can lead to different situations since the film or drops are continuously cooled at the water-air interface, with the liquid water emissivity being close to unity. There is thus no thermal resistance in the film since condensation and cooling occur at the same place and the Nusselt model no longer applies. Concerning dropwise condensation, (the most usual case encountered with natural dew), the substrate and drop exhibit the same emissivity close to unity, and cooling is preserved throughout the process.

The original version of the book was revised: Belated corrections have been updated. The correction to the book is available at https://doi.org/10.1007/978-3-030-90442-5_16

© Springer Nature Switzerland AG 2022, corrected publication 2023
D. Beysens, *The Physics of Dew, Breath Figures and Dropwise Condensation*,
Lecture Notes in Physics 994, https://doi.org/10.1007/978-3-030-90442-5_13

13.1 Heat Transfer

13.1.1 Heat Transfer Coefficients

When a material is submitted to a heat flux Q (or q per unit surface area), corresponding to a temperature difference ΔT, one can define a heat transfer coefficient a by:

$$q = a\Delta T \tag{13.1}$$

The coefficient a can be related to a thermal resistance R:

$$a = \frac{1}{R} \tag{13.2}$$

For example, when the heat flow is related to conduction (heat diffusion) through a material of thickness e, one can write, with λ, the material thermal conductivity:

$$a = \frac{\lambda}{e} \tag{13.3}$$

Figure 13.1 presents typical filmwise and dropwise condensation designs where cooling is ensured either by contact with a thermostat (quite often provided by a flow of coolant) or by radiation deficit at the coating/air and condensate/air interfaces. Heat flux Q being conserved at the interface, the temperature differences between interfaces and thermal resistances can thus be added. One can write, with j the different materials listed in Fig. 13.1 and $\Delta T = T_a - T_0$:

$$Q = \text{const.}; \ \Delta T = \sum \Delta T_j; \ R = \sum R_j; \ \frac{1}{a} = \sum \frac{1}{a_j} \tag{13.4}$$

In different situations such as in Sect. 13.1.2, the temperature difference between two interfaces is constant, and heat fluxes must be added. In this case, one obtains:

$$\Delta T = \text{const.}; \ Q = \sum Q_j; \ \frac{1}{R} = \sum \frac{1}{R_j}; \ a = \sum a_j \tag{13.5}$$

The heat transfer coefficients due to conduction in the solid substrate and coating are determined through Eq. 13.3 from solid thicknesses and thermal conductivities. The heat transfer in humid air (or vapor plus non-condensable gases), q_a, can be also deduced from the air thermal conductivity λ_a and the thermal boundary layer thickness δ_T (see Sect. 2.2) corresponding to a convective heat transfer coefficient $a_a = \frac{\lambda_a}{\delta_T}$:

$$q_a = \frac{\lambda_a}{\delta_T}(T_a - T_i) \tag{13.6}$$

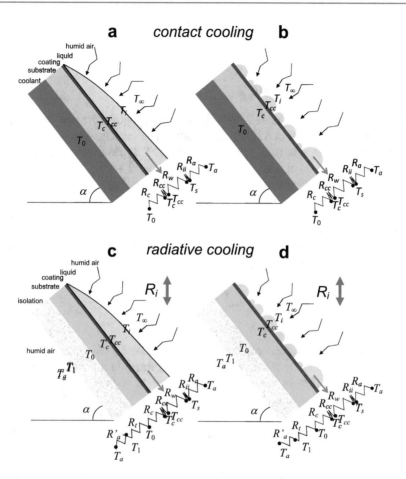

Fig. 13.1 **a–b** Schematics of contact cooling for **a** filmwise and **b** dropwise condensation when non-condensable gas is present in the vapor (e.g. humid air), showing thermostat (T_0), substrate (thermal resistance R_c), coating (R_{cc}), condensed liquid (R_w), air/liquid interface (R_{ii}), humid air (R_a) with the corresponding interface temperatures. **c–d** Radiative cooling (power R_i) for **a** filmwise and **b** dropwise condensation showing thermal isolation (thermal resistance R_l), substrate (R_c), coating (R_{cc}), condensed liquid (R_w), air/liquid interface (R_{ii}) and humid air (R_a) with the corresponding interface temperatures

For usual δ_T values (~ mm, see Sect. 2.2) and difference $T_a - T_i \sim$ K between the condensed liquid interface temperature, T_i and air temperature, T_a, data from Table 1.2 give $q_a \sim 26$ W·m^{-2}.

The interface heat coefficient $a_i = 1/R_i$ corresponds to the release of the latent heat by the condensation process at the gas-liquid interface. Its value depends upon whether non-condensable gases are present or not in the vapor. This coefficient is considered independent of the interface location (Umur and Griffith 1965) and is detailed below in Sects. 13.3.1 and 13.3.2.

Below, one will assume for the sake of simplicity that vapor is at saturation at temperature T_∞.

13.1.2 Substrate Coating

The use of patterned substrates can lead to non-negligible thermal resistance as outlined by Miljkovic et al. (2012), especially when coatings are superhydrophobic. On such substrates (see Appendix E, Sect. E.2) drops can either be in a Cassie–Baxter air-pocket state (Fig. 13.2a) or a Wenzel impaled state (Fig. 13.2b). The heat transfer drop/substrate will be different according to these two situations.

When in the Cassie–Baxter state, conduction in air can be neglected with respect to conduction through the pillars with height h_p, surface fraction ϕ_s, thermal conductivity λ_p, hydrophobic coating with thickness h_c and thermal conductivity λ_c (Fig. 13.2a). The heat transfer coefficient through the whole coating, from the conduction Eqs. 13.3 and 13.4, thus, becomes:

$$a_{cc}^{CB} = \frac{\phi_s}{\frac{h_p}{\lambda_p} + \frac{h_c}{\lambda_c}} \tag{13.7}$$

When a Wenzel state is concerned (Fig. 13.2b), one has to add conduction through the pillar and conduction through the coating structure in a parallel heat transfer pathway from the base of the droplet to the substrate surface. It readily becomes, using Eq. 13.5 and adding the flux through pillars + coating (a_{cc}^{CB}) and water interpillars + coating (a_{cc}^{CB} with surface fraction $1 - \phi_s$ and pillar thermal conductivity λ_c replaced by water conductivity λ_w). The heat transfer coefficient for the Wenzel state becomes:

$$a_{cc}^W = a_{cc}^{CB} + (1 - \phi_s)a_{cc}^{CB} = \frac{\phi_s}{\frac{h_p}{\lambda_p} + \frac{h_c}{\lambda_c}} + \frac{1 - \phi_s}{\frac{h_p}{\lambda_w} + \frac{h_c}{\lambda_c}} \tag{13.8}$$

Note that this formulation can also represent the Cassie–Baxter case when one writes $\lambda_w = 0$. Later, the heat transfer coefficients of Eqs. 13.7 or 13.8 will be simply denoted by the unique $a_{cc} = a_{cc}^{CB}\text{or}a_{cc}^W$.

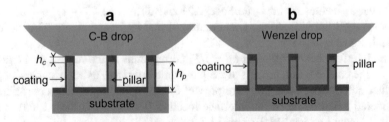

Fig. 13.2 Droplet in **a** the CB state and **b** the Wenzel state (schematic)

13.2 Filmwise Condensation

In this condensation mode, coatings are generally not used and $T_c = T_{cc}$. The liquid condensate forms a continuous film over the substrate and the film flows down under the actions of gravity, shear force due to vapor flow or other forces. The latent heat released at the vapor-liquid interface is transferred through the condensate film and then through the solid wall and is eventually removed by a coolant flowing on the other side of the condensing surface (Fig. 13.1a) or by radiative exchange at the condensate and substrate interface with air (Fig. 13.1c). (One considers below both cooling processes).

A steady state is established when the rate of condensation is balanced by the rate of flow of the condensate. The heat transfer through the condensed film forms, in the conduction cooling mode, the greatest fraction of the thermal resistance. The film thickness varies under the influence of condensation and drainage (Fig. 13.1a) and depends upon whether cooling is performed by substrate contact underneath or by a radiative deficit at the interface film/air.

A sketch of the condensation process on a smooth square planar substrate of surface $S_c = L \times L$, making an angle α from horizontal, is depicted in Fig. 13.3a. One assumes that condensation occurs as a film with a non-uniform thickness $h(\xi, t)$. Flow in the film is supposed to be laminar with flow velocity $u(\xi, \zeta)$ and stick boundary conditions. The equation of motion can be deduced from the Navier–Stokes equations (see Appendix C). With η the dynamic (shear) viscosity, g the earth acceleration constant, it comes in the limit of negligible fluid inertia:

$$\eta \left(\frac{\partial^2 u}{\partial \xi^2} + \frac{\partial^2 u}{\partial \zeta^2} \right) + \rho_l g \sin \alpha = 0. \tag{13.9}$$

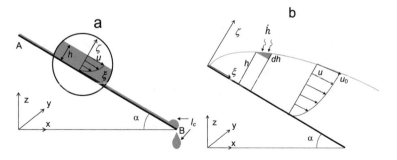

Fig. 13.3 Filmwise condensation on a smooth planar surface and drop detachment (notations: see text). AB $= L$. **a** General view. **b** Film profile

As the thickness h being very small compared with the length of the flow, the lubrication flow approximation can be used. Equation 13.9 thus becomes:

$$\eta \frac{\partial^2 u}{\partial \zeta^2} + \rho_l g \sin \alpha = 0. \tag{13.10}$$

Integration gives the classical parabolic velocity profile. With boundary conditions $u(0) = 0$ (no slip); $\left(\frac{\partial u}{\partial \zeta} \right)_{\zeta = h} = 0$ (continuity of the shear stress at the interface), one obtains:

$$u(\zeta) = \left(\frac{\rho_l g \sin \alpha}{\eta} \right) \zeta \left(h - \frac{\zeta}{2} \right) \tag{13.11}$$

By definition F, the volumic flux per unit length dy, can be written as:

$$F(\xi) = \int_0^{h(\xi)} u(\zeta) d\zeta \tag{13.12}$$

Making use of Eq. 13.11 to express u as a function of h, it becomes:

$$F(\xi) = \left(\frac{\rho_l g \sin \alpha}{3\eta} \right) h^3(\xi) \tag{13.13}$$

13.3 Interface Heat Transfer

The interface heat coefficient a_i corresponds to the release of the latent heat by the condensation process at the gas-liquid interface. This coefficient is considered independent of the interface location (Umur and Griffith 1965). It is detailed below for filmwise condensation (Sect. 13.3.1) and dropwise condensation (Sect. 13.3.2). With L_v, the latent heat of condensation, ρ_l the liquid density, S_c the condensing surface area, dh the local variation of liquid condensate height corresponding to volume variation dV, then the corresponding heat flux q_i reads as:

$$q_i = \rho_l L_v \frac{1}{S_c} \frac{dV}{dt} = \rho_l L_v \frac{dh}{dt} \tag{13.14}$$

13.3.1 Pure Vapor

When pure vapor is concerned (e.g. steam, see Fig. 13.4), the rate $dh/dt = \dot{h}$ is classically expressed from the kinetic theory of gases. Incorporation of molecules at the liquid-vapor interface and molecule evaporation occurs in the Knudsen layer, corresponding to a few mean free paths and thus a few molecular sizes thick (see Sect. 4.1). It leads, for relatively weak subcooling, to the following expression for the interfacial heat transfer coefficient (Eq. 4.10)

$$\frac{q_i}{T_\infty - T_i} = a_i = \left(\frac{2\sigma}{2-\sigma}\right)\frac{\rho_v L_v^2}{T_\infty^{3/2}\sqrt{2\pi r_g}} \tag{13.15}$$

For steam at $T_\infty = 373$ K under atmospheric pressure where $\rho_v = 0.6$ kg·m^{-3} and $L_v = 2.26 \times 10^6$ J·kg^{-1}, the transfer coefficient $a_i = 16$ MW·m^{-2}·K^{-1} for $\sigma = 1$ and 0.63 MW·m^{-2}·K^{-1} for $\sigma = 0.04$. For steam at $T_\infty = 297$ K under 3 kPa pressure where $\rho_v = 0.022$ kg·m^{-3}, $L_v = 2.4 \times 10^6$ J·kg^{-1}, one finds $a_i = 0.92$ MW·m^{-2}·K^{-1} for $\sigma = 1$ and 36 kW·m^{-2}·K^{-1} for $\sigma = 0.04$.

Fig. 13.4 Filmwise condensation with a non-condensable gas (black/interrupted curves) and without (blue/continuous curves). Vapor is assumed at saturation pressure p_s. C: substrate coating; L: liquid; G: gas; ζ: diffuse boundary layer; K: Knudsen layer. Other notations: see text

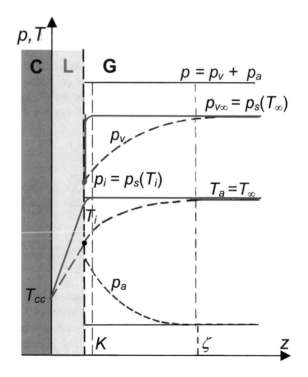

13.3.2 Vapor with Non-condensable Gases

Let us now consider a non-condensable gas with partial pressure p_a. The total pressure $p = p_v + p_a$ is uniform. In addition to the Knudsen layer at the immediate proximity of the interface, the accommodation of molecules at the gas-liquid interface induces a diffuse concentration gradient (Fig. 13.4) and then a vapor pressure gradient. The probability for a molecule to hit the surface is indeed greater close to it than from far away. This gradient is thus inherent to liquid growth. The deposition of latent heat at the liquid/gas surface in Eq. 13.14 must be expressed in the growth rate (Peterson et al. 1993; Frymyer et al. 2020). Note that quite often the thermal resistance $1/a_i$ associated to the kinetic process in the Knudsen layer (Eq. 13.15) is neglected with respect to the diffuse thermal resistance coming from the vapor concentration gradient.

13.3.2.1 Filmwise Condensation

The approach, in this case, is similar to the equivalent film, Eq. 6.13 and the rate of condensation can be seen as the rate of condensation of the equivalent film (Eq. 6.14):

$$\dot{h} = \frac{D(c_\infty - c_s)}{\rho_l \zeta} \tag{13.16}$$

In this equation, D is the mutual diffusion coefficient for water molecules in air, ζ is the length of the diffuse boundary layer (see Chap. 2), c_∞ is the water molecule concentration in mass per volume at distances larger than than ζ and c_s is the concentration at the liquid-air interface. From Eq. 1.46:

$$c = p_v / r_v T \tag{13.17}$$

where r_v (= 462 J·kg^{-1}·K^{-1} for water) is the specific gas constant. The relationship Eq. 13.16 can then be evaluated from Eq. 13.17 in terms of vapor pressure:

$$\dot{h} = \frac{D(p_{v\infty} - p_s)}{\rho_l \zeta r_v T} \tag{13.18}$$

In order to make apparent the diffusive heat transfer coefficient $a^*_{i,f}$ for vapor with non-condensable gases, one expresses Eq. 13.18 in terms of temperature difference from the Clausius–Clapeyron relation $p_v - p_i \approx \frac{\rho_v L_v}{T_\infty}(T_\infty - T_i)$ (Appendix A, Eq. A.5 with ρ_v the vapor density). Substituting Eq. 13.18 in Eq. 13.16, one gets, with $T = T_\infty$:

$$a^*_{i,f} = \frac{q_i}{T_\infty - T_i} = \frac{\rho_v L_v^2 D}{\zeta r_v T_\infty^2} \tag{13.19}$$

For water vapor in air at $T_d = 297$ K under 3 kPa partial pressure where $\rho_v = 0.022$ kg·m^{-3} and $L_v = 2.4 \times 10^6$ J·kg^{-1}, and taking a usual value for the boundary layer thickness $\zeta \approx 10^{-3}$ m (see Sect. 2.2), one obtains $a^*_{i,f} \approx 2.7$ kW·m^{-2}·K^{-1}. This value is obviously sensitive to the value of ζ.

13.3.2.2 Dropwise Condensation
This case is treated in Sect. 13.5.3.

13.4 Conductive Cooling: The Nusselt Film

As conduction is made through the condensed film, one cannot express independently the conduction through the condensed film and its thickness evolution. In the steady state, the energy balance is concerned with q_a, the convective heat flux with the surrounding gas, which involves only vapor plus non-condensable gases (e.g. humid air) (Eq. 13.6), q_i, the latent heat release at the interface (Eq. 13.14), and q_l the conduction through the liquid film (Fig. 13.1a):

$$q_l = q_a + q_i \approx q_i \qquad (13.20)$$

In pure vapor, obviously q_a is not defined. In vapor plus non-condensable gases, the approximation where q_a is neglected is valid when one compares the respective values of the heat flux Eqs. 13.6 and 13.15.

In order to evaluate q_l, one must determine the film thickness $h(\xi)$. Equation 13.20 becomes, making use of Eq. 13.14:

$$q_l(\xi) = \frac{\lambda_l}{h(\xi)}(T_i - T_{cc}) = \rho_l L_v \dot{h} \qquad (13.21)$$

where λ_l is the liquid thermal conductivity.

The shape of the film thickness $h(\xi)$ along the substrate can be deduced from the balance of volumic flux between times t and $t + dt$, through a fixed volume of control located between ξ and $\xi + d\xi$. With $dh_c = \dot{h} dt$, the condensed volume per unit area, one obtains the volume conservation relation:

$$dh d\xi = F(\xi) - F(\xi + d\xi) + d\xi dh_c \qquad (13.22)$$

Equation 13.22 can also be written as:

$$\frac{\partial h}{\partial t} = -\frac{\partial F}{\partial \xi} + \dot{h} \qquad (13.23)$$

After calculating $dF/d\xi$ by deriving Eq. 13.13, the above Eq. 13.23 becomes:

$$\frac{\partial h}{\partial t} = -\left(\frac{\rho_l g \sin\alpha}{\eta}\right) h^2 \frac{\partial h}{\partial \xi} + \dot{h} \qquad (13.24)$$

Let us consider first the stationary state $\frac{\partial h}{\partial t} = 0$ where condensation compensates drainage. Equation 13.24 becomes, expressing $\dot{h}$ from Eq. 13.21:

$$h^2 dh = \left(\frac{\eta}{\rho_l g \sin\alpha}\right)\left(\frac{\lambda_w}{h(\xi)}\frac{T_i - T_{cc}}{\rho_l L_v}\right) d\xi \qquad (13.25)$$

Integration gives:

$$\int_0^h h^3 dh = \int_0^\xi \left(\frac{\eta}{\rho_l g \sin\alpha}\right)\left(\frac{\lambda_l}{h(\xi)}\frac{T_i - T_{cc}}{\rho_l L_v}\right) d\xi \qquad (13.26)$$

With the boundary condition $h(\xi = 0) = 0$, it becomes:

$$h = \left(\frac{4\eta\lambda_l(T_i - T_c)}{\varrho_l^2 L_v g \sin\alpha}\right)^{1/4} \xi^{1/4} \qquad (13.27)$$

In the same way as Eq. 13.3, the heat transfer coefficient, a_l, due to film thickness $h(\xi = L)$ (L is the sample length in the gravity direction) can now be evaluated as:

$$a_l = \frac{\lambda_l}{h(L)} = \lambda_l \left[\frac{4\eta\lambda_l(T_i - T_{cc})}{\varrho_l^2 L_v g \sin\alpha}\right]^{-1/4} L^{-1/4} \approx \lambda_l \left[\frac{4\eta\lambda_l(T_\infty - T_c)}{\varrho_l^2 L_v g \sin\alpha}\right]^{-1/4} L^{-1/4} \qquad (13.28)$$

The approximation $T_c \approx T_{cc}$ neglects the effect of coating, which is valid as film-wise condensation corresponds to perfect liquid wetting (see Sect. 13.1.2). The approximation $T_i \approx T_\infty$ is valid for pure vapor and for vapor with non-condensable gases where the heat flux q_a can be neglected (Fig. 13.4 and Eq. 13.20). In addition, the difference $T_i - T_{cc}$ operates in Eq. 13.28 at a very low power ($-1/4$).

For gas (vapor or vapor with non-condensable gases) at 373 K with supersaturation $T_\infty - T_c = 10$ K, $L = 5$ cm and the numerical values from Table 1.2 except dynamic viscosity η, which has a strong temperature dependence ($\eta = 0.25 \times 10^{-3}$ Pa·s), $a_l \approx 10$ kW·m^{-2}·K^{-1}. Larger (50 K) supersaturation gives $a_l = 6.7$ kW·m^{-2}·K^{-1}.

For gas at $T_d = 297$ K with supersaturation $T_\infty - T_c = 10$ K, $L = 5$ cm and the numerical values from Table 1.2 ($\eta = 1.5 \times 10^{-3}$ Pa·s) $a_l = 3.0$ kW·m^{-2}·K^{-1}.

Rohsenow (1956) and Rohsenow et al. (1956) accounted in the determination of the liquid film for the effect of the buoyancy force, the transfer of the sensible heat from vapor in the condensate and the deviation to linearity in the distribution of the temperature in the condensate layer (deviation to Eq. 13.3). These effects, however, are generally low. The simple Nusselt analysis agrees usually well with experimental data obtained for film condensation, except for metal vapors (see Tanasawa 1991).

13.4.1 Pure Vapor

From Eq. 13.4, one can write $T_\infty - T_c = (T_\infty - T_i) + (T_i - T_{cc}) + (T_{cc} - T_c)$. The total condensation heat transfer coefficient, a_f, can then be written as, supposing negligible the effect of coating, if any $(T_{cc} = T_c)$:

$$\frac{q}{T_\infty - T_c} = a_f = \frac{1}{\frac{1}{a_l} + \frac{1}{a_i}} \approx a_l \qquad (13.29)$$

The coefficients a_i and a_l are Eqs. 13.15 and 13.28. The approximation $a_f \approx a_l$ is justified when one compares the numerical values calculated in Sects. 13.3.1 and 13.4 with Eqs. 13.15 and 13.28. It means that the film thickness represents the main thermal resistance.

13.4.2 Vapor with Non-condensable Gas

The same calculation as in Sect. 13.4.1, using the heat exchange coefficients a_i, $a_{i,f}^*$ and a_l (Eqs. 13.15, 13.21 and 13.28) and neglecting the effect of coating, leads to the total condensation heat transfer coefficient, a_f^*:

$$a_f^* = \frac{1}{\frac{1}{a_l} + \frac{1}{a_i} + \frac{1}{a_{i,f}^*}} \approx \frac{1}{\frac{1}{a_l} + \frac{1}{a_{i,f}^*}} \qquad (13.30)$$

The approximation accounts for the high value of a_i ($\mathrm{MW \cdot m^{-2} \cdot K^{-1}}$, see Eq. 13.15) $\gg a_{i,f}^*, a_l$ ($\mathrm{kW \cdot m^{-2} \cdot K^{-1}}$, see Eqs. 13.21 and 13.28).

13.5 Dropwise Condensation

Many studies have dealt with heat transfer by dropwise condensation on a substrate cooled by contact. This configuration indeed gives higher heat transfer rates because the process leaves free a fraction of the substrate and is thus not limited by the thermal resistance of a condensed film. Coatings are mandatory to make the substrate hydrophobic or super-hydrophobic in order to maintain a high contact angle during the condensation process. The thermal resistance of the coating can thus be non-negligible, especially when the drop is in the Cassie–Baxter configuration where the drop sits on air pockets with only limited contact with the substrate (see Appendix E, Sect. E.2.2).

One first considers a single hemispherical droplet (Fig. 13.5). The drop has a radius of curvature ρ, makes a contact angle θ_c with the substrate and has a contact radius $R = \rho \sin \theta_c$. Similar to filmwise condensation in Sect. 13.4 above, one neglects the flux due to air convection when compared with the flux due to the latent heat of condensation through the interface.

Fig. 13.5 Conduction heat transfer between two neighboring isothermal surfaces

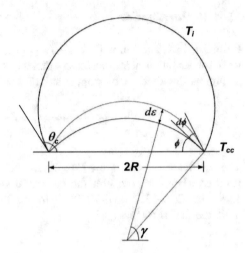

13.5.1 Effect of Drop Curvature

The drop curvature imposes a Laplace capillary pressure difference Δp_c between the fluids at both sides of the gas-liquid interface. The pressure difference between the liquid and the gas, $p_g - p_l$, can be written (with $\rho > 0$ if convex seen from the gas phase):

$$p_g - p_l = \Delta p_c = \frac{2\sigma_{LV}}{\rho} \tag{13.31}$$

This pressure difference modifies the vapor pressure of the gas phase above the interface, p_g, which then deviates from the saturation pressure. This is the so-called Kelvin equation (see Appendix P). For a weak difference of vapor pressure, it can be written, with ρ_l the vapor density, as (Eq. P.9), as:

$$\frac{\Delta p_c}{p_s} = \frac{2\sigma_{LV}}{\rho_l r_v T_\infty \rho} \tag{13.32}$$

Converting now the pressure difference into temperature difference ΔT_c from the Clausius–Clapeyron relation (Appendix A, Eq. A.5), it becomes:

$$\Delta T_c \approx \frac{2T_\infty \sigma_{LV}}{\rho_l L_v \rho} \tag{13.33}$$

When looking into the formulation of the critical radius for nucleation, (Eqs. 13.13 and 13.36) $\rho^* = 2\frac{\sigma_{LG}T}{\rho_l L_v \Delta T}$, where $\Delta T = T_\infty - T_{cc}$, one can express Eq. 13.35 as a function of ρ^*:

$$\Delta T_c = \frac{\rho^*}{\rho}(T_\infty - T_{cc}) \tag{13.34}$$

This additional temperature difference must be taken into account for small drops close to the critical nucleation radius. For example, when considering a water drop growing from humid air saturated at $T_\infty = 300$ K, $\Delta T_c = 2$ K for $\rho = 5$ nm and becomes negligible (< 0.01 K) for $\rho > 2$ μm.

13.5.2 Interface Heat Transfer—Pure Vapor

The heat transfer coefficient through the interface is given by Eq. 13.15. The rate q_i in Eq. 13.14 can be deduced from the drop volume V expression (Appendix D, Eq. D.5), $V = \pi\, F(\theta_c)\rho^3$ with $F(\theta_c) = \frac{2-3\cos\theta_c+\cos\theta_c{}^3}{3}$ (Eq. D.6).

The gas/liquid interface area of the drop reads as:

$$S_d = 2\pi\rho^2(1 - \cos\theta_c) \tag{13.35}$$

The interfacial heat flux, Q_i, over the entire drop interface of area S_d is:

$$Q_i = q_i S_d = 2\pi\rho^2(1 - \cos\theta_c)a_i \tag{13.36}$$

In the above expression, a_i is the interfacial heat transfer coefficient given by Eq. 13.15.

13.5.3 Interface Heat Transfer—Vapor with Non-condensable Gases

The heat transfer coefficient through the interface is also given by Eq. 13.15.

13.5.3.1 Early Stage of Drop Growth

For an isolated single drop, the growth law is $R = \rho\sin\theta_c = (A_s t)^{1/2}$ where $A_s = \frac{4D(c_\infty - c_s)f_0(\theta_c)}{3\rho_w\sin\theta_c F(\theta_c)}$ (Sect. 4.3, Eqs. 4.45 and 4.46). The function $f_0(\theta_c)$ is Eq. 4.47 (for a hemispherical drop, $f_0(\theta_c) = 1$) and the function $F(\theta_c)$ relates drop volume and drop radius (Eq. D.6). Expressing the concentration difference in terms of vapor pressure and temperature difference following the Clausius–Clapeyron relation $p_v - p_i \approx \frac{\rho_v L_v}{T_\infty}(T_\infty - T_i)$ (Appendix A, Eq. A.5 with ρ_v the vapor density), with $T_i \approx T_\infty$, one obtains the drop growth rate in units of volume per surface area, $\dot{h}_{d,1}$:

$$\dot{h}_{d,1} = \frac{\dot{V}}{S_d} = \frac{1}{1 - \cos\theta_c}\frac{f_0(\theta_c)F(\theta_c)D\rho_v L_v}{\rho_l r_v T_\infty^2 \rho}(T_\infty - T_i) \tag{13.37}$$

It results in the diffuse heat transfer coefficient for the early stage of drop growth:

$$a_{i,d1}^* = \frac{1}{1 - \cos\theta_c}\frac{f_0(\theta_c)F(\theta_c)D\rho_v L_v^2}{r_v T_\infty^2 \rho} \tag{13.38}$$

13.5.3.2 Late Stages of Drop Growth

When drops start to coalesce, the concentration profiles around each drop overlap to form a gradient perpendicular to the substrate and droplets coalesce to rescale the pattern (Sect. 6.3). It makes the drop surface coverage $\varepsilon_2 = $ constant and the equivalent film approach is valid. The drop contact radius R_i between coalescences grows as (Eq. 6.15):

$$R_i = \left(A_{pi}t\right)^{1/3} \tag{13.39}$$

Here:

$$A_{pi} = \frac{D(c_\infty - c_s)}{\pi G(\theta_c)\rho_w\zeta}A_i \tag{13.40}$$

$G(\theta_c)$ relates the drop cap volume to the cap radius (Eq. D.8) and A_i is the influence area of each drop, which can be deduced from the definition of ε_2 (Eq. 6.29):

$$\varepsilon_2 = \frac{\pi \sum R_i^2}{\sum A_i} = \frac{\pi \langle R_i \rangle^2}{\langle A_i \rangle} \tag{13.41}$$

Following the same procedure as in Sect. 13.5.3.1, one obtains the growth rate:

$$\dot{h}_{d,2} = \frac{\dot{V}}{S_d} = \frac{\sin\theta_c^2}{1-\cos\theta_c}\frac{\varepsilon_2 D\rho_v L_v}{\zeta\rho_l r_v T_\infty^2}(T_\infty - T_s) \tag{13.42}$$

It follows that the interface heat transfer coefficient, due to the use of the equivalent film approach, is independent of the drop radius in the late stage of growth. In this approach, the concentration gradient perpendicular to the surface is indeed constant (Sect. 6.1.2). One finds:

$$a_{i,d2}^* = \frac{\sin\theta_c^2}{1-\cos\theta_c}\frac{\varepsilon_2 D\rho_v L_v^2}{\zeta r_v T_\infty^2} \tag{13.43}$$

13.5.4 Drop Conduction and Convection Modes

Heat transfer inside the liquid drops can be performed by solid conduction and/or convection. The latter is triggered by the temperature gradients, which form inside the drop and, for condensation with non-condensable gases, at the gas-liquid drop interface.

13.5.4.1 Pure Conduction

Kim and Kim (2011) calculated the heat flux through a liquid drop assuming uniform temperature of the gas-liquid interface and heat diffusion as in a solid (Fig. 13.5). Heat is assumed to be conducted from one isothermal surface having a contact angle ϕ to another isothermal surface with angle $\phi + d\phi$. The surface area of the lower isothermal surface, A_s, is calculated by integration with respect to the angle γ as defined in Fig. 13.5:

$$dA_s = 2\pi \left(\frac{R}{\sin\phi}\right)^2 \cos\gamma \, d\gamma \tag{13.44}$$

It becomes:

$$A_s = 2\pi \int_{\frac{\pi}{2}-\phi}^{\frac{\pi}{2}} \left(\frac{R}{\sin\phi}\right)^2 \cos\gamma \, d\gamma = 2\pi R^2 \frac{1-\cos\phi}{\sin^2\phi} \tag{13.45}$$

The distance $d\varepsilon$ between two isothermal surfaces is the largest at $\gamma = 90°$. The mean distance, $\langle d\varepsilon \rangle$, can be taken as half the maximum distance at $\gamma = 90°$, that is:

$$\langle d\varepsilon \rangle = \frac{R}{2}\left\{\left[\sin^{-1}(\phi + d\phi) - \sin^{-1}\phi\right] - \left[\cot(\phi + d\phi) - \cot\phi\right]\right\}$$

$$\langle d\varepsilon \rangle = \frac{R}{2}\frac{1-\cos\phi}{\sin^2\phi}d\phi \tag{13.46}$$

The heat flux through the drop, Q_d, is related to $\langle d\varepsilon \rangle$ corresponding to the temperature difference $\langle dT \rangle$. With λ_w the liquid (water) thermal conductivity, it becomes:

$$Q_d = \frac{\lambda_w}{\langle d\varepsilon \rangle} A_s \langle dT \rangle \tag{13.47}$$

The temperature difference $T_i - T_{cc}$ is obtained by integration, with A_s evaluated by Eq. 13.45:

$$T_i - T_{cc} = \frac{Q_d}{\lambda_w} \int_0^{\theta_c} \frac{\langle d\varepsilon \rangle}{A_s} = \frac{Q_d}{4\pi R\lambda_w} \int_0^{\theta_c} d\phi = \frac{Q_d\theta}{4\pi R\lambda_w} \tag{13.48}$$

It becomes, making use of $R = \rho \sin\theta_c$ (Appendix D, Eq. D.1) and $T_i - T_{cc} = \frac{Q_d\theta}{4\pi R\lambda_w} = \frac{Q_d\theta}{4\pi\rho\lambda_w \sin\theta_c}$, or

$$Q_d = \frac{4\pi\rho\lambda_s \sin\theta_c}{\theta}(T_i - T_{cc}) \tag{13.49}$$

This corresponds to the coefficient of drop heat transfer by conduction:

$$\frac{Q_d}{T_i - T_{cc}} = a_d = \frac{4\pi\rho\lambda_s \sin\theta_c}{\theta} \tag{13.50}$$

13.5.4.2 Conduction and Convections

The assumption of isothermal gas-liquid interface made in the preceding Sect. 13.5.4.1 is not generally encountered. Near the contact line, high heat flux indeed occurs due to the presence of large thermal gradients (Chavan et al. 2016), the same phenomenon observed earlier in the ebullition process (Nikolayev and Beysens 1999). The existence of temperature gradients can trigger buoyancy and/or thermocapillary convections inside the drop as analyzed by Guadarrama-Cetina et al. (2014).

The first source of convection is the expanding volume of the drop due to condensation, leading to an expansion of the gas/liquid interface and the motion of the contact line. These effects were included in a simulation by Xu et al. (2018) of condensing droplets and appear to be the dominant process for large droplets.

Buoyancy can also induce convection flows, triggered by density inhomogeneities produced by temperature gradients. In order to evaluate them, let us consider the Rayleigh number:

$$\mathrm{Ra} = \frac{g\beta_p \Delta T R^3}{\nu D_{Tl}} \tag{13.51}$$

Here $\beta_p = (1/V)(\partial V/\partial T)_p$ is the volumetric liquid thermal expansion at constant pressure p, g is the earth acceleration constant, ΔT is the temperature difference that acts typically on lengthscale R to generate the instability, ν is the liquid kinematic viscosity and D_{Tl} is the liquid thermal diffusivity. The instability threshold in the classical Rayleigh–Bénard configuration (two infinite parallel plates with temperature difference ΔT separated by distance R) corresponds to a critical Rayleigh number $\mathrm{Ra}_c \approx 1700$, above which convection starts. For a water drop of radius $R = 500$ µm, convection would start only with an enormous temperature difference in the drop (1200 K with Table 1.2 numerical data). For smaller drops, the temperature difference would be even larger. Although the temperature gradients in a drop do not fit the classical Rayleigh–Bénard configuration, one can nevertheless conclude that buoyancy flows alone cannot efficiently contribute to the heat exchange in a water drop.

Another possible source of convection is thermocapillary flows originating because of temperature differences at the drop air-liquid interface. When pure vapor is involved, the kinetics of evaporation and condensation is so effective that the interfacial temperature should be evenly distributed at saturation temperature. Any temperature variation will immediately be canceled by evaporation or condensation. As a matter of fact, no thermocapillary convection in boiling vapor at saturation has ever been observed (Straub 2001). The origin of observed thermocapillary convection in the similar process of evaporation (boiling) under weightlessness, where buoyancy is suppressed, was thus attributed to the presence of non-condensable gases, even at very low concentration as evaluated by Barthes et al. (2007). These gases cause local differences in the partial vapor pressure, thus resulting in temperature gradients along the interface.

The governing parameter for thermocapillary flows is the Marangoni number:

$$\text{Ma} = \left(-\frac{d\sigma_{GL}}{dT} \right) \frac{R\Delta T}{\eta D_T} \tag{13.52}$$

In this expression, $(d\sigma_{GL}/dT)$ is the thermal variation of the gas-liquid surface tension σ_{GL} and η is the liquid dynamic viscosity. For the same water drop radius as above (500 μm), $\text{Ma} \approx 4 \times 10^2 \, \Delta T$. With a critical Marangoni number $\text{Ma} \approx 60$ above which thermocapillary flows can start, convections should occur for temperature differences as small as 150 mK. Note that $\text{Ma} \sim R$, such that smaller drops can be free of convection.

However, the ideal configuration corresponding to having no buoyancy convection below Rac is not encountered in the drop geometry. Convection in a drop will, therefore, be the result of a mixing of buoyancy forces F_b and Marangoni forces F_M, the ratio of which being:

$$\frac{F_M}{F_b} = \frac{\left(-\frac{d\sigma_{GL}}{dT} \right) R\Delta T}{\rho_l g \beta_p R^3 \Delta T} = \frac{\left(-\frac{d\sigma_{GL}}{dT} \right)}{\rho_l g \beta_p R^2} \tag{13.53}$$

Marangoni flows predominate for drops smaller than a typical size R_0 such as:

$$R_0 \sim \frac{\left(-\frac{d\sigma_{GL}}{dT} \right)}{\rho_l g \beta_p} \tag{13.54}$$

It corresponds for water to $R_0 \sim 9$ mm when using Table 1.2 data. This value is only approximate in this scaling approach. Only the precise determination of the flow field in the exact drop geometry, which depends on the drop contact angle, can give a precise value, as numerically studied by Al-Sharafi et al. (2016) who mimicked condensation by considering a $R \sim 1$ mm drop on a colder substrate.

The simulations by Xu et al. (2018) (Fig. 13.6) take into account all effects cited above: conduction, volume expansion and contact line motion, temperature dependence of surface tension for Marangoni flow and density for buoyancy. The curvature radius ρ of a drop from was varied from 50 nm to 500 μm for a contact angle ranging from 90° to 140°. The commercial code COMSOL was used to solve in a 2D axisymmetric configuration the Navier–Stokes equations (see Appendix C). Condensation of pure vapor (steam) was considered with condensation heat flux imposed through the interfacial heat transfer coefficient Eq. 13.15.

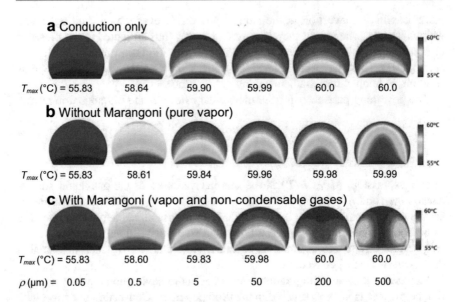

a Conduction only

T_{max} (°C) = 55.83   58.64   59.90   59.99   60.0   60.0

b Without Marangoni (pure vapor)

T_{max} (°C) = 55.83   58.61   59.84   59.96   59.98   59.99

c With Marangoni (vapor and non-condensable gases)

T_{max} (°C) = 55.83   58.60   59.83   59.98   60.0   60.0

ρ (µm) =   0.05   0.5   5   50   200   500

Fig. 13.6 Temperature distribution inside a growing water drop with 120° contact angle whose radius of curvature ρ ranges from 50 nm to 500 µm. **a** Assuming only conduction, **b** adding conduction and volume change (condensation from pure vapor), **c** adding Marangoni convection (condensation from vapor with non-condensable gases). Convection becomes visible for radius above 5 µm and convection dominates the heat transport above 200 µm. (Substrate temperature: 55 °C; vapor saturation temperature: 60 °C). (Adapted with permission from Xu et al. 2018)

Strictly speaking, no temperature gradient along the interface can form in a pure vapor configuration where the substrate has to accommodate its temperature around the contact line. The above simulation leads to a temperature gradient from the contact line to the top of the drop. The results will thus be concerned with condensation in the presence of non-condensable gases, and nevertheless an interfacial heat transfer coefficient calculated for the case of pure vapor. The results are interesting in terms of how conduction and convections induced by Marangoni, buoyancy and drop plus contact line expansion interact during drop growth. In particular, it was found for substrate temperature 55 °C and vapor saturation temperature 60 °C that the conduction mode was still valid for droplet sizes smaller than about 5 µm. Below a curvature radius $\rho = 5$ µm, conduction is only visible. Between 5 and 200 µm, strong interactions between Marangoni flow and interfacial mass flow increase the heat transfer in the center of the droplet bottom. Above 200 µm, the Marangoni flow disappears and the expansion of interfacial mass flow induces convection in the entire droplets. One notes that the temperature distribution near the liquid-solid interface is not uniform, with the zone near the contact line being characterized by a larger heat flux gradient.

Such simulations can be compared with infrared pictures that detect drop surface temperature. Figure 13.7a displays a picture of a condensing water drop with 1.75 mm diameter. The surface temperature distribution (Fig. 13.7b) looks similar

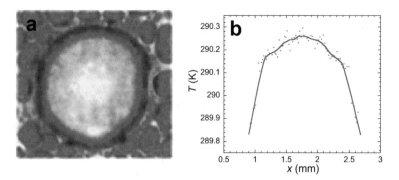

Fig. 13.7 **a** Top view by infra-red microscopy of a 1.75 mm diameter droplet condensing on polyvinyl chloride (PVC) with 87° contact angle (air temperature: 21 °C; dew point temperature: 17.6 °C). Smaller condensing drops are also visible. Temperature rise from the contact line (dark blue) to the top of the drop (orange-yellow). **b** Corresponding temperature distribution. (From Daoud et al., 2019)

to the simulation with Marangoni effects (Fig. 13.6c), with the largest temperature gradient near the contact line.

Average heat transfer through the drop surface of contact is reported in Fig. 13.8 as a function of its radius. The cases where Marangoni flows are absent or present are compared. Marangoni flows are seen to matter only in the range 20–200 μm and *decrease* the heat transfer. This effect is due to flow circulations in the opposite direction for Marangoni and drop expansion.

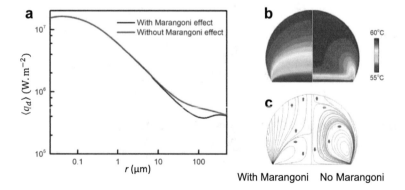

Fig. 13.8 Contribution of Marangoni convections (substrate temperature: 55 °C; vapor saturation temperature: 60 °C). **a** Average heat flux as a function of droplet radius of curvature. **b** Temperature distribution of a 100 μm radius droplet. Left: with Marangoni effect. Right: without. **c** Corresponding velocity field. Left: with Marangoni effect, the maximum velocity is 0.764 m·s^{-1}. Right: without Marangoni effect, the maximum velocity is 0.769 m·s^{-1}. (From Xu et al. 2018, Adapted with permission)

Simulation of flow field and heat transfer was also numerically studied by Phadnis and Rykaczewski (2017) in a 2D configuration. Condensing water droplets in the presence of non-condensable gas (air) were considered, using however as in the above the Xu et al. (2018) simulation of the interfacial heat transfer coefficient for pure vapor, Eq. 13.15, with condensation coefficient $\sigma \approx 0.004$). They considered only the effect of conduction and Marangoni convection in drops whose contact angles were varied (values $\theta_c = 90°$, $12°$, $150°$) and radius of curvature ranging from 0.1 μm to 1 mm. Their result compares well with the study by Xu et al. (2018) in the sense that the effect of convection on heat transfer becomes important only for a droplet radius above 1–10 μm.

13.5.5 One Drop Overall Heat Transfer

The overall heat transfer, Q, through one drop, corresponding to temperature T_∞ and substrate temperature T_c, can be expressed by adding the temperature difference between each interface, Eq. 13.4:

$$T_\infty - T_c = \Delta T_i + \Delta T_d + \Delta T_{cc} + \Delta T_c \tag{13.55}$$

Here $\Delta T_i = T_\infty - T_i = \frac{Q}{a_i}$ (pure vapor, Eq. 13.15) or $Q\left(a_i^{-1} + a_i^{*-1}\right)$ (vapor with non-condensable gases, Eqs. 13.15 and 13.19) is due to the interface. $\Delta T_d = T_i - T_{cc}$ (Eq. 13.50 for pure conduction) corresponds to the heat transfer through the drop. ΔT_c represents the drop curvature effect (Eq. 13.34). $\Delta T_{cc} = T_{cc} - T_c$ (Eqs. 13.7 and 13.8) is due to heat conduction through the substrate coating.

One thus obtains, for solid conduction through the drop:

13.5.5.1 Pure Vapor

$$Q_d(\rho, \theta_c) = \frac{\pi \rho^2 (1 - \rho^*/\rho)}{\frac{1}{2a_i(1-\cos\theta_c)} + \frac{\rho\theta_c}{4\lambda_l \sin\theta_c} + \frac{1}{a_{cc}\sin^2\theta_c}} (T_\infty - T_c) \tag{13.56}$$

13.5.5.2 Vapor with Non-condensable Gases

$$Q_d(\rho, \theta_c) = \frac{\pi \rho^2 (1 - \rho^*/\rho)}{\frac{1}{2a_i(1-\cos\theta_c)} + \frac{1}{2a_{i,d1}^*(1-\cos\theta_c)} + \frac{\rho\theta_c}{4\lambda_l \sin\theta_c} + \frac{1}{a_{cc}\sin^2\theta_c}} (T_\infty - T_c)$$

$$\tag{13.57}$$

13.5.6 Heat Transfer with Drop Size Distribution (Pure Vapor)

The heat transfer in the steady state must be integrated over the whole distribution of drop sizes with the distribution depending on the growth mode. In the pure vapor growth mode, the distribution of droplet size, as evaluated in Sect. 6.7, is different according to the radius regions where drops do or do not undergo coalescence.

Droplets initially grow on nucleation sites with surface density n_s. They are separated on average, for uniform distribution of nucleation sites, by the distance $\langle d_0 \rangle \sim 1/\sqrt{n_s}$, assuming the sites are distributed on a square lattice (Eq. 6.96).

The region where *no coalescences* occur between drops corresponds to having a drop radius of curvature $\rho < \rho_c = \langle d_0 \rangle/2$ (contact angle $\theta_c \geq \pi/2$) or cap radius $R = \rho \sin\theta_c < R_c = \langle d_0 \rangle/2$ (contact angle $\theta_c < \pi/2$). The distribution in this no-coalescence region is (Sect. 6.7, Eq. 6.110):

$$n(\rho) = N(\rho_c) \frac{\rho(\rho_c - \rho^*)}{\rho_c(\rho_c - \rho^*)} \frac{A_2\rho + A_3}{A_2\rho_c + A_3} \exp(B_1 + B_2) \tag{13.58}$$

In this equation:

$$B_1 = \frac{A_2}{\tau A_1}\left[\frac{\rho_c^2 - \rho^2}{2} + \rho^*(\rho_c - \rho) - \rho^{*2}\ln\left(\frac{\rho - \rho^*}{\rho_c - \rho^*}\right)\right] \tag{13.59}$$

$$B_2 = \frac{A_3}{\tau A_1}\left[\rho_c - \rho - \rho^*\ln\left(\frac{\rho - \rho^*}{\rho_c - \rho^*}\right)\right] \tag{13.60}$$

$$A_1 = \frac{T_\infty - T_c}{2\rho_l L_v}, \; A_2 = \frac{\theta_c(1 - \cos\theta_c)}{4\lambda_l \sin\theta_c} \tag{13.61}$$

$$A_3 = \frac{1}{2a_i} + \frac{1(1 - \cos\theta_c)}{a_{cc}\sin^2\theta_c} \tag{13.62}$$

In the *coalescence* region, the distribution is adequately described by Eq. 6.92 where drops are assumed hemispherical ($R = \rho$):

$$n_c(\rho) = \frac{1}{3\pi\rho_0^3}\left(\frac{\rho}{\rho_0}\right)^{-8/3} \tag{13.63}$$

Here, ρ_0 is the radius of the departing drop by gravity (see Sect. 14.1.2 and Eq. 14.17 for evaluation).

The total heat flux Q follows by integrating the heat flux in one drop (Q_d) with radius ρ over the drop size distributions, Eq. 13.58 (no coalescence) and Eq. 13.63 (coalescences):

$$Q = \int_{\rho^*}^{\rho_c} Q_d n(\rho) d\rho + \int_{\rho_c}^{\rho} Q_d N(\rho) d\rho \tag{13.64}$$

Figure 13.9 shows the heat flux variation as a function of the supersaturation temperature $\Delta T = T_\infty - T_c$ as calculated by Kim and Kim (2011) for pure vapor,

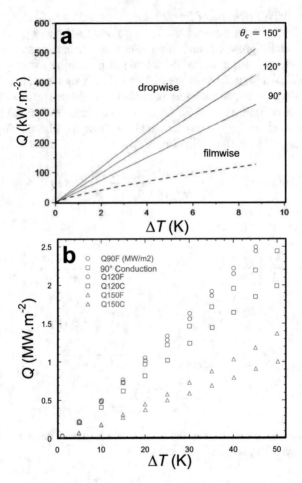

Fig. 13.9 Overall heat transfer rate per unit area versus pure vapor to solid surface subcooling ($\Delta T = T_\infty - T_c$) for contact angles $\theta_c = 90°$, $120°$, $150°$, in comparison with filmwise condensation (Nusselt film on 5 cm long substrate). (Adapted with permission from Kim and Kim 2011). **b** Thermocapillary enhancement of dropwise condensation heat transfer: Simulated heat flux with respect to subcooling ($\Delta T = T_\infty - T_c$) with (F) and without (C) Marangoni flow for dropwise condensation of water onto surfaces with various wetting properties. (Data from Phadnis and Rykaczewski 2017)

with $T_\infty = 373$ K for different contact angles and using a substrate of $L = 5$ cm length. For the sake of comparison, the result for filmwise conditions (Nusselt film Eqs. 13.28 and 13.29) is also shown. Filmwise condensation (see e.g. Tanasawa 1991) exhibits much lower heat transfer than filmwise condensation.

The effect of (Marangoni) convection was investigated numerically by Phadnis and Rykaczewski (2017). The main result is that conduction only slightly underestimates the heat transfer by a minor value of 10% or less. The reason is that convection matters only for large drops, which is due to the drop distribution,

Eq. 13.64 and Fig. 6.23 contribute weakly to the overall heat transfer. The important point is that, in most cases, flow contribution to the dropwise condensation heat transfer coefficient is in the order of typical experimental uncertainties and can be neglected.

13.5.7 Heat Transfer with Drop Size Distribution (Vapor with Non-condensable Gases)

When non-condensable gases are present with vapor, drop growth obeys the different stages summarized in Sect. 3.3 and detailed in Chaps. 3, 4, 6: (i) Nucleation, (ii) Individual drop growth in a radial vapor concentration profile, $\rho_i \sim t^{1/2}$, (iii) Drop growth with overlapping concentration profiles giving a mean concentration profile perpendicular to the surface with $\rho_i \sim t^{1/3}$ and droplet coalescence rescaling the growth as $\langle \rho_i \rangle \sim t$, (iv) Nucleation of new drop pattern on the surface left free by coalescence and (v) Drop sliding by gravity when the drop radius exceeds the critical value ρ_0 as determined by the balance between weight and pinning forces (Eq. 14.26 with $\rho_0 = R_0/\sin\theta_c$). ρ_0 thus corresponds to the largest drop present on the surface.

13.5.7.1 General
Heat transfer in each drop corresponds to Eq. 13.57:

$$Q_d(\rho, \theta_c) = \frac{\pi\rho^2(1 - \rho^*/\rho)}{\frac{1}{2a_i(1-\cos\theta_c)} + \frac{1}{2a_i^*(1-\cos\theta_c)} + \frac{\rho\theta_c}{4\lambda_l\sin\theta_c} + \frac{1}{a_{cc}\sin^2\theta_c}}(T_d - T_c) \quad (13.65)$$

13.5.7.2 Individual Growth (Stage ii)
In the above Eq. 13.65, one uses the coefficient of interface heat transfer (Eq. 13.38) $a_i^* = a_{i,d1}^* = \frac{1}{1-\cos\theta_c}\frac{f_0(\theta_c)F(\theta_c)D\rho_v L_v^2}{r_v T_d^2 \rho}$.

13.5.7.3 Growth in a Pattern (Stage iii)
One now considers the coefficient $a_i^* = a_{i,d2}^* = \frac{\sin\theta_c^2}{1-\cos\theta_c}\frac{\varepsilon_2 D\rho_v L_v^2}{\zeta r_v T_d^2}$ (Eq. 13.43).

Frymyer et al. (2020) developed the heat transfer expression by using a similar approach. They were able to evaluate numerically the instantaneous heat transfer rate within some assumptions on the expression and thickness of the boundary layer.

13.6 Radiative Cooling

Here, we consider only radiative cooling of vapor with non-condensable gases like air (e.g. dew formation). Conduction through the substrate is supposedly negligible (adiabatic conditions) and only heat exchange with air above the substrate is considered (Fig. 13.1c, d). A surface exposed to atmosphere can be cooled down below the surrounding air temperature T_a in the absence of solar radiation (e.g. at night) because it radiates more energy to the atmosphere than it receives from it (see Appendix O). The power radiated by unit surface area of a condensing surface with emissivity ε_c and temperature $T_c \approx T_a$ can be written as (Eq. A1.10):

$$P_r = \varepsilon_c \sigma T_a^4 \tag{13.66}$$

Here $\sigma = \frac{2k_B^4 \pi^5}{15c^2 h^3} = 5.670 \ 10^{-8} \ \mathrm{W \cdot m^{-2} \cdot K^{-4}}$ is the Stefan–Boltzmann constant.

The power radiated on the condensing surface by the atmosphere with emissivity ε_s, which increases with the water vapor content and is thus a function of dew point temperature T_d (see Appendix O) is (Eq. O.20; per unit surface area):

$$P_i = \varepsilon_s(T_d) \sigma T_a^4 \tag{13.67}$$

The radiation deficit $R_i = P_i - P_r$ (the available cooling energy) is thus:

$$R_i = \varepsilon_c \sigma [1 - \varepsilon_s(T_d)] T_a^4 \tag{13.68}$$

The radiation deficit is a function of T_a and T_d (from $\varepsilon_s(T_d)$) or, equivalently, of T_a and air relative humidity RH. The radiation deficit increases when RH or T_d decreases at constant T_a (the air vapor content decreases). For typical nocturnal conditions where dew forms ($T_a = 288$ K and $RH = 85\%$, $T_d = 285.5$ K), $R_i \approx 60$ W/m^2 (Bliss 1961) assuming $\varepsilon_c \approx 1$.

It must be noted that for filmwise condensation, as soon as the film is nucleated, ε_c will be the emissivity of liquid water ($\varepsilon_c \approx 0.98$ in the atmospheric window, from the imaginary part of the refractive index in Downing and Williams 1975). In case of dropwise condensation, since the substrate remains partly dry, one must consider an effective emissivity, which accounts for the substrate and water droplet properties. From Trosseille et al. (2021b), the effective emissivity remains close to the above liquid water emissivity.

13.6.1 Filmwise Condensation

In this case, the condensation rate $\dot{h}$ is determined by the radiative deficit surface/air per unit surface area, $R_i < 0$. In contrast to the Nusselt film (Eq. 13.21), this cooling power is no more dependent on h. The energy balance is written as:

$$R_i = -(q_a + q_i) \tag{13.69}$$

Here q_a and q_i are given by Eqs. 13.6 and 13.14, giving:

$$\dot{h} = -\frac{q_a + R_i}{\rho_l L_v} \tag{13.70}$$

The same calculation as done with the Nusselt film in Sect. 13.4 leads to:

$$h^2 dh = -\left(\frac{\eta}{\rho_l g \sin \alpha}\right)\left(\frac{q_a + R_i}{\rho_l L_v}\right) d\xi \tag{13.71}$$

Integration of Eq. 13.71 gives:

$$\int_0^h h^2 dh = -\int_0^\xi \left(\frac{\eta}{\rho_l g \sin \alpha}\right)\left(\frac{q_a + R_i}{\rho_l L_v}\right) d\xi \tag{13.72}$$

With the boundary condition $h(\xi = 0) = 0$, one obtains:

$$h = \left[-\frac{3\eta(R_i + q_a)}{\rho_l^2 L_v g \sin \alpha}\right]^{1/3} \xi^{1/3} \tag{13.73}$$

This film thickness variation is similar to the Nusselt formulation, Eq. 13.27, but with a different power law dependence (1/3 instead of 1/4). This change is because condensation in the present case is not limited by the conductive heat flux in the film. In contrast to the Nusselt film, there is no experimental study of such a condensation film under radiative cooling.

The available heat transfer for condensation, q_i, can be evaluated as (Eq. 13.69):

$$q_i = -R_i - q_a \tag{13.74}$$

For humid air at $T_a = 288$ K, $T_d = 285.5$ K or $RH = 85\%$ (typical nocturnal conditions where dew condenses), $R_i \approx 60$ W/m^2 (Bliss 1961). Using $\zeta \sim$ mm (see Sect. 2.2), $T_a - T_i \sim$ K and numerical values from Table 1.2, one finds, using Eq. 13.6 where $\delta_T \approx \zeta$, $q_i = 34$ W·m^{-2}. Air heat losses correspond to more than 50% of the radiative cooling power.

13.6.2 Dropwise Condensation

In this mode, one can simplify the equivalent film approach of Sect. 6.1.2. Each drop is thus a fraction of the film studied in Sect. 13.6.1. Then, in units of condensed surface area, the heat flux, Eq. 13.74, is still valid. In this simplification, however, one neglects the cooling effect of the dry substrate by conduction along the substrate and the drop motion due to coalescence, which allow drops to sit on new, fresh areas of the substrate.

Gravity Effects

14

Water collection by gravity has been well studied for condensing surfaces cooled by contact, but only recently for surfaces cooled by radiation deficit. In addition, it appears that gravity-induced flows are different according to whether condensation is filmwise or dropwise.

Below these different processes are addressed for both smooth and patterned substrates, including liquid infused substrates and soft substrates. In addition to the steady states of condensation, the lag time where the condensed volume accumulates on the substrate before flowing down is discussed. This lag time is often the most important parameter when the condensation rate is weak; therefore, condensation proceeds only during the period of time where transient phenomena dominate.

14.1 Smooth Substrates

14.1.1 Filmwise

14.1.1.1 Contact Cooling

The film thickness gives the most important thermal resistance as the thermal exchange with air is negligible in general. This is the so-called *Nusselt film*, which was outlined in Sect. 13.4. When condensation occurs from saturated vapor at temperature T_∞ on a smooth square planar substrate of surface $S_c = L \times L$ whose temperature is T_c, with axis ξ on the plane in the direction of gravity (Fig. 14.1), making an angle α with horizontal, the film thickness $h(\xi, t)$ is non-uniform and obeys the steady state (see Sect. 13.4 and Eq. 13.27 with the definition of parameters). With T_i the liquid-gas interface temperature:

$$h = \left(\frac{4\eta\lambda_l(T_i - T_c)}{\varrho_l^2 L_v g \sin\alpha} \right)^{1/4} \xi^{1/4} \approx \left(\frac{4\eta\lambda_l(T_\infty - T_c)}{\varrho_l^2 L_v g \sin\alpha} \right)^{1/4} \xi^{1/4} \tag{14.1}$$

© Springer Nature Switzerland AG 2022

D. Beysens, *The Physics of Dew, Breath Figures and Dropwise Condensation*, Lecture Notes in Physics 994, https://doi.org/10.1007/978-3-030-90442-5_14

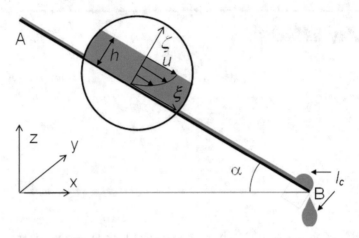

Fig. 14.1 Filmwise condensation on a smooth planar surface and drop detachment (notations: see text). AB = L

The thickness h_L of the film at the lower end of the plane corresponds to $\xi = L$. In the situation (Lee et al. 2012b) where $T_\infty - T_c = 12.5$ K, $\alpha = 90°$ and $L = 3$ cm and numerical values from Table 1.2, one finds $h_L \approx 85$ μm.

At the lower end of the film, a puddle forms. Droplets detach and water can be collected in a gutter when the volume of the puddle becomes in the order of $s_c L = \pi l_c^2 L$ (s_c is the puddle cross-section; l_c is the capillary length (see Appendix E, Sect. E.1.4). Special shapes (drip edge) at the lower end can also gather the puddle into a single drop of volume $v_c = (2\pi/3)l_c^3$.

14.1.1.2 Radiative Cooling

The substrate can be considered adiabatic below its surface (see Fig. 13.1c, d), and the condensation rate is determined solely by the radiative deficit surface/air per unit surface, $R_i < 0$, and the heat exchange with air (Eq. 13.6). With λ_a the air thermal conductivity:

$$q_a = \frac{\lambda_a}{\delta_T}(T_\infty - T_i) \tag{14.2}$$

Here δ_T is the thermal boundary layer ($\delta_T \approx \zeta$, the diffuse boundary layer, see Sect. 2.2). Heat flux q_a now becomes the limiting thermal resistance. The film thickness can be evaluated (Eq. 13.73) as:

$$h = \left[-\frac{3\eta\left(R_i + \frac{\lambda_a}{\delta_T}(T_\infty - T_i) \right)}{\rho_l^2 L_v g \sin \alpha} \right]^{1/3} \xi^{1/3} \tag{14.3}$$

The thickness h_L of the film at the lower end of the plane corresponds to $\xi = L$. Let us consider a typical situation (Beysens 2018) where $R_i \approx -60$ W.m^{-2},

$T_\infty - T_i \approx 2$ K, $\alpha = 30°$ and $L = 1$ m. Using $\delta_T = \zeta \sim$ mm (see Sect. 2.2) and numerical values from Table 1.2, one obtains $h_L \approx 35$ μm.

14.1.1.3 Flow Onset

The above calculations assume a stationary state but a precise description of the earlier transient states has not been investigated yet. However, one can give some characteristics of this state. A stationary state is reached only after (i) a continuous film has formed above the plate (time t_0) and (ii) a stationary profile is obtained (time t_1). Assuming the plate roughness to be Gaussian distributed with mean arithmetic amplitude (see Appendix L):

$$R_a \equiv \overline{\delta\zeta} = \frac{1}{n}\sum_{i=1}^{n}|\delta\zeta_i| \qquad (14.4)$$

a film forms when:

$$t_0 \approx \frac{\overline{\delta\zeta}}{\dot{h}} \qquad (14.5)$$

With $\overline{\delta\zeta} \sim 0.5$ μm and numerical values concerning weak condensation as encountered with natural dew ($\dot{h} \approx 0.2$ mm/10 h. night or 5.6 nm.s^{-1}), a film will form after $t_0 \approx 90$ s. It corresponds to a dead volume per surface area $\overline{\delta\zeta} = 0.5$ μm.

After time t_0, water starts to flow. Since the unstationary states of the film evolution are not known, one can obtain only a time limit (t_1) to obtain the stationary state. Time t_1 corresponds to the condensation of the volume stored in the stationary film, neglecting the flow. The stored water volume V_1 for plate width dy is:

$$V_1 = dy \int_0^L h(\xi)d\xi = nh_L L dy \qquad (14.6)$$

In case of contact cooling, $n = 4/5$ (Eq. 14.1) and for radiative cooling $n = 3/4$ (Eq. 14.3), it then becomes for time:

$$t_1 = n\frac{h_L}{\dot{h}}0.775\frac{h_L}{\dot{h}} \qquad (14.7)$$

With the numerical values above, one obtains $t_1 \sim 2000$s, which is the maximum time to obtain a stationary flow.

After this time, a steady water flow is thus present at the lower edge of the plane. Water accumulates at the plane end as a puddle. A lag time t_c is then needed for the first water drop to detach from the edge. It corresponds to a limiting s_c for the section along y where the gravitational force on the puddle, $\rho g s_c dy$, balances the capillary force, $2\sigma dy$ (Lee et al. 2012b). It follows $s_c \approx 2l_c^2 = 1.5 \times 10^{-5}$ m^3. From Eq. 14.6, the exit flux F per unit length dy at $\xi = L$ and the definition of velocity u (Fig. 14.1) can be written as:

$$F(\xi = L) = \int_0^{h_L} u(\zeta)d\zeta = \int_0^{L} \dot{h}d\xi = L\dot{h} \tag{14.8}$$

which is a predictable result. The time t_c to fill the puddle is then:

$$t_c = \frac{s_c}{L\dot{h}} = \frac{2l_c^2}{L\dot{h}} \tag{14.9}$$

With the numerical values above, one finds $t_c \approx 2600$ s. In case a drip edge device is used to concentrate the cylindrical puddle in a unique drop of volume $v_c \sim (4\pi/3)l_c^3 = 8.2 \times 10^{-8}$m^3, t_c reduces to:

$$t_{c0} = \frac{v_c}{L^2\dot{h}} \tag{14.10}$$

that is, a very small value $t_{c0} \approx 15$ s. The actual time to reach the above puddle values will be nevertheless larger because of the duration of the non-stationary state, where condensation provides the volume to build up the stationary film in addition to the formation of the puddle volume.

The final time then corresponds to the time $t_0 + t_1 + t_c$. Assuming that the time to reach the stationary state is smaller than forming a cylindrical puddle (it is the case here where $t_1 < t_c$), the final time t_f to get the first drop at the plane end will thus be:

$$t_f = t_0 + t_1 + t_c = \frac{1}{\dot{h}}\left(\frac{1}{3}\overline{\delta\zeta} + nh_L + \frac{s_c}{L}\right) \tag{14.11}$$

In the case where a drip edge device is used, the time to form a unique drop is so short that the stationary state is not reached. It is only by solving Eq. 14.8 in the non-stationary state that a value can be estimated. One may state that:

$$t_f < t_0 + t_1 + t_{c0} = \frac{1}{\dot{h}}\left(\frac{\overline{\delta\zeta}}{3} + nh_L + \frac{v_c}{L^2}\right) \tag{14.12}$$

The values with the numerical values above give $t_f \approx 4700$ s (linear puddle) or lower than 2100s (drip edge device).

The above description of film drainage is in qualitative agreement with the observation of Lee et al. (2012b) using cooling contact (Fig. 14.2).

Fig. 14.2 Vertical
hydrophilic surface cooled by
contact after 1800 s.
Advancing/receding contact
angles are 4° and 0°,
respectively. Surface area is
$L \times L = 30\text{mm} \times 30\text{mm}$.
Condensation rate is
70 nm.s^{-1}. The white bar is
10 mm. (From Lee et al.
2012b, with permission)

14.1.2   Dropwise

The forces acting on a drop set on an inclined substrate (Fig. 14.3) are the gravity
force F_g and the contact line pinning forces (F_s), whose sum is non-zero because
of the difference between the advancing (θ_a) and receding (θ_r) contact angles.

With V_d the drop volume, the gravity force in the direction ξ becomes:

$$F_g = \rho V_d g \sin \alpha \tag{14.13}$$

Fig. 14.3 Drop detachment.
(Notations: see text; adapted
from ElSherbini and Jacobi
2006)

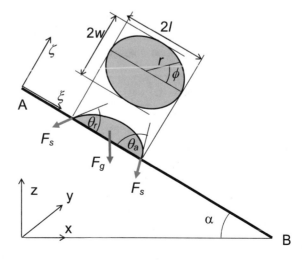

The pinning or retention forces in the direction ξ can be related to a drop length scale R representing the radius of the drop contour and the advancing and receding contact angle (Extrand and Gent 1990; Extrand and Kumagai 1995):

$$F_{s,\xi} = k^* \sigma R(\cos \theta_r - \cos \theta_a) \tag{14.14}$$

Here $k^* \sim 1$ is a numerical constant that depends on the precise shape of the drop. ElSherbini and Jacobi (2006) calculated the sum of pinning forces on an ellipsoidal contour (Fig. 14.3). Due to symmetry, the components of the surface tension force in the y-direction cancel. The resulting surface tension force acts only in the ξ-direction and can be calculated from:

$$F_{s,\xi} = -2\sigma \int_0^\pi r(\phi) \cos \theta \cos \phi d\phi \tag{14.15}$$

Here, the function $r(\phi)$ describes the ellipse through:

$$r(\phi) = \frac{l}{\sqrt{\cos^2 \phi + \beta^2 \sin^2 \phi}} \tag{14.16}$$

The value of k^* in Eq. 14.14 depends on the shape of the drop contour in the integration of Eq. 14.15. The parameter β expresses the deformation under the action of gravity. With Bo the Bond number (see Appendix H, Eq. H.2), $\beta = 1 + 0.096\text{Bo}$. When the drop contour can be approximated by a circle, it corresponds to $\beta = 1$ and $k^* = \frac{48}{\pi^3} = 1.548$. For drops of an ellipse-like shape, Eq. 14.15 with $\beta = 1$ and Eq. 14.14 with $k^* = 1.548$ still holds if one considers the effective drop contour radius found by representing the contour by a circle having the same area as the ellipse.

During condensation, R increases and the drop slides when R reaches a sliding critical radius R_0 such as $F_g = F_s$. Expressing the volume as a function of R with $G(\theta_c)$ from Eqs. D.7 and D.8 in Appendix C and using the capillary length l_c (Appendix E, Eq. E.6), one obtains:

$$R_0 = l_c \left(\frac{k^*}{\pi G(\theta_c)} \right)^{1/2} \left(\frac{\cos \theta_r - \cos \theta_a}{\sin \alpha} \right)^{1/2} \tag{14.17}$$

For a substrate tilt angle $\alpha = 30°$, contact angles $\theta_r \approx 40°$ and $\theta_a \approx 70°$ ($\theta_c \approx 55°$ and $G(\theta_c) = 0.29$), one finds $R_0 \approx 1.2 l_c \approx 3.2$ mm, that is a value slightly larger than the capillary length $l_c = 2.7$ mm.

The time t_f for a drop with mean radius $< R >$ to reach R_0 can be evaluated for the same condensation rate $\dot{h} \approx 5.6$ nm.s^{-1} as in the preceding Sect. 14.1.1 for filmwise condensation. Using the mean film approximation in Sect. 6.1.2 and Eq. 6.40, and recognizing that the radius distribution is narrow (Figs. 6.6 and 6.8):

$$h = \dot{h} t_f = G(\theta)\varepsilon_2 \langle R \rangle = G(\theta)\varepsilon_2 R_0 \tag{14.18}$$

where ε_2 is the drop surface coverage, one deduces the time:

$$t_f = R_0 \frac{G(\theta)\varepsilon_2}{\dot{h}} \qquad (14.19)$$

The dead volume is $\dot{h}t_f$. With the values used above ($R_0 = 3.2$ mm, $\theta \approx 55°$ giving $\varepsilon_2 \sim 0.7$ from Fig. 6.78), one obtains $t_f \sim 1.2 \times 10^5$ s (30 h.). It must be noted that the model in Eq. 14.14 is static. However, drop detachment always occurs owing to a coalescence event (Trosseille et al. 2019a). The reason is two-fold. First, coalescence makes the drop radius grow instantaneously, enabling the critical sliding radius to be surpassed. Second, during coalescence the drop contact line moves, thus lowering the pinning forces. Gao et al. (2018) indeed highlighted the fact that, as in the case of solid-solid friction, the forces of adhesion of a drop on a solid are increasingly weaker in a dynamic mode than in a static mode. The experimental critical radius is consequently weaker than calculated by Eq. 14.14.

As shown in Fig. 14.4a, some drops are much larger than the average and detach sooner, sweeping drops on their way down. Some drops are larger because (i) the distribution in drop size is not a Dirac function and some drops can indeed be larger than the average (see Sects. 6.3.2 and 6.8), (ii) the presence of geometrical and/or chemical defects and pollution effects favor earlier nucleation and further growth; they can also modify the sliding criteria by changing advancing and receding contact angles. The large dispersion of the t_c values in Fig. 14.4b (from 2500 to 5000 s) found in different experiments corresponds to these uncontrolled defects.

Another process that speeds up drop detachment concerns edge effects, which gives 3 to 5 times growth acceleration (see Chap. 8). Such effects on drop collection are detailed below.

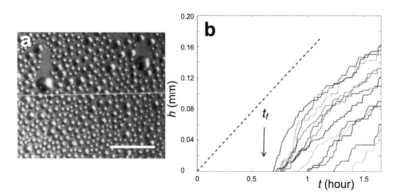

Fig. 14.4 Condensation on a vertical Duralumin smooth surface (roughness $Ra = 0.4$ μm, advancing/receding angles 120° and 40°, respectively, 14.7 cm long, 3 cm wide, condensation rate 48 nm.s^{-1}) cooled by conduction. Edge effects have been canceled by absorbers. **a** Picture showing some droplets much larger than average at the sliding onset. The white bar is 10 mm. **b** Condensed mass with lag time t_c. The dispersion in lag times corresponds to uncontrolled defects and pollution effects. (From Trosseille et al. 2019a, with permission)

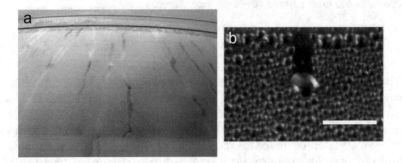

Fig. 14.5 Dew condensing on **a** a car rear windshield (photograph courtesy of Medici 2014) and **b** on a vertical Duralumin substrate cooled from below, under the same conditions as Fig. 14.4. Drops on the upper edge are larger and detach sooner than drops on the surface

14.2   Edge Effects

When looking into dew forming on a kitchen window (contact cooling) or on the external side of a car windshield (radiative cooling), large edge effects can be frequently observed. Drops are seen to detach primarily from the top or the side of the condensing surface; they fuse with surface drops, acting as natural wipers (Fig. 14.5).

At the top horizontal edge, the drops indeed grow faster as discussed in Chap. 8. They thus reach the critical size for detachment sooner and slide down, collecting the other drops. On a fresh bare area, a new generation of drops nucleates and grows. From Table 8.1, the gain in drop growth rate is between ~ 3 (edge) and ~ 5 (corner), resulting in a large reduction in lag time t_c. With the numerical values of Sect. 14.1.2, instead of 30 h on the plane surface it occurs at about 10 h for linear edges and 6 h for corners.

This border effect explains why condensers with edges can collect more water than similar condensers without edges. The gain for small yields can reach up to 400–450% of the yield of the reference plane (Beysens et al. 2013).

14.3   Textured Substrates

Special surface patterning can improve either film flow or drop sliding and/or drop coalescence, to provide a few large drops in place of a myriad of tiny droplets. Many kinds of micropatterning can be envisaged. They have in common microroughness, which can increase either the hydrophobicity when the smooth substrate is hydrophobic, or hydrophilicity when the substrate is hydrophilic (see Appendix E, Sect. E.2).

14.3.1 Filmwise

14.3.1.1 Posts

When considering microtextured substrates, pillars are quite often used. Seiwert et al. (2011a) studied the drainage of liquid films on cylindrical pillars of diameter $a = 3$ μm separated by distance $b = 7$ or 17 μm (corresponding to periodicity $d = a + b = 10$ and 20 μm) and height c between 1 and 35 μm (similar to Fig. 10.2). The pillar density $\varphi = \pi a^2/4d^2$ was low, below 10%. The experiment was carried out by pulling the microtextured substrate from silicon oil. Lee et al. (2012b) carried out condensation studies on denser cylindrical pillars with $\varphi \approx 20\%$ ($a = 0.5$ mm diameter with periodicity $d = 1$ mm).

When the film is formed by condensation, one distinguishes several regimes depending on the film thickness, h, with respect to the pillars' height, c:

(i) When $h < c$, the condensed liquid invades the structure (Texier et al., 2016 and references therein) and the film is trapped by capillary forces.
(ii) When $h \approx c$ (zone 1), flow occurs as if the pillars increased fluid viscosity, with no-slip conditions at the solid surface ($u(\zeta) = 0$).
(iii) When $h > c$ (zone 2), the boundary conditions at the pillar surface ($\zeta = c$) are no longer zero velocity thus the film moves more easily. Two parabolic profiles are present, depending on both zones as sketched in Fig. 14.6.

In zone 1, the friction in the trapped layer takes place on both the bottom surface and pillar walls. In a cell of size $d \times d \times c$ with the plot in the middle, the viscous force F_v will scale as:

$$F_v = \eta u \frac{d^2}{c} + \eta \beta \int_0^c (u/c)\zeta \, d\zeta = \eta u \frac{d^2}{c}\left(1 + \beta \frac{c^2}{d^2}\right) \qquad (14.20)$$

Fig. 14.6 Filmwise condensation on microtextured planar surface and drop detachment. AB = L; other notations: see text. Zones 1 and 2 correspond to flow inside and outside, respectively, of the microstructure

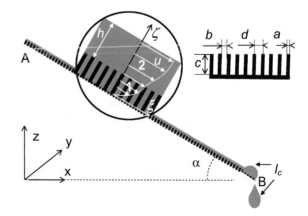

The first term $\eta u d^2/c$ represents the viscous force in a film of thickness c on a flat surface. The second term $\eta \beta \int_0^c (u/c)\zeta d\zeta = \eta \beta u c/2$ is the supplementary force due to the plot dissipation. The factor β depends on the surface fraction of the plot $\sim a^2/d^2$ and on plots mutual interactions (Hasimoto 1959; Seiwert et al. 2011b): $\beta = \frac{4\pi}{\ln(2a/d)-1.31}$. The presence of plots eventually leads to an increase of viscous dissipation by the factor:

$$k = 1 + \beta c^2/d^2 \tag{14.21}$$

when compared with a film of same thickness on a flat surface. The approach of Sect. 14.1 with Eqs. 14.1 or 14.3 is thus still valid if replacing viscosity η by the enhanced viscosity η^*:

$$\eta^* = k\eta = \left(1 + \beta \frac{c^2}{d^2}\right)\eta \tag{14.22}$$

In zones $1 + 2$, the two velocity profiles u_1 in zone 1 and u_2 in zone 2 correspond to four boundary conditions concerning no-slip at the materials wall, no stress at the free interface and continuity of velocity and stress at $\zeta = c$:

$$u_1(\zeta = 0) = 0; \quad \frac{\partial u_2(\zeta = h + c)}{\partial \zeta} = 0;$$

$$u_1(\zeta = c) = u_2(\zeta = c); \quad k\eta \frac{\partial u_1(\zeta = c)}{\partial \zeta} = \eta \frac{\partial u_2(\zeta = c)}{\partial \zeta} \tag{14.23}$$

After some algebra, the velocity profile in zone 1 becomes:

$$u_1(\zeta) = \left(\frac{\rho_l g \sin \alpha}{k\eta}\right)\zeta\left[(h+c) - \frac{\zeta}{2}\right] \tag{14.24}$$

and in zone 2:

$$u_2(\zeta) = \left(\frac{\rho_l g \sin \alpha}{\eta}\right)\left\{\zeta\left[(h+c) - \frac{\zeta}{2}\right] - (1-k^{-1})c\left(h + \frac{c}{2}\right)\right\} \tag{14.25}$$

The flow velocity can be considered as the sum of a flow velocity on a smooth plane at the post surface $\left(\frac{\rho_l g \sin \alpha}{\eta}\right)\zeta\left[h - \frac{\zeta}{2}\right]$ and a supplementary velocity corresponding to fluid sliding above the posts,

$$\left(\frac{\rho_l g \sin \alpha}{\eta}\right)\left[c\zeta - (1-k^{-1})c\left(h + \frac{c}{2}\right)\right].$$

The volumic flux F per unit length y is expressed as:

$$F = \int_0^c u_1(\zeta)d\zeta + \int_c^{c+h} u_2(\zeta)d\zeta \qquad (14.26)$$

Making use of Eqs. 14.24 and 14.25, it becomes:

$$F = \left(\frac{\rho_l g \sin\alpha}{3k\eta}\right)(kh^3 + 3h^2 c + 3hc^2 + c^3) \qquad (14.27)$$

In the presence of condensation with constant rate $\dot{h}$ per unit surface, the flux conservation gives Eq. 13.23, $\frac{\partial h}{\partial t} = -\frac{\partial F}{\partial \xi} + \dot{h}$. It follows making use of the flux expression Eq. 14.27:

$$\frac{\partial h}{\partial t} = -\left(\frac{\partial h}{\partial \xi}\right)\left(\frac{\rho_l g \sin\alpha}{k\eta}\right)(kh^2 + 2ch + c^2) + \dot{h} \qquad (14.28)$$

In the stationary state $\frac{\partial h}{\partial t} = 0$ where condensation compensates drainage, Eq. 14.28 becomes:

$$\left(\frac{\rho_l g \sin\alpha}{k\eta}\right)(kh^2 + 2ch + c^2)dh = \dot{h}d\xi \qquad (14.29)$$

(i) In case of *conduction cooling*, the condensation rate is limited by the thermal exchange through the condensed film (Nusselt film, Sect. 13.4) where (Eq. 13.21), with $T_{cc} \approx T_c$:

$$\dot{h} = \frac{\lambda_l}{\rho_l L_v h(\xi)}(T_i - T_c) \qquad (14.30)$$

Here L_v is latent heat of condensation and λ_l is liquid thermal conductivity. With the expression of $\dot{h}$ from Eq. 14.30, integration of Eq. 14.29 from 0 to h (left hand) and 0 to ξ (right hand) gives the stationary film profile:

$$\xi = \left(\frac{\rho_l^2 L_v g \sin\alpha}{k\eta\lambda_l(T_d - T_c)}\right)\left(c^2\frac{h^2}{2} + 2\frac{ch^3}{3} + k\frac{h^4}{4}\right) \qquad (14.31)$$

(ii) For *radiative cooling*, $\dot{h}$ is a constant $\dot{h}_0$ and depends on the radiative power, R_i and air thermal exchange q_a (Eq. 13.69):

$$\dot{h} = -\frac{q_a + R_i}{\rho_l L_v} \qquad (14.32)$$

The same calculation as above for the conduction case, starting from Eq. 4.29, gives:

$$\xi = \left(\frac{\rho_l g \sin \alpha}{k \eta \dot{h}_0} \right) \left(c^2 h + ch^2 + k \frac{h^3}{3} \right) \tag{14.33}$$

In the expressions, Eqs. 14.31 or 14.33, making $c = 0$ or $k \gg 1$, give Eqs. 14.1 or 14.3, valid for a smooth substrate. It is also valid for a condensed film thick enough ($h > 3c/k$) where the third term at the right hand side is dominant. For a thinner film, the second term in the bracket may prevail, but Eqs. 14.31 or 14.3 also include the flow inside the texture (first term), which becomes dominant.

The scenario for a condensation process is thus the following. First, the microstructure fills with condensed water, which remains trapped by capillarity (as in Sect. 14.1.1 due to the substrate roughness). Then the film above the posts starts to flow, which also implies a secondary flow in the microstructure. When the film above the posts becomes thick enough, the film flow is like a flow above a smooth plane. Water condensation experiments on vertical micropatterned substrates by Lee et al. (2012b) led to the same conclusion: flow starts only when the microstructures are filled with condensed liquid.

From Eq. 14.33, radiative cooling case, one can obtain the thickness h_L of the film at the plate end $\xi = L$. With the numerical values above ($\dot{h} \approx 5.6$ nm s^{-1}, $\alpha = 30°$, $L = 1$ m) and $a = 3$ μm, $b = 17$ μm, $c = 5$ μm giving $k = 2.8$ from Eq. 14.21), one finds $h_L = 14$ μm, a value comparable with the value obtained on a smooth surface (Sect. 14.1.1: $h_L = 17$ μm).

Similar to Eq. 14.11, the actual time to reach a puddle of Section s_c (or volume v_c if using a drip edge) will be the sum of the time t_0 to fill the volume between pillars, the time t_1 to reach the stationary state and the time t_c to fill the puddle.

The time t_0 to fill the posts can be evaluated as (no flow in the microstructures):

$$t_0 \approx \frac{c}{\dot{h}_0} \tag{14.34}$$

Using the numerical values used above ($\dot{h} \approx 5.6$ nm.s^{-1} and $c = 5$ μm), one gets $t_0 = 900$ s.

The volume V_1 of the stationary film above the post is obtained by integrating $h(\xi)$ from 0 to L. The algebra is rather complex, thus one linearizes the profile by sake of simplification:

$$h \sim \frac{h_L}{L} \xi \tag{14.35}$$

giving:

$$V_1 = L \int_0^L h(\xi) d\xi = \frac{1}{2} h_L L^2 \tag{14.36}$$

The time t_1 to condense V_1 is thus:

$$t_1 \sim \frac{h_L}{2\dot{h}_0} \qquad (14.37)$$

With the numerical values used above and $h_L = 14$ μm, one gets $t_1 = 1100$ s. The time to fill the puddle is the same as for a smooth surface since $\dot{h}$ is the same (Eqs. 14.9 or 14.10: $t_c = 2600$ s or $t_{c0} = 15$ s). The total time t_f can thus be written as:

$$t_f = t_0 + t_1 + (t_c \, \text{or} \, t_{c0})\frac{1}{\bar{h}}\left[c + \frac{1}{2}\frac{h_L}{L} + \left(\frac{s_c}{L} \, \text{or} \, \frac{v_c}{L^2}\right)\right] \qquad (14.38)$$

The values with the numerical values above give $t_f = 4600$ s (linear puddle) or 2000s (drip edge device). The volume in the post is lost because it is trapped by capillarity and corresponds to volume per surface area $c = 0.005$ mm. Deeper posts (15–50 μm) will increase the lag time t_1 (3300–1100 s) and the dead volume (0.015–0.05 mm).

When compared with water collection using smooth hydrophilic substrates (Sect. 14.1.1: $t_f = 4700$ s for linear puddle or lower than 2100s for drip edge device), superhydrophilic substrates are at best as efficient to collect condensed water. Although flow is larger in the film above the post, one needs a supplementary time to fill the volume between the posts, a volume that will be lost for collection because it is trapped by capillarity.

14.3.1.2 Grooves

A particularly simple microtextured surface is microgroove as described in Sect. 10.3.1. Grooves are directed in the gravity direction. As soon as condensation starts, a film forms, which fills the grooves by capillarity effects and water starts to flow downwards.

The limiting times are the time t_0 to form a film considering the surface mean arithmetic roughness $R_a = \overline{\delta\zeta}$ (Eq. 14.4) and the time t_1 to fill the $L/(a + b)$ grooves of section bc and length L (assuming no flow in the grooves where water is trapped by capillarity):

$$t_1 = \frac{\left(\frac{L}{a+b}\right)}{L^2 \dot{h}_0}Lbc = \frac{1}{\bar{h}}\left(\frac{bc}{a+b}\right) \qquad (14.39)$$

and adding the time to fill a cylindrical puddle (t_c) or a drip edge (t_{c0}). It, therefore, becomes comes for the total time t_f:

$$t_f = t_0 + t_1 + (t_c \, \text{or} \, t_{c0})\frac{1}{\bar{h}}\left[\frac{\overline{\delta\zeta}}{3} + \frac{bc}{a+b} + \left(\frac{s_c}{L} \, \text{or} \, \frac{v_c}{L^2}\right)\right] \qquad (14.40)$$

Fig. 14.7 Left: Microgrooved vertical hydrophilic surface ($a = b\ 100\,\mu$m, $c = 65\,\mu$m, $L = 14.5$ cm, collection area 20 cm², tilt angle $\alpha = 60°$) at the onset of puddle detachment from the bottom end. Right: Corresponding smooth surface. Contact angles (left) $\theta_r = 0°$ and $\theta_a < 4°$ and (right), $\theta_c = 90°$. Time is 800 s after condensation has started by contact cooling (condensation rate 18 nm.s^{-1}). The bar is 5 mm. (Adapted from Lhuissier et al. 2015)

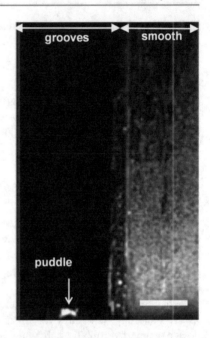

The total time is reduced with respect to microposts because the microstructure volume is less. The dead volume expressed per spatial period, $bc/(a + b)$, is also smaller, by the ratio $b/(a + b)$. Using the above values ($\overline{\delta\zeta} = 0.5\,\mu$m, $a = 3\,\mu$m, $b = 17\,\mu$m, $c = 5\,\mu$m, $L = 1$ m, $\dot{h} \approx 5.6$ nm s^{-1}) one obtains $t_0 = 90$ s, $t_1 = 760$ s, $t_c = 2600$ s or $t_{c0} = 15$ s (same values as for a smooth surface). The total time becomes 3500 s (cylindrical puddle) or 900 s (drip edge). When compared with posts ($t_f = 4600$ s for linear puddle or 2000s for drip edge device), there is less film trapped in the grooves and the lag time is reduced in proportion. Lag time is also reduced when compared with a smooth surface (Sect. 14.1.1: $t_f = 4700$ s for linear puddle or lower than 2100s for a drip edge device), because the film thickness remains limited to the groove depth (Fig. 14.7).

14.3.1.3   Superposed Biphilic Surfaces

Thin superposed biphilic surfaces were considered by Oh et al. (2018). A thin hydrophilic porous matrix of nickel inverse-opal (NIO) nanostructures (pore radius ≈ 250 nm and ratio void-space volume/total volume, porosity ≈ 0.8) is coated with a hydrophobic decomposed hydrophobic polyimide, where dropwise condensation can take place (Fig. 14.8; see Sect. 10.3.2.2). It results in a superhydrophilic sublayer and superhydrophobic top layer. Nucleated drops grow on the outside hydrophobic layer in the Cassie-Baxter state, then transition to the Wenzel state when in contact with the hydrophilic regions, to be eventually sucked by the porous hydrophilic structure (Figs. 10.10 and 14.8) due to their larger capillary pressure.

When the substrate is set vertical, liquids in the hydrophilic layer can be efficiently collected by gravity. The flow in a porous structure is described by Darcy's

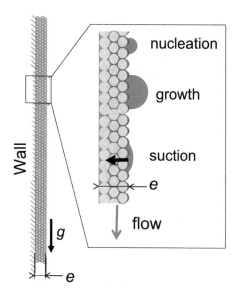

Fig. 14.8 Porous sublayers (thickness e) with upmost hydrophobic coating (red color)

law: the viscous pressure drop, Δp_η, through a porous medium for a flowing fluid is related to the volumetric flow rate, F, dynamic viscosity, η, coefficient of permeability, $\kappa \sim r_p^2$ with r_p the pore radius, flow length, L and flow cross-sectional area, $A_c = Le$, by:

$$F = -\frac{\kappa A_c \, \Delta p_\eta}{\eta L} \tag{14.41}$$

The process by which liquid flows in the structure is dominated by the comparison between the viscous pressure Δp_η and the capillary pressure confining the condensate in the porous structure, Δp_c. If $\Delta p_\eta > \Delta p_c$, the condensate will flow out of the top pores, eventually resulting in flooded condensation similar to the Nusselt film (see Sect. 13.4). Assuming that the Darcy flow is only related to the mean condensation rate through the permeable layer (thus not accounting for the gradual volume addition along the propagation path), one can write, with $\dot h$ the condensation volume per unit of time and surface area:

$$F = A_c \dot h \tag{14.42}$$

Equaling F in Eqs. 14.41 and 14.42, it becomes:

$$\Delta p_\eta = -\frac{\eta L \dot h}{r_p^2} \tag{14.43}$$

The Laplace capillary pressure barrier confining the condensate in pores with radius r_p can be estimated assuming a hemispherical liquid-vapor interface shape

Fig. 14.9 Capillary pressure difference (Δp_c, black solid line) and viscous pressure difference ($\Delta \eta_h$, colored dotted lines) as a function of pore radius (r_p) for different volumic condensation rates $\dot{h}$ (log-log plot). The region (pink area) above Δp_c corresponds to flooded condensation, while below (green area), it corresponds to film condensation. Film thickness is $e = 10\ \mu m$, which makes the region (red) $r_p > e$ inaccessible. (Adapted from Oh et al. 2018, with permission)

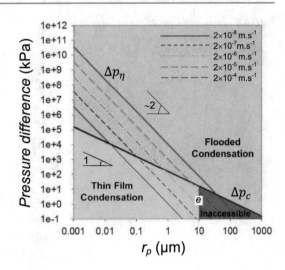

expanding through the hydrophobic pores as:

$$\Delta p_c = \frac{2\sigma_{LV}}{r_p} \tag{14.44}$$

Figure 14.9 contains the variation of Δp_c with respect to r_p, for pressure regions $\Delta p > \Delta p_c$ where water overflows the thin porous film and $\Delta p < \Delta p_c$ where condensed water is confined and propagatse inside the structure. Also shown are the lines of constant condensation rates. The region $r_p > e$ remains obviously inaccessible. As $\Delta p_c \sim r_p^{-1}$ and $\Delta p_\eta \sim r_p^{-2}$, flooded condensation will proceed even for small condensation rates $\dot{h}$ at small r_p.

Note that the above results are only approximate. The linear dependence of Δp_η with the condensation rate $\dot{h}$ is due to the dependence on condensate flow rate in the superhydrophilic layer and its accumulation in the superhydrophilic layer and in the structures. It was assumed to be 1D Darcy flow through a permeable layer without accounting for the steady mass addition along the propagation length. In reality, condensate addition occurs over the whole sample surface area, making Eq. 14.43 a conservative estimate. When Oh and al. (2018) computed the non-linear pressure drop along the sample length by discretizing the condensation domains, they found data approximately 10 times larger than the evaluation from Eq. 14.43. In addition, the effect of gravity to drain the liquid condensate in the thin film should be taken into account and estimated.

14.3.2 Dropwise

14.3.2.1 Posts

When condensing on micropatterned surfaces with hydrophobic posts, water primarily condenses as droplets on pillars and between pillars (see Sect. 10.2). Then,

Fig. 14.10 Picture (left) of square micropost surface in the vertical direction at the onset of drop sliding and the corresponding smooth surface (right) after 10,000 s of condensation by contact cooling (bar: 10 mm). Conditions are the same as in Fig. 14.7 with (left) contact angles $\theta_r < 5°$ and $\theta_a = 34°$ and (right), $\theta_r = 34°$ and $\theta_a = 90°$. (From Lhuissier et al. 2015)

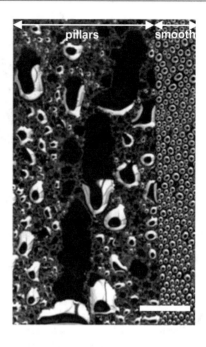

depending on the roughness and hydrophobicity, droplets larger than the typical post length scale come into Wenzel (W) or Cassie-Baxter (CB) states. Drops behave as on a smooth planar surface with large hysteresis of contact angle (W state) or small hysteresis (CB state). W-droplets exhibit a large critical sliding radius R_0, while for CB-droplets, it is much smaller. Condensation experiments by Lee et al. (2012b) clearly demonstrate this behavior. However, as shown by Lhuissier et al. (2015) when compared with the same smooth substrate, droplet growth on such micropatterned surfaces with hydrophobic posts is still accelerated and lag time to obtain drop sliding occurs after the shortest time (Fig. 14.10).

The onset of shedding occurs at time t_f when the first drop with radius R_0 detach, corresponding to Eq. 14.19, $t_f = \frac{R_0 G(\theta_c)\varepsilon_2}{h}$.

14.3.2.2   Drop Jumping

On *superhydrophobic surfaces*, the coalescence of two microdroplets may provide sufficient energy to make them jump out of the surface (see Sect. 10.4). Mouterde et al. (2017) and Liu et al. (2021) used microcones instead of cylindrical pillars, with improved efficiency.

Such drops can fall further on the surface or out of it, depending on the tilt angle α with horizontal. Mukherjee et al. (2019) studied the corresponding change in drop size distribution and condensation rate, which can be significantly increased when α varies from 0 to 90°, increasing with respect to horizontal the condensation rate and thus heat transfer by 40% for $\alpha = 45°$ and by 100% for the vertical orientation (Fig. 14.11). Indeed, for a horizontal surface, jumping droplets return

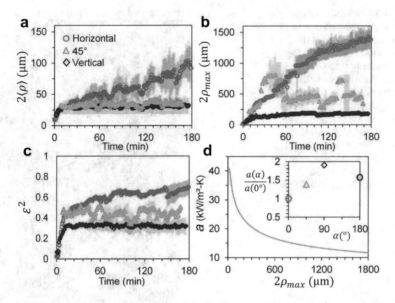

Fig. 14.11 Surface orientation dependence of droplet size and heat transfer in a condensation experiment from humid air. **a–c** Variation of (**a**) mean droplet radius of curvature ρ, **b** maximum droplet radius, **c** droplet surface coverage ε^2. Error bars represent one standard deviation. **d** Exponential decay of the theoretical steady-state condensation heat transfer coefficient (**a**) with maximum droplet radius on the surface with condensation coefficient $\sigma = 0.35$ (Eq. 4.5) and supersaturation temperature $\Delta T = 5$ K. Inset: heat transfer coefficients at different surface inclinations with respect to horizontal orientation. (Adapted from Mukherjee et al. 2019, with permission)

to the surface and eventually become stuck. When the surface is vertical, droplets can either return to it or jump out of it. Stuck droplets can shed by the action of gravity, which also serves to sweep away neighboring drops. At high heat fluxes, drops jumping from upside-down surfaces $\alpha = 180°$) can return to the surface against gravity due to the effect of vapor flow entrainment (see Sect. 10.4). Gas flow can be due to buoyancy effects on vapor near the surface and flow required for mass conservation of the condensing vapor (Miljkovic et al. 2013b).

The use of Black Silicon (conical microstructures on silicon, see Sect. 10.1.2.7) made hydrophobic by deposition of PTFE leads to quite efficient drop shedding. As seen in Fig. 14.12, which compares the collected volume at the bottom of a Black Silicon plate coated or not with PTFE, the sliding radius is so small on the coated materials that it leads to an undetectable t_f onset value.

Hydrophobic posts with hydrophilic upper part have been tested following the pioneering work by Parker (2001). Drops nucleate preferentially at the post top and grow there. However, the large hysteresis of contact angle at the post top prevents efficient shedding (Lee et al. 2012b).

Fig. 14.12 Evolution of the collected shedding droplets volume per unit area, h, by plain silicon (S) and Black Silicon PTFE-coated (BS + PTFE, see inset). The steps for S correspond to sliding droplets, starting at time t_f. On BS + PTFE, surface droplets present immediate departure characteristics, $t_f \approx 0$. (Adapted from Liu et al. 2020)

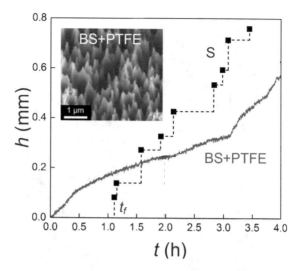

14.3.2.3 Grooves

Using grooves parallel to gravity can markedly increase the collected amount of water (Bintein et al. 2019). The presence of grooves indeed promotes drop coalescence from plateaus to water inside the grooves and long-range coalescences for drops that share the same channels, favoring the rapid emergence of a few large drops (see Sect. 10.3.1). In addition, these drops can easily slide down because the contact line pinning is canceled parallel to the grooves and reduced to the plateau surface perpendicular to them. Such groove-induced coalescence phenomena thus favor the emergence of only a small number of large, mobile droplets instead of many small pinned droplets as found on a smooth surface with the same wetting properties (Figs. 14.13 and 14.14). Note that the occurrence of imbibition strings of water in the grooves is linked to the relation between the ratio c/b (depth/spacing) and the advancing contact angles θ_A (water/A bottom material) and θ_B (water/B wall material) (Eq. 10.4):

$$\frac{c}{b} > \left(\frac{c}{b}\right)_c = \frac{1 - \cos\theta_A}{2\cos\theta_B} \tag{14.45}$$

The critical run-off start time to obtain the first drop sliding on the surface depends on the groove geometrical properties. The latter are length L (plate dimension), thickness a, spacing b, depth c and periodicity $d = a + b$, see Figs. 10.6, 10.7, 14.13f, Sect. 10.3.1 and Fig. 10.7. Growth was discussed in the case of a horizontal substrate. In the present case, the substrate makes an angle α with horizontal. In Fig. 14.13, a pattern is shown of droplets colored by a fluorescent line growing on a grooved substrate at $\alpha = 45°$ from horizontal. Figures 14.13a, b corresponds to the formation of drops overlapping channels and drying plateaus (which look dark) on both sides over length $L^* = L/p$ with $p \geq 1$ (see Sect. 10.3.1). Figure 14.13c shows when such drops start to slide down.

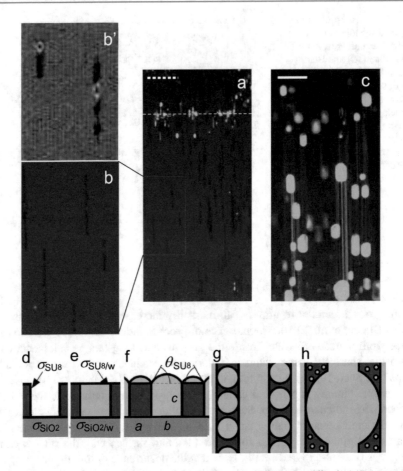

Fig. 14.13 Drops colored by fluorescein (initially placed at the interrupted line in (**a**)) on a rectangular grooved plate (area 20 cm², $L = 14.7$ cm parallel to grooves, inclined at $\alpha = 45°$). Grooves have thickness a, spacing b, depth c. In the bottom of the channel, the advancing/receding contact angles are $\theta_a = 34°/\theta_r < 5°$, making a mean contact angle $\theta_c \approx 20°$. On the plateau coated with SU8, the advancing/receding contact angles are $\theta_a = 80°/\theta_r = 66°$, making the mean contact angle $\theta_c \approx 73°$. The groove characteristics are $a = 100$ μm, $b = c = 65$ μm. Condensation rate is $h = 55$nm.s^{-1} The bar is 10 mm. **a** Time t_1. First drops overlapping channels and drying plateaus on both sides appear dark. **b** Zoom of the window in (**a**) (white interrupted rectangle) and (**b'**), similar view at larger magnification without fluorescein. **c** Time t_2 : Drops at sliding onset. Some coalescence between drops sharing the same channels can be observed, with the smallest drops (larger capillary pressure) draining out from large drops (lower pressure). **d–e** Schematics of the texture. **f–h** Schematics of saturation and overflow considered for t_1: **f** cross-section, **g** Top views. (Adapted from Bintein et al. 2019, with permission)

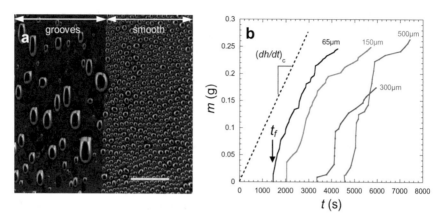

Fig. 14.14 **a** Picture of grooved surface ($a = b = 100 \ \mu$m, $c = 65 \ \mu$m, collection area 20 cm^2, length $L = 14.7$ cm, sliding critical radius ≈ 3 mm, tilt angle $\alpha = 45°$, condensation rate $\dot{h} = 27$ n ms^{-1} (adapted from Royon et al. 2016). Contact angles on SU8 walls and silicon in the channel bottom are in Fig. 14.13. The yellow bar is 10 mm. **b** Evolution of the condensed mass m (g) for substrate (**a**) at different values of plateau thickness a for $b = 100 \ \mu$m and $c = 56 \ \mu$m. The first drop is collected at time t_f. The dashed line is the effective condensation on the substrate. (Adapted from Bintein et al. 2019, with permission)

The time t_f when large drops start to slide down from a given surface is the sum of (i) the time t_1 to create the first large drop and (ii) the time t_2 to grow a drop large enough to depin from the plateaus. Time t_1 is the time where water in the channel coalesces with drops on two adjacent plateaus. It is the time t_{1g} for water to fill a groove and t_{1p} for plateau drop perimeters to reach the plateau groove edges. Quite generally, $t_{1p} < t_{1g}$ as the plateau volume involved is less than the groove volume. One can thus assume that the resulting water volume in the overflowing channel is in the order of the volume whose meniscus touches the plateau edge, v_c. With water making the advancing contact angle $\theta = \theta_a$ on the walls during condensation, the channel volume is evaluated as in Appendix Q, Eq. Q.8:

$$v_c = \left[bc + b^2 g(\theta_a)\right]L \tag{14.46}$$

where (Eq. Q.6):

$$g(\theta_a) = \frac{2\theta - \sin 2\theta_a}{8 \sin^2 \theta_a} \tag{14.47}$$

The channel volume results from vapor condensation on the surface $(2a + b)L$. One obtains:

$$t_1 = \frac{1}{\dot{h}} \frac{\left[bc + b^2 g(\theta_a)\right]L}{(2a + b)L} = \frac{1}{\dot{h}} \frac{b[c + bg(\theta_a)]}{2a + b} \tag{14.48}$$

The second time t_2 corresponds to Wenzel drops growing until they slide down. Drops are only pinned by the contact line on the dry plateaus, giving them an elongated form (Figs. 14.13 and 14.14). With N_0 being the number of plateaus covered by the drop, the pinning force can be expressed from Eq. 14.14 where $k \sim 1$ since large drops are pinned only on the dry plateaus:

$$F_s = \sigma N_0 a (\cos \theta_r - \cos \theta_a) \tag{14.49}$$

At the sliding threshold, F_s is equal to the gravity force such as:

$$\sigma N_0 a (\cos \theta_r - \cos \theta_a) = \rho V_c g \sin \alpha \tag{14.50}$$

Here, V_c is the critical drop volume at the sliding onset. It corresponds to the volume that has condensed on surface $V_c = L^* N_0 (a + b)$. The length $L^* = L/p$, with $p > 1$, corresponds to the mean distance between drops at the sliding onset and is the result of a complex growth law where drop coalescence occurs through flows in the channels. The smallest drop with the highest pressure fills the largest drop with the lowest pressure. It thus becomes:

$$V_c \approx L^* N_0 (a + b) \dot{h} t_2 \tag{14.51}$$

Using Eq. 14.50 to express V_c and neglect the effect of film thermal resistance in the channel when cooling is ensured by contact, the time t_2 can be written as:

$$t_2 = \frac{1}{\dot{h}} \left(\frac{a}{a+b} \right) \frac{p l_c^2 (\cos \theta_r - \cos \theta_a)}{L \sin \alpha} \tag{14.52}$$

It is interesting to note that this time does not depend on N_0 because this parameter enters both in drop growth and contact line pinning.

Since water in the channels remains trapped, the final time t_f is the sum of t_1 and t_2. It follows, from Eqs. 14.48 and 14.52:

$$t_f = t_1 + t_2 = \frac{1}{\dot{h}} \left[\frac{b[c - bg(\theta_a)]}{2a + b} + \left(\frac{a}{a+b} \right) \frac{p l_c^2 (\cos \theta_r - \cos \theta_a)}{L \sin \alpha} \right] \tag{14.53}$$

A typical pattern is reported in Fig. 14.14a with corresponding shedding times t_f in Fig. 14.14b. There is a strong decrease when the plateau thickness a is reduced.

The variations of t_f with the groove parameters a, b, c are shown in Fig. 14.14. Compared with groove spacing b and depth c, the variation with thickness a is very strong, and t_f is diminished with a until reaching a plateau for small a. The time t_f corresponds to a volume per surface area, $h_f = \dot{h} t_f$, where $\dot{h}$ is the condensation rate. A good fit of the experimental data with Eq. 14.53, when all groove parameters and condensation rates are varying, is found with $p \approx 2.7$ (Fig. 14.15d) (Bintein et al. 2019).

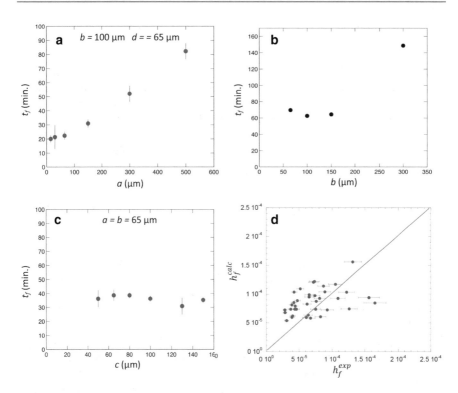

Fig. 14.15 Variations of the sliding threshold time t_f with **a** groove thickness a, **b** spacing b, **c** depth c under constant condensation rate condensation rate $\dot{h} = 27$ nm.s^{-1}. (Adapted from Royon et al. 2016). **d** Correlation of the calculated threshold volumes $h_f^{calc} = h_f^{calc}$ (Eq. 14.53) with the experimentally measured threshold volumes $h_f^{exp} = \dot{h} t_f^{exp}$. (Data from Bintein et al. 2019)

Water dead volume that remains trapped in the grooves is in the order of bcL_c, with $L_c \approx \frac{2l_c^2}{b \sin \alpha}$ the capillary rise in the channels (Jurin's law, see e.g. Batchelor 2000). The dead volume can then be minimized by reducing the b and c values.

Using the numerical values of Fig. 14.13, one obtains for $a = b = 100$ μm and $c = 65$ μm, $t_1 = 450$ s, $t_2 = 1260$ s and $t_f = 1710$ s. These values should be compared with the corresponding smooth surface applying Eq. 14.19, where $R_0 = 1.4$ mm giving $t_0 = 10{,}000$ s. The sliding time is reduced by a factor of about six when using the grooved surface.

14.3.2.4 Biphilic Stripe Patterns

Alternate hydrophilic-hydrophobic stripes exhibit behavior similar to grooves. In particular (see Sect. 10.3.2.1), Chaudhury et al. (2014) and Peng et al. (2014) identified that, similar to what occurs on groove plateaus (see the preceding Sect. 14.3.2.3), coalescences on hydrophobic region could efficiently move the drops (see Chap. 9), which eventually coalesce and are sucked by contact with a hydrophilic stripe. However, in contrast to the study with grooves, which is concerned with transient effects, it is only the steady state that has been studied so far.

Wu et al. (2017) considered the steady-state collection of *fog droplets* by biphilic stripes on a surface tilted at 45° with horizontal (Fig. 14.16). Phenomena similar to what has been observed with grooves are found with enhanced water collection efficiency observed on all hybrid surfaces when compared with smooth superhydrophobic and superhydrophilic surfaces. One can define a super-hydrophilic surface area ratio, S^*, with S_{phil} and S_{phob} the hydrophilic and hydrophobic surface areas, respectively:

$$S^* = \frac{S_{phil}}{S_{phil} + S_{phob}} = \frac{b_{phil}}{a_{phob} + b_{phil}} \tag{14.54}$$

Here a_{phob} and b_{phil} are the width of the hydrophobic and hydrophilic stripes, respectively (corresponding to the plateau width a and groove width b in the groove pattern, see Figs. 10.7 and 14.16). The micropatterned surface with $S^* = S_M^* = 0.09$ was seen to give the highest efficiency. However, they paradoxically reported that, although the minimum departure volume is smaller when stripes are parallel to gravity, higher fog harvesting efficiency is found with stripes *perpendicular* to gravity, probably a result of the accumulation area that should dominate the fog harvesting efficiency rather than the minimum departure volume. Contact angles are indeed quite different in the directions perpendicular and orthogonal to the tracks (Fig. 14.16b, c). It shows a very strong anisotropic wettability along these two directions leading, therefore, to different drop collection performances.

Nevertheless, fog is formed of water droplets of a given size that hit the surface. This process is different from nucleation and growth of droplets, making uncertain

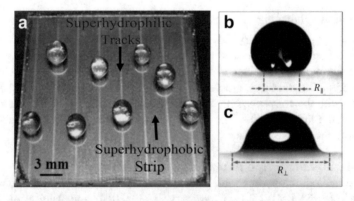

Fig. 14.16 **a** Superhydrophobic substrate with superhydrophilic tracks. Fog droplets roll on the superhydrophilic tracks. **b, c** 4 μL water droplet on a horizontal 300 μm wide superhydrophilic track taken **b** parallel and **c** perpendicular to the hydrophilic tracks direction. $R_\parallel$ and $R_\perp$ are contact width of the water droplet with a superhydrophilic track in the parallel and orthogonal track directions, respectively. (Adapted from Wu et al. 2017, with permission)

the comparison with condensation even when drop shedding is involved. Lee and al. (2020) considered steady condensation on a striped patterned biphilic surface. Droplets of the near super-hydrophobic area are quickly removed and drained to the nearby super-hydrophilic region. Rapid drainage of droplets from the near super-hydrophobic surface results in the renewal of new dropwise condensation, which leads to an increased collection rate. The authors varied the stripe width of the biphilic surface and, similar to the grooved substrate analyzed in the preceding Sect. 14.3.2.3, the drainage rate improved. The maximum amount of condensate on the biphilic surface occurred at a super-hydrophilic area ratio $S_M^* = 0.75$ or more. These values also correspond to the better performance obtained in grooves when the plateau width a is the smallest, with the filled channels playing the role of a hydrophilic stripe.

However, S_M^* is much larger than the value 0.09 found for fog collection as discussed above. The reason lies in the different processes of drop formation, with fog drops hitting both hydrophilic and hydrophobic stripes at the same rate. This is not the case for steady condensation where dropwise condensation on the hydrophobic stripe is more efficient than filmwise condensation on the hydrophilic stripe due to enhanced thermal resistance (Nusselt film, see Sects. 13.4 and 14.1.1).

Condensation on biphilic stripes is shown for different ratios S^* ranging from 0.5 to 0.85. Dropwise condensation occurs on the hydrophobic stripes and filmwise condensation on the hydrophilic areas. Droplets grown on the hydrophobic region are absorbed into the hydrophilic region before reaching the critical sliding size. Even if the droplets of the hydrophobic region reach the critical size, instead of sliding by gravity they are rather absorbed into the hydrophilic region owing to the strong wettability gradient between the hydrophilic and hydrophobic regions. When the hydrophilic stripe width is much smaller than the critical droplet size (Fig. 14.17c), droplets merge onto the superhydrophobic region, resulting in droplets covering the adjacent hydrophobic regions and forming a 'bridge droplet'. These bridge droplets are ultimately drained, but they reduce drainage and heat transfer performance because they cover a part of the super-hydrophobic region.

The condensate collected in the hydrophilic stripes flows downward by gravity, producing an edge effect at the end of the surface where water accumulates before being completely drained. Some of the liquid films at these ends form bridges again and coalesce, reducing the condensation heat transfer in the hydrophobic regions (Fig. 14.17c, d). When the width of the hydrophilic stripe is wide (Fig. 14.17e), the bridging phenomenon of such films is reduced (Fig. 14.18).

The condensate recovery of the biphilic surface can be expressed as the sum of filmwise condensation (on the hydrophilic stripes) and dropwise condensation, with better performance (on the hydrophobic stripes). The contribution of the hydrophilic regions $\dot{m}_{phil} = \dot{h}_{phil} S_{phil}$ follows from the latent heat flux

$$q_i = \rho_l L_v \dot{h}_{phil} = \frac{\rho_v L_v^2 D}{\zeta r_v T_\infty^2}(T_\infty - T_i) \text{ (Eqs. 13.14 and 13.19)},$$ with T_i the interface temperature and T_∞ the temperature of humid air supposed at saturation.

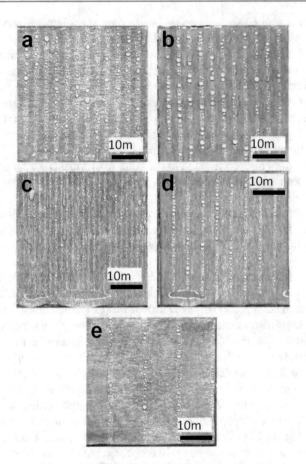

Figure. 14.17 Comparison on biphilic surface 210 min after condensation starts. **a** Comparison on biphilic surface 210 min after condensation starts. **a** $a_{phob} = 1.5$ mm, $b_{phil} = 1.5$ mm, $S^* = 0.5$, **b** $a_{phob} = 1.3$ mm, $b_{phil} = 2$ mm, $S^* = 0.6$, **c** $a_{phob} = 0.64$ mm, $b_{phil} = 0.96$ mm, $S^* = 0.6$ **d** $a_{phob} = 1.1$ mm, $b_{phil} = 3.4$ mm, $S^* = 0.75$, **e** $a_{phob} = 1.5$ mm, $b_{phil} = 8.5$ mm, $S^* = 0.85$. (Air temperature: 14 °C; relative humidity: 100%; substrate temperature: 5 °C). (Adapted from Lee and al. 2020, with permission)

Making apparent the hydrophilic ratio S^* and the total condensing surface $S_c = S_{phil} + S_{phob}$, it comes:

$$\dot{m}_{phil} = \frac{q_i}{L_v} S_{phil} = \frac{q_i}{L_v} S^* S_c \tag{14.55}$$

Lee et al. (2020) assumed that the contribution of dropwise condensation could be expressed by one single length, the length L_i of the interface between the hydrophilic and hydrophobic regions, which obviously changes with the different stripes characteristics. Figure 14.15a reports on the condensation yield from the only hydrophobic area as a function of L_i. Condensate recovery was measured

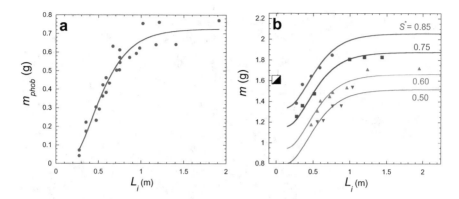

Fig. 14.18 **a** Condensation recovered from the only hydrophobic stripes. The curve is Eq. 14.56. **b** Total condensation rate from the entire biphilic surfaces. The curves correspond to Eq. 14.57. Half-filled black square corresponds to the yield if the entire surface was hydrophilic. Condensate recovery was measured for 4 h from 1 h after condensation started. (Data from Lee et al. 2020)

for 4 h from 1 h after condensation started. An empirical function can fit this behavior, with m_{phob} the condensed mass on the hydrophobic regions:

$$m_{phob}(g) = 0.695\left(1 - e^{-L_i/0.2354}\right)^{6.744} \tag{14.56}$$

With the help of Eqs. 14.55 and 14.56, the total recovery from both hydrophilic and hydrophobic regions, m can thus be expressed as, with τ the condensation time and m_{phil} the condensed mass on the hydrophilic regions:

$$m = m_{phil} + m_{phob} = \frac{q_i \tau}{L_v} S^* S_c + 0.695\left(1 - e^{-4.2488L_i}\right)^{6.744} \tag{14.57}$$

14.3.2.5 Biphilic Triangular Pattern

Superhydrophilic (S-philic) triangular patterns on superhydrophobic (S-phobic) background have been studied by Hou et al. (2020). In Fig. 14.16, triangles and spacing are of mm length scale. The triangular structures have the particularity to foster water displacement by the Laplace pressure gradient induced by the surface structural gradient from the tip to the base of the triangle even in the horizontal position, in a manner similar to what is observed on cactus spines (Ju et al. 2012).

The S-philic triangles on the copper S-phobic substrate are isosceles, with single triangle base length of 0.5 mm and height 5.5 mm. In one direction (vertical), the tip of the lower triangle is connected to the base of the upper triangle. In the horizontal direction, the triangles are spaced with base edge-to-edge spacing varying from 0.1 to 4 mm (see Fig. 14.19).

The water droplets preferentially nucleate on the S-hydrophilic triangular patterns at the onset of condensation because the nucleation energy barrier is the

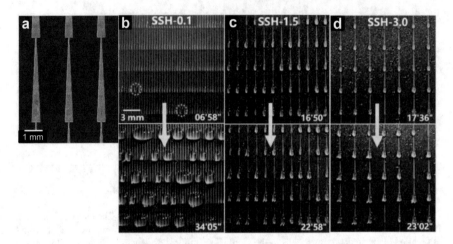

Fig. 14.19 **a** Photo of a biphilic triangular substrate. **b–d** Water collecting dynamics with triangle spacing **b** 0.1 mm, **c** 1.5 mm and **d** 3.0 mm. The upper image on panel (**b**) shows the time when the droplets start to coalesce between rows (dashed green **c**). The corresponding lower images on each panel (yellow arrows) indicate the time when droplets start to slide down. (Adapted from Hou et al. 2020, with permission)

lowest (see Sect. 3.2). The condensed droplets then grow and coalescence with each other, forming a thin layer of water film along the triangle (filmwise condensation). Between the triangles, on the S-phobic surface, near spherical water droplets nucleate and grow (dropwise condensation). They eventually coalesce with each other and are incorporated in the water film on the stripes due to their random displacement as induced by coalescence (see Chap. 9). Thick water films accumulate as big irregular droplets at the base of each triangle (Fig. 14.19 b, c, d), which can coalesce together when the spacing is small and detach when their size exceeds the critical sliding size (see Sect. 14.1.2). The critical size is determined by the contact angle hysteresis (Eq. 14.17).

Although S-philic substrates are favorable for droplet nucleation, the large pinned droplets on their surface can reduce the area available for favorable continuous dropwise nucleation. When the triangle spacing is varied between 0.1 and 4 mm, the water collection rate exhibits a bell-shaped behavior, with a maximum at $\approx$ 1.5 mm spacing (Fig. 14.20). Smaller triangle spacing indeed leads to earlier coalescence between adjacent triangles (Fig. 14.19) leading to larger contact angle hysteresis since the triangular pattern acts as pinning points. When the pattern spacing is equal or larger than 1.5 mm (Fig. 14.19), the droplets are confined to single triangles and no coalescence between rows happens, decreasing the contact angle hysteresis. However, the S-philic region that captures drops from the S-phobic background is reduced, leading to the degradation of the water collection performance.

When compared with pure S-phobic and S-philic substrates (Fig. 14.20), the collection rate $\dot{h}_c$ on the biphilic surface with smaller triangle spacing (0.1 mm)

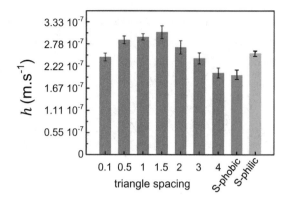

Fig. 14.20 Water collection rate per surface area $\dot{h}_c$ at different triangle spacings. Pure superhydrphobic and superhydrophilic surfaces are shown for comparison. (Adapted from Hou et al. 2020, with permission)

is comparable to the S-philic substrate, due to a large wettable region. The same remark holds for the S-phobic substrate when compared with the 4 mm biphilic surface, making apparent the role of the S-phobic region in collecting water. Optimizing the triangle spacing on the biphilic surface makes the collection rate larger than pure S-philic and S-phobic surfaces (20% with respect to pure S-philic and 57% with respect to pure S-phobic), due to the fast water capturing rate on the S-philic region and enhanced water removal efficiency on the S-phobic region. To this effect adds a continuous directional water transport induced by the triangular patterns.

14.4 Other Rough and Porous Substrates

14.4.1 Sand Blasting

The utilization of controlled, enhanced random roughness on initially smooth surfaces has several benefits. It averages the substrate defects and increases the hydrophilic or hydrophobic character of the substrate. With θ_c the angle on the smooth substrate and θ_W the contact angle on the rough substrate with roughness r (ratio of drop actual contact are / projected area), one gets $\cos\theta_W = r\cos\theta_c$ (see Eq. E.9 in Appendix E).

Technically speaking, it is easy to produce large rough surfaces by sand blasting. Experiments by Trosseille et al. (2019a) were carried out with a Duralumin substrate whose initial mean arithmetic roughness perpendicular to the substrate ($R_a = \overline{\delta\zeta} = 0.4$ μm) was increased by sand blasting with $\rho = 12.5$ μm radius silica beads at jet pressure $p = 8$ bar (see Appendix L for details). Roughness was increased to $\overline{\delta\zeta} = 6.6$ μm, with lateral roughness parallel to the substrate $\overline{\delta\xi} = 2\sqrt{2}\left(\varrho\overline{\delta\zeta}\right)^{1/2} = 25$ μm (Appendix L, Eq. L.2).

Such sand blasting gives interesting properties to the surface. First, the Wenzel roughness factor r (see Appendix E, Sect. E.2.2) is increased by roughness. The calculation for sand blasting as reported in Appendix L, Eq. L.9 gives $r = 1 + \frac{\overline{\delta\zeta}}{2\varrho}$,

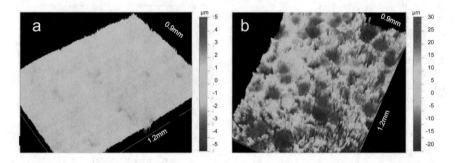

Fig. 14.21 Scanned portion (0.9 mm × 1.2 mm) of the surface of a 175 mm × 175 mm × 5 mm Duralumin plate. **a** "Smooth surface", with mean roughness $R_a = \overline{\delta\zeta} = 0.4$ μm. **b** Same after being hit perpendicularly by silica beads of 25 μm diameter under air pressure of 8 bars during approximately 3 min, with a density of impact of about 45 mm^{-2}. The jet was scanned parallel and perpendicular to one side of the square. Roughness due to the bead impacts is in the order $R_a = \overline{\delta\zeta}$ = 14.6 μm, corresponding to $\overline{\delta\xi} = 25$ μm (see text). (Adapted from Trosseille et al. 2019, with permission)

where $\overline{\delta\zeta}$ is the mean amplitude roughness. The latter increases linearly with the sand blast jet pressure p (Fig. L.2.b). For smooth Duralumin materials ($p = 0$), Fig. 14.21a with mean $\overline{\delta\zeta} = 0.4$ μm roughness corresponds to $r = 1$. When sand is blasted at $p = 8$ bar, giving $\overline{\delta\zeta} = 6.6$ μm (Fig. 14.21b), r increases up to 1.14. This increase has little effect on the drop pinning forces. The critical drop radius at the sliding onset thus remains nearly the same than using the corresponding smooth materials.

Enhanced roughness, however, has secondary strong effects on nucleation sites, leading to a much increased number of sites. When compared with the smooth materials (Fig. 14.4b), the time t_f for drop sliding onset remains nearly the same for all condensation experiments (Fig. 14.22b), in contrast with the same smooth materials. In the latter case, a dispersion of 200% is observed because of defects, acting as uncontrolled nucleation sites (see Sect. 14.1.2 above, Fig. 14.4b). The roughness-enhanced materials, in addition to canceling this dispersion, give a sliding onset time less than the minimum time found for the corresponding smooth surface. An enhanced number of nucleation sites indeed reduces the initial periods of drop slow growth (radius $\sim t^{1/2}$ and $t^{1/3}$), which more quickly enter the fast growth stage (radius $\sim t$), see Sect. 6.3.2. Early nucleation thus overcompensates the delay increase due to enhanced roughness.

As noted above in Sect. 14.1.2, the drops are always seen to detach during a coalescence event. Coalescence accelerates drop radius growth and moves the contact line, thus lowering the pinning forces (Gao and Geyer 2018). Drop shedding then occurs for a radius smaller than the critical radius of the static case Eq. 14.17 (Trosseille et al. 2019).

14.4.2   Porous Substrates (Fibrocement)

Fibrocement, a cheap material, is widely used for roofing. It is made of cement (e.g. Portland), organic fibers (e.g. polyvinylalcohol) and mineral additives. Its superficial structure is highly porous (Fig. 14.22a, b) and the question naturally arises how much condensed water can be trapped on such materials. The condensation process starts by filling the pores (Fig. 14.22c), and then an irregular film forms and water starts to flow (Fig. 14.22d).

By comparing collected dew water on radiatively cooled smooth steel coated surfaces with paint made hydrophilic by mineral additives (OPUR 2021), Doppelt and Beysens (2013) found that water starts to be collected on Fibrocement after h_0 ≈ 0.1 mm water has been collected by the painted surface. This is in the order of the Fibrocement surface roughness $\overline{\delta \zeta} \sim 0.1$ mm (Fig. 14.22a, b), in accord with the discussion of Sect. 14.1.1. Dew condensation and light rain collection tests (Doppelt and Beysens 2013) show that 0.1 mm corresponds indeed to the mean value that remains trapped in the Fibrocement structure. Spray experiments where microdroplets are projected on the surface show the same value (Fig. 14.23).

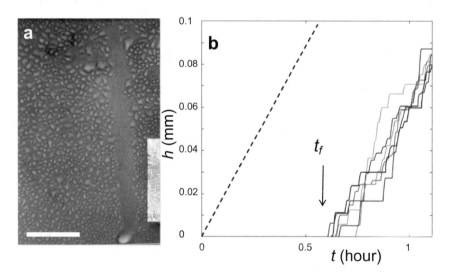

Fig. 14.22 Condensation on a vertical Duralumin sand-blasted surface with 25 μm silica beads (same as Fig. 14.13: roughness $Ra = 14.6$ μm, advancing/receding angles 105°/23°, plate 14.7 cm long, 3 cm wide, condensation rate 45 nm.s^{-1}) cooled by conduction. Edge effects have been canceled by absorbers. **a** Picture showing droplets at the sliding onset. The white bar is 10 mm. **b** Condensed mass with lag time t_f. Note the lack of dispersion in lag times as compared with the smooth materials in Fig. 14.4b. (Adapted from Trosseille et al. 2019, with permission)

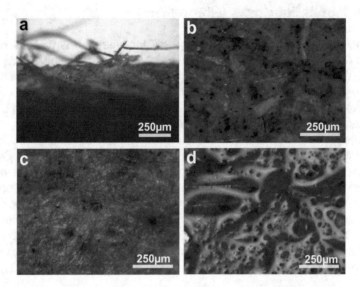

Fig. 14.23 Dry EDILFIBRO Fibrocement: **a** Side view, **b** front view. Under steady condensation (front view): **c** Wet, **d** imbibed. (Adapted from Doppelt and Beysens 2013)

14.5   Effect of Tilt Angle and Condensation Rate

The time t_f at which the first drop detaches from the condensing surface is a function of both the condensation substrate tilt angle,α, with horizontal and condensation rate per surface area, $\dot{h}$. Gerasopoulos et al. (2018) experimentally investigated the influence of these two parameters on t_f and $\dot{h}_c$, the average rate of collected water per condensing surface area during the steady state. Small $\dot{h}_c$ corresponds to dew water collection cooled by radiation deficit whereas high $\dot{h}_c$ corresponds to industrial devices cooled by contact.

Four typical substrates were investigated; hydrophilic, hydrophobic, superhydrophobic and superbiphilic (hydrophilic islands on a square array of diameter $a = 0.5$ mm separated by hydrophobic regions of $b = 2$ mm). The wetting characteristics (contact angle with hysteresis) are listed in Table 14.1.

Two typical subcooling temperatures were considered, with the same humid air characteristics ($RH = 98\%$, temperature$T_\infty \approx T_d = 37°C$). Low subcooling corresponds to Breath Figure formation with substrate temperature at $T_c = 30 °C$ ($\Delta T = T_\infty - T_c = 7°C$) leading to condensation rate $\dot{h} \approx 200$ nm.s^{-1} (10 times larger than the natural dew rate, see Beysens 2018) and large subcooling, with $T_c = 10 °C$ ($\Delta T = T_\infty - T_d = 27°C$), giving $\dot{h} \approx 670$ nm.s^{-1}, a value closer to industrial devices.

Water collection from substrates tilted from horizontal was studied at 4 angles α ($= 10°, 30°, 60°, 90°$). Because gravity affects sliding as $g\sin\alpha$, two characteristic

Table 14.1 Substrate wetting characteristics. The surface area is 4 cm × 4.2 cm.

Substrate		Contact angle (°)		
		Mean	Advancing	Receding
Hydrophilic		66	77	55
Hydrophobic		116	121	111
Superhydrophobic		157	160	154
Superbiphilic (square array of hydrophilic islands separated by b = 2 mm)	Superhydrophilic (diameter a = 0.5 mm)	≈0	≈0	≈0
	Superhydrophobic	157	160	154

Fig. 14.24 Typical pictures of substrates 60 min after condensation onset for **a** hydrophilic, **b** hydrophobic, **c** superhydrophobic and **d** superbiphilic samples at $\alpha = 10°$ angle and $\Delta T = 7\,°C$ subcooling. Images are similar for large subcooling. (Adapted from Gerasopoulos et al. 2018, with permission)

main domains correspond to (I) null or weak gravity shedding with small $\alpha = 0$–$10°$ ($\sin\alpha \approx 0$–0.2) and (II) strong gravity shedding with large $\alpha = 30$–$90°$ ($\sin\alpha \approx 0.5$–1).

Typical pictures are shown in Fig. 14.24. Representative cumulated mass collected for both small and large subcoolings are is in Fig. 14.25. All evolutions exhibit a more and less pronounced delay t_f in water collection, followed by steady condensation with mean rate $\dot{h}$. It must be noted that, with the same rate $\dot{h}$, the total collected volume will be the largest for the shortest t_f, the "dead" volume corresponding to the delay being $t_f\dot{h}$.

14.5.1 Low Supercooling

Figure 14.26 shows for supercooling $\Delta T = 7°C$ the delay in water collection, t_f, and the subsequent mean condensation rate, $\dot{h}$. Superhydrophobic substrate shows the shortest delay as water is rapidly shed to the collector. In the small angle region (I), t_f can be extremely long, especially for the superhydrophilic and superbiphilic substrates. It corresponds to pooled water, which adds a thermal resistance and reduces the condensation rate as seen in Fig. 14.26b.

In the large angle region, t_f is considerably reduced. For vertical positions, nearly all substrates exhibit the same small values, with the superhydrophobic

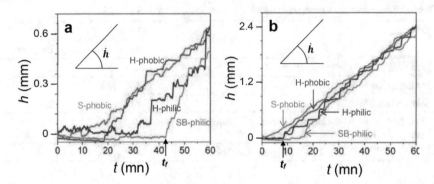

Fig. 14.25 Comparison of cumulated collected volume evolution per unit area $h(t)$ for four different substrates tilted at 30° from horizontal showing the critical shedding time t_f (arrow) and the mean collection rate (slope $\dot{h}$). H-philic: Hydrophilic; H-phobic: Hydrophobic; S-phobic: Superhydrophobic; SB-philic: Superbiphilic. **a** Low subcooling $\Delta T = 7°$. **b** Large subcooling $\Delta T = 27°$. (Adapted from Gerasopoulos et al. 2018, with permission)

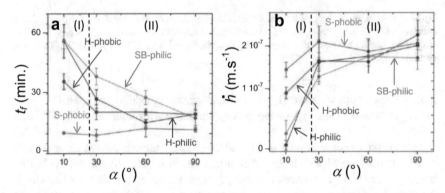

Fig. 14.26 **a** Onset of water collection t_f at different tilt angles α for $\Delta T = 7°C$ subcooling. **b** Corresponding mean water collection rate $\dot{h}$ for times $t > t_f$. Regions (I) and (II) separated by the vertical interrupted line correspond to weak and large gravity shedding regions (see text). (Adapted from Gerasopoulos et al. 2018, with permission)

substrate still giving better results. The condensation rate also exhibits nearly the same values for both substrates. As noted above at the beginning of Sect. 14.5, with the same $\dot{h}$, the total collected volume will be largest for the shortest t_f. It is worthy to note that superbiphilic surfaces, although designed to increase nucleation and favor drop rolling, are the least efficient to collect water. This is because the border between the hydrophilic area and the hydrophobic region around necessarily causes large differences between the advancing and receding contact angles, increasing drop pinning.

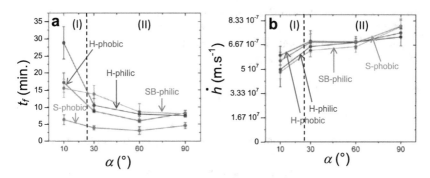

Fig. 14.27 **a** Onset of water collection t_f at different tilt angles α for $\Delta T = 27\,°C$ subcooling. **b** Corresponding mean water collection rate for times $t > t_f$. Regions (I) and (II) separated by the vertical interrupted line corresponds to weak and large gravity shedding regions (see text). (Adapted from Gerasopoulos et al. 2018, with permission)

14.5.2   Large Supercooling

The results for large supercooling $\Delta T = 27\,°C$ are shown in Fig. 14.27. As expected, the large condensation rate promotes faster growth to the drop critical sliding radius when compared with low supercooling. Similar trends as in low supercooling condensation are shown, however with much reduced variability between the samples. This is especially noticeable for zone I of small angles, where a comparable volume of water is collected for all substrates.

14.6   Liquid-Imbibed Substrate (LIS)

The different situations where a water droplet can sit on an oil-imbibed microsubstrate are depicted in Sect. 11.2, Fig. 11.13. Depending on whether the lubricant has a low or moderate surface energy, oil can engulf the water droplet or form an annular ring around it. In addition, oil and water can also partially or completely wet the substrate. It results in 12 possible different situations (see Fig. 11.13) where only four situations, where an oil wetting layer isolates the drop from the surface, corresponds to non-pinning drops. Although water droplets can nucleate and grow on liquids and liquid impregnated surfaces (see Sect. 11.1.1), encapsulation of water by oil is not desirable as some oil will be removed from the impregnated surface and collected with the water.

The lubricant encapsulating the texture is stable only if it wets the texture completely, as otherwise portions of the textures dewet and emerge from the lubricant. Although complete wetting of the texture is most desirable in order to eliminate pinning, texture geometry can be exploited to reduce the emergent areas and obtain a very small sliding drop radius, which increases the range of lubricants that can be used. In case drop pinning occurs (Fig. 11.13, cases A2,3-W1,2), the critical radius for sliding can be deduced by considering the water drop and oil pinning forces on the surface pattern (Eq. 14.14). With ϕ_s the area fraction of the liquid–solid contact (see Appendix E, Eq. E.2.2), $\phi_s^{1/2}$ will be the fraction of the droplet perimeter making contact with the emergent features of the textured substrate. The pinning force Eq. 14.14 thus becomes:

$$F_s = k^* \phi_s^{1/2} R \left[\sigma_{ow} \left(\cos \theta_{r,os/w} - \cos \theta_{a,os/w} \right) + \sigma_{oa} \left(\cos \theta_{r,os/a} - \cos \theta_{a,os/a} \right) \right] \tag{14.58}$$

The length of the contact line over which pinning occurs is indeed expected to scale as $\phi_s^{1/2} R$. σ_{wo} is the surface tension water–oil and σ_{oa} is the surface tension oil–air. The receding and advancing contact angles are denoted by the subscripts r and a, respectively, $\theta_{os/w}$ is the water drop oil-solid contact angle and $\theta_{os/a}$ is the oil-solid-air contact angle. Equaling the pinning force Eq. 14.14 with water drop weight and expressing the drop volume as a function of its radius (Appendix D, Eq. D.7) gives, after some algebra where one has made apparent the capillary length (Appendix E, Eq. E.6), the critical radius R_0 at sliding onset is:

$$R_0^2 = \phi^{1/2} \frac{l_c^2}{\pi \sigma G(\theta) \sin \alpha}$$
$$\left[\sigma_{ow} \left(\cos \theta_{r,os/w} - \cos \theta_{a,os/w} \right) + \sigma_{oa} \left(\cos \theta_{r,os/a} - \cos \theta_{a,os/a} \right) \right] \tag{14.59}$$

A detailed study has been reported by Smith et al. (2013).

In order to visualize the effect, Anand et al. (2012) compared in Fig. 14.28f the shedding velocity of condensing droplets on the same microtexture (cubical microposts of silicon $a = b = c = 10$ μm with nanograss), with and without a lubricant. Nanograss ensures that the posts remain filled with oil while keeping the pinning contact very small (Fig. 14.28a, b, c). Although the critical sliding radius is in the order of 1.8 mm on the bare posts, it becomes as small as 50 μm when lubricant is added. This small critical radius is due to the weak pinning by nanograss.

This method to reduce water drop pinning is very appealing. It suffers, however, from some drawbacks. Lubricants indeed slowly evaporate and some quantity is always removed from the substrate by the sliding drops. It means that the impregnated surface progressively dries out and some (small) quantities of oil are found in the collected water. These problems can be somewhat overcome by using low vapor pressure, food-secure lubricants and periodically refilling the structures.

14.7   Soft Substrates

Droplet condensation on a soft or deformable substrate, that is on a substrate whose shear modulus G is below typically 500 kPa, shows some aspects of condensation on liquid surfaces. Indeed, capillary force and disjoining pressure during dropwise condensation can deform the substrate around the drops (see Sect. 11.3).

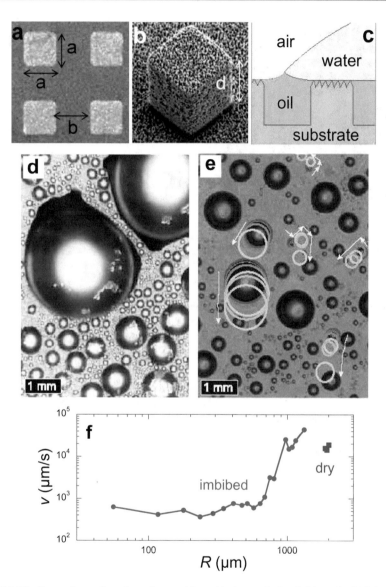

Fig. 14.28 Comparison of condensation on hierarchical superhydrophobic and oil (1-butyl-3-methylimidazolium bis(trifluoromethylsulfonyl)imide, [BMIm⁺][Tf2N⁻]) impregnated surfaces with identical microtexture. **a, b, c** Microposts with nanograss. $a = b = c = 10 \, \mu$m. **d** Condensation on dry nanograss microposts near the shedding threshold. **e** Condensation on oil-imbibed nanograss microposts. Both small and large droplets are highly mobile (see colored circles to follow the droplet motion). **f** Median speed of droplets, V, undergoing gravity shedding as a function of drop radius, R, for unimpregnated (squares) and impregnated surfaces (circles) (log–log plot). (Adapted from Anand et al. 2012, with permission)

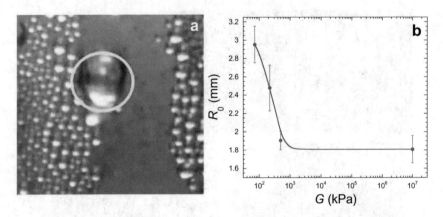

Fig. 14.29 **a** Condensation on a vertical soft substrate with a departing droplet in the yellow circle. **b** Departure radius R_0 with respect to shear modulus G. (Adapted from Phadnis et al. 2017, with permission)

Highly irregular-shaped drops are formed during coalescence on the softer substrates. In particular, droplets become non-circular after coalescence because of the presence of a wetting ridge deformation at the meeting point of the two constituent drops and around their receding contact lines. This ridge decays only very slowly. As a result, when condensation occurs on a substrate tilted with horizontal, the critical sliding radius is increased and the sliding velocity decreases due to the irregular drop shape (Fig. 14.29a). In addition, sliding involves the motion of ridges, which during drop coalescence increases contact line friction and lowers the shedding velocity. The effect increases with decreasing shear modulus. In Fig. 14.29b, the decrease of the modulus below 500 kPa is seen to increase the drop departure radius (from 1.8 mm at 500 kPa to 3.1 mm at 75 kPa).

Optical Effects

15

The most common and ubiquitous effect of dew condensing on glass, windows, mirrors is to blur the vision. To be more precise, dropwise condensation scatters light and makes the substrate look "milky" (while filmwise condensation leaves it transparent). The ability of tiny droplets to make the substrate translucent when breathing on it has long been used to determine whether glass optical components were clean or contaminated by fat substances. When glass is clean, its surface is hydrophilic and an invisible transparent water film forms. However, when only a monolayer of fat molecules is present on glass, the water contact angle becomes non-zero, droplets condense and scatter light. This phenomenon is at the origin of the term "Breath Figures" by Aitken (1893) and later Rayleigh (1911). Breathing was thus a simple and quite efficient manner to check the property of optical surfaces at a molecular level.

Scattered light, however, contains much more information than making glass look "white". This chapter outlines how visible and infra-red scattered and transmitted light by the droplet pattern can reveal its statistical properties. Infra-red light matters because radiative cooling is at the origin of natural dew condensation and condensation modifies the substrate emissivity.

15.1 Transmitted or Reflected Light

15.1.1 Description

In the analysis below (Nikolayev et al., 1998), drops are considered as spherical caps of base radius R and radius of curvature ρ (Fig. 15.1), the influence of gravity is being neglected. Gravity effects (Sect. 6.6.2 and Chap. 14) occur when $R \geq l_c \approx$ 2.7 mm, which is the capillary length (see Eq. E.6 in Appendix E) and gives the scale of the approximation.

© Springer Nature Switzerland AG 2022
D. Beysens, *The Physics of Dew, Breath Figures and Dropwise Condensation*, Lecture Notes in Physics 994, https://doi.org/10.1007/978-3-030-90442-5_15

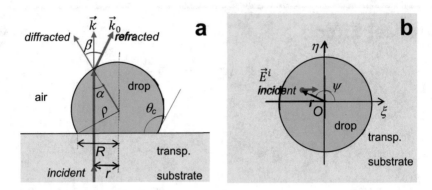

Fig. 15.1 Definitions of the quantities used in the text

The pattern is assumed to evolve on a transparent substrate. It is illuminated by a monochromatic linearly polarized plane wave with amplitude E_s, which travels through the substrate and falls normally on the interface substrate/water drops (Fig. 15.1).

15.1.2   Optical Properties of a Single Drop

The incident wave is partially reflected, and the amplitude of the wave just after hitting the plane substrate–water interface is (see e.g. Born and Wolf, 1999):

$$E^i = \frac{2n_w}{n_w + n_s} E^s e^{ik_0 n_w l} \tag{15.1}$$

Here n_w and n_s are the refractive indices of water and the substrate, respectively. The phase shift in the exponential contains the incident wavevector of light, $k_0 = 2\pi/\lambda_0$ with λ_0 the light wavelength in a vacuum, and the geometrical path of the ray inside the water drop, l, (see Fig. 15.1 for the definition of the other parameters):

$$l = \sqrt{\rho^2 - r^2} - \rho\cos\theta_c \tag{15.2}$$

After the ray has been refracted by the drop, its polarization changes according to the Fresnel formulae (see e.g. Born and Wolf, 1999). One obtains for the component of the transmitted field *parallel* to the plane of incidence:

$$E_\parallel^t = t_\parallel E_\parallel^i \tag{15.3}$$

with the factor of transmission:

$$t_\parallel = \frac{2\sin\beta\cos\alpha}{\sin(\alpha + \beta)\cos(\alpha - \beta)} \tag{15.4}$$

Here, α is the angle of incidence and β is the angle of refraction (Fig. 15.1a). They are connected by the Descartes-Snellius law, which reads, with n_a the air refractive index, as:

$$n_w \sin\alpha = n_a \sin\beta \tag{15.5}$$

By geometry (Fig. 15.1), α is measured by r:

$$\sin\alpha = \frac{r}{\rho} \tag{15.6}$$

Concerning the field component *perpendicular* to the plane of incidence, one obtains:

$$E_\perp^t = t_\perp E_\perp^i \tag{15.7}$$

with the transmission factor:

$$t_\perp = \frac{2\sin\beta\cos\alpha}{\sin(\alpha + \beta)} \tag{15.8}$$

Le us consider a polar coordinate system in a plane parallel to the substrate (see Fig. 15.1b) where the reference point is the drop center O and the reference direction, ξ, is the direction of polarization of the incoming light wave. For a point with coordinates r, ψ, the incident field writes as:

$$E_\parallel^i = E^i \cos\psi; \quad E_\perp^i = E^i \sin\psi \tag{15.9}$$

In Fig. 15.1b, Cartesian axes $\xi O\eta$, Eqs. 15.3, 15.4, 15.7, and 15.8 together with Eq. 15.9 shows that the transmitted field components in directions ξ and η can be written as:

$$E_\xi^t = E^i \left(t_\parallel \cos^2\psi + t_\perp \sin^2\psi \right) \tag{15.10}$$

$$E_\eta^t = E^i \left(t_\perp - t_\parallel \right) \sin\psi \, \cos\psi \tag{15.11}$$

Since waves of different polarization do not interfere, they can be analyzed separately. The η-components of the waves emitted by two mirror symmetrical points of the surface of the drop at r, ψ and $r, -\psi$ therefore obey the relation:

$$E_\eta^t(r, \psi) + E_\eta^t(r, -\psi) = 0 \tag{15.12}$$

The above relation can also be deduced from the fact that the ray incidence being normal to a spherical cup, the angles α and β as defined by Eqs. 15.5 and 15.6 are independent of the ψ—coordinate of the incident ray (Fig. 15.1). Thus

$t_\parallel$ and $t_\perp$ as defined in Eqs. 15.4, 15.8 are also independent of ψ and Eq. 15.12 directly comes from Eq. 15.11.

As far as only the zero order of diffraction is concerned, the η-polarized components of the waves that come from two of these points annihilate each other. This argument is applicable to every two symmetrical points, thus the η—polarized components of the waves give no contribution. Therefore, the wave in the zero order of diffraction, in contrast to any other point in the image plane, is polarized in the ξ-direction, which is the direction of polarization of the incident wave. The index ξ will be, therefore, omitted in the following and the scalar wave theory for the ξ-component will be applied.

The key parameter of this theory is the complex transparency of the drop. One notes that the cap radius of the drop, R, is related to the contact angle θ_c through the factor:

$$\zeta = \frac{R}{\rho} = \begin{cases} sin\theta_c & \text{if } \theta_c < \pi/2 \\ 1 & \text{otherwise} \end{cases} \tag{15.13}$$

The contribution of the rays that are multiply reflected or refracted before leaving the drop will be neglected. Indeed, each reflection or refraction greatly decreases the amplitude of the corresponding fields according to the Fresnel formulae, the angles of reflection or refraction being large. Another reason for amplitude decrease is the angle factor K, which is small for large refraction angles (see below Eq. 15.20).

Full internal reflection occurs when the incidence α is larger than α_{cr}, defined by:

$$sin\alpha_{cr} = \gamma = \frac{n_a}{n_w} \tag{15.14}$$

Two regions can thus be defined in the drop: (i) an inner transparent region of radius $R' < R$ and (ii) an outer annular opaque, "black" region. The ratio $\chi = R'/R$ can be evaluated from Fig. 15.1:

$$\chi = \frac{R'}{R} = \begin{cases} \gamma & \text{if } \alpha_{cr} < \theta_c < \pi - \alpha_{cr} \\ sin\theta_c & \text{otherwise} \end{cases} \tag{15.15}$$

The contribution of the rays that must cross an air layer before entering the drop when $\theta_c > \pi/2$ is neglected following the above "single-refraction" assumption.

Considering Eqs. 15.10 and 15.1 and division by the amplitude of the wave unperturbed by the presence of the drop, one obtains the complex transmittance of the drop:

$$\tau_R(\rho) = \begin{cases} \frac{n_w(n_a+n_s)}{n_a(n_w+n_s)} C(\psi)e^{ik_0(n_w-n_a)l-\mu l}, & if \, 0 \leq r < R' \\ 0, & if \, R' < r < R \\ 1, & if \, r \geq R \end{cases} \tag{15.16}$$

Here, μ is the coefficient of light absorption, negligible for visible light. In the following, we will make $\mu = 0$. The function $C(\psi)$ reads as:

$$(\psi) = t_\| \cos^2 \psi + t_\perp \sin^2 \psi \tag{15.17}$$

15.1.3 Diffraction by a Single Drop

The Kirchhoff's formula, reduced for the diffraction in the far zone (see e.g. Born and Wolf, 1999), gives the following expression for the field amplitude, depending on the direction to the point of observation as determined by the wavevector $\overrightarrow{k}$:

$$E\left(\overrightarrow{k}\right) = \int \tau_R(\rho) K e^{-i n_a \overrightarrow{r} \cdot (\overrightarrow{k}_0 + \overrightarrow{k})} d\overrightarrow{r} \tag{15.18}$$

Here, K is an angle factor given by:

$$K = \frac{1}{2k_0} \cdot \left(\overrightarrow{k_0} + \overrightarrow{k}\right) \cdot \overrightarrow{n} \tag{15.19}$$

$\overrightarrow{n}$ is a vector unit normal to the reference plane parallel to the substrate (Fig. 15.1). The integration in Eq. 15.18 is performed over the reference plane parallel to the substrate. $\overrightarrow{k_0}$ (modulus $\left|\overrightarrow{k_0}\right| = \left|\overrightarrow{k}\right| = k_0$) is a normal vector to the wavefront that falls on the reference plane. Equation 15.18 is valid for small scattering angles where $\overrightarrow{k} \approx k_0 \overrightarrow{n}$ (here $\overrightarrow{k} = k_0 \overrightarrow{n}$ exactly). One can thus rewrite Eq. 15.19 as:

$$K = \frac{1}{2} \cdot \left(\frac{\overrightarrow{k_0} \cdot \overrightarrow{n}}{k_0} + 1\right) = \frac{1}{2}[\cos(\beta - \alpha) + 1] \tag{15.20}$$

In the common case of the Fraunhofer diffraction by a hole or by a flat phase object, k_0 is not a function of r. $E(\overrightarrow{k})$ is in this case the Fourier transform of the transmittance $\tau_R(\rho)$ (Eq. 15.16). In the case of a drop, the situation is more complex as the direction of $\overrightarrow{k}_0$ is a function of r. $E(\overrightarrow{k})$ is now the Fourier transform of the function:

$$\widetilde{\tau}_R(\rho) = \tau_R(\rho) K e^{-i n_a \overrightarrow{r} \cdot \overrightarrow{k_0}} \tag{15.21}$$

Note that $\widetilde{\tau}_R(\rho) = \tau_R(\rho) = 1$ outside the drop.

15.1.4 Zero-Order Diffraction by a Dew Pattern

The theory of diffraction by an assembly of objects was analyzed by Hosemann and Bagchi (1962). The assumptions of applicability are the conditions of Fraunhofer diffraction. The amplitude in the image plane is the Fourier transform of the object transparency. The intensity in this zero diffraction order, I_0, is defined as:

$$I_0 = \left| E\left(\vec{k} = 0\right)\right|^2 = |\langle \tau \rangle|^2 \tag{15.22}$$

where the brackets $\langle \tau \rangle$ denote an average of the drop size distribution. With $s_1 = S_1/N$, the total illuminated area of the substrate, S_1, divided by N, the total number of drops, one obtains:

$$tau = \frac{1}{s_1} \left\langle \int_{s_1} \tau_R(r)dr \right\rangle \tag{15.23}$$

Note that the invariance with respect to the spatial distribution of the drops is an interesting feature of Eq. 15.22.

Hosemann and Bagchi (1962) developed the theory for a real transmittance $\tau_R(r)$; however, it can be generalized for the complex case, using $\tilde{\tau}_R(r)$ from Eq. 15.21 (this function is radially symmetric, a necessary condition to apply Eq. 15.23). Equation 15.23 thus becomes:

$$\tau = 1 - \frac{\pi R^2}{s_1} + \frac{1}{s_1} \frac{n_w(n_a + n_s)}{n_a(n_w + n_s)} \int_0^{R'} J K r e^{i k_0 (n_w - n_a)l + n_a r \sin(\beta - \alpha)} dr \tag{15.24}$$

Here:

$$J = \int_0^{2\pi} C(\psi)d\psi = \pi\left(t_\parallel + t_\perp\right) \tag{15.25}$$

When averaging Eq. 15.24, one needs to introduce a drop distribution $H(R)$, which defines the probability to find a drop with the visible radius R. One considers the Maxwellian distribution following Hosemann and Bagchi (1962), which depends on only two free parameters, the mean drop size $\langle R \rangle$ and the drop polydispersity $g = \left(\langle R^2 \rangle - \langle R \rangle^2\right)^{1/2}/\langle R \rangle$ (see Eqs. 6.36, 6.37). This distribution is given by:

$$H(R) = \frac{B(m)}{R_0}\left(\frac{R}{R_0}\right)^m e^{-(R/R_0)^2} \tag{15.26}$$

In this expression, $B(m)$ is a normalization constant. The parameter $m(\geq -1)$ determines g and $\langle R \rangle$ by:

$$g \approx (2m + 2)^{-1/2}$$

$$\langle R \rangle \approx R_0 \left(\frac{m + 1/2}{2} \right)^{1/2} \tag{15.27}$$

The approximations are valid within less than 1% when $m > 4$.

The apparent mean surface coverage, ε_2^*, of drops seen from above the substrate, is related to the actual drop surface coverage, ε_2, by:

$$\varepsilon_2^* = \begin{cases} \varepsilon_2 = \pi \frac{\langle R^2 \rangle}{S_1} & \text{if } \theta_c \leq \pi/2 \\ \frac{\varepsilon_2}{\sin^2 \theta_c} & \text{if } \theta_c > \pi/2 \end{cases} \tag{15.28}$$

It involves the projected area covered by the drops, S_1, which differs from the actual area in contact with water when the contact angle $\theta_c > \pi/2$ (Fig. 15.1).

Equations 15.24, 15.25, 15.26, 15.27 and 15.28 give $\langle \tau \rangle$ as a function of the parameters R_0, g (or equivalently, R_0, m) and ε_2^*:

$$
\begin{aligned}
\tau = 1 - \varepsilon_2^* & + \frac{\varepsilon_2^*}{\zeta^2 \gamma^2} \frac{n_a + n_s}{n_w + n_s} B(m + 2) \\
& \times \int_{\sqrt{1-\chi^2}}^{1} dx \frac{x^2 (1 + \gamma + vx - x^2)}{(v + x)(vx + 1 - x^2)} \\
& \times \int_0^\infty dy \, y^{m+2} \exp \left\{ iyn_w \frac{2\pi R_0}{\zeta \lambda_0} \left[(1 - \gamma)(x - \cos \theta_c) \right. \right. \\
& \left. \left. - (1 - x^2)(x - v) \right] - y^2 \right\}
\end{aligned} \tag{15.29}
$$

In this equation, $v = \sqrt{x^2 + y^2 - 1}$ and ζ, γ, χ are given by Eqs. 15.13, 15.14 and 15.15, respectively.

The intensity of the zero diffraction order $I_0(\langle R \rangle)$ (Fig. 15.2) follows from Eq. 15.22, where g and ε_2^* (or ε^2 by using Eq. 15.28) are the parameters. Oscillations in I_0 occur due to the exponential dependence on iR_0/λ_0 in Eq. 15.29.

The asymptotic values for small and large R/λ_0 can be obtained explicitly from:

$$
\begin{aligned}
I_0(R \to 0) = & \left[1 - \varepsilon_2^* + \frac{\varepsilon_2^*}{\zeta^2 \gamma^2} \frac{n_a + n_s}{n_w + n_s} \right. \\
& \left. \int_{\sqrt{1-\chi^2}}^{1} dx \frac{x^2 (1 + \gamma + vx - x^2)^2}{(v + x)(vx + 1 - x^2)} \right]^2
\end{aligned} \tag{15.30}
$$

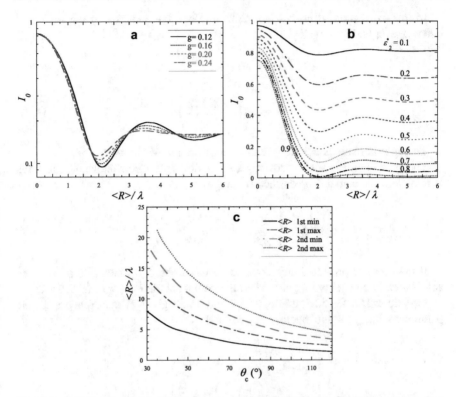

Fig. 15.2 (**a,b**) Variations of $I_0(\langle R/\lambda \rangle)$ for **a** several polydispersity values (g) at constant surface coverage $\varepsilon_2^* = 0.6$ for contact angle $\theta_c = 90°$ (semi-log plot) and **b** for several surface coverages ε_2^* at constant polydispersity $g = 0.16$ and contact angle $\theta_c = 90°$. **c** Positions of the four first extrema of $I_0(\langle R/\lambda \rangle)$ with respect to contact angle θ_c for $g = 0.16$ and $\varepsilon_2^* = 0.6$. (Adapted from Nikolayev et al., 1998, with permission)

where one has used $\langle R \rangle \to 0$, $\varepsilon_2^* \to 0$, $I_0(\langle R \rangle \to 0) = 1$. When ($\langle R \rangle \to \infty$), the asymptotic value becomes:

$$I_0(\langle R \rangle \to \infty) = \left[1 - \varepsilon_2^*\right]^2 \tag{15.31}$$

One notes that these expressions do not depend on polydispersity. Equation (15.31) relates $I_0(\langle R \rangle \to \infty)$ with the surface coverage ε_2^*. One also notes that Eq. 15.31 represents the transparency of an assembly of black disks in the limit of geometrical optics where $\langle \tau \rangle = 1 - \varepsilon_2^*$. In this case, I_0 depends either on the shape or on the size of the spots through ε_2^* only. When oscillations are present, the latter are thus connected to the phase shift of the transmitted wave induced by the drops.

Figure 15.2 presents the results of calculations of I_0 when the different parameters (θ_c, ε_2^*, g) are varied. Note that the curves do not correspond to the evolution of dropwise condensation since g and ε_2^* are not constant in the early times of growth (see Chaps. 4, , 6). In Fig. 15.2, one notes that the amplitude of the oscillations of I_0 is controlled mainly by the polydispersity, while the mean level, around which I_0 oscillates, depends strongly on ε_2^*. Large polydispersity can even suppress the oscillations. The positions of the I_0 extrema are nearly independent of g and ε_2^* over a wide range of parameter values (Fig. 15.2a, b). However, the dependence of the positions of the extrema on contact angle θ_c is stronger (Fig. 15.2c).

An experimental evolution of I_0 from Nikolayev et al. (1998) is shown in Fig. 15.3. The characteristics of I_0 allow the values of mean radius $\langle R \rangle$ and surface coverage ε_2^* to be determined as follows.

(1) The value of I_0 at long time corresponds to the self-similar regime where surface coverage $\varepsilon_2 = \varepsilon_{2,\infty}$ is constant (see Sects. 6.3 and 6.4). Figure 15.3a gives $I_0 = 0.16$, leading from Eq. 15.31 to $\varepsilon_{2,\infty} = 0.6$, to be compared to 0.55 (Fig. 6.6) and 0.57 (Fig. 6.8).
(2) From the $\varepsilon_{2,\infty}$ dependence on contact angle (Eq. 6.78), one obtains $\theta_c \approx 90°$, validating that the apparent surface coverage $\varepsilon_2^* = \varepsilon_2$ (see Eq. 15.28).
(3) Because the positions of the extrema of I_0 are nearly independent of ε_2^* and g (Fig. 15.2), the value of $\langle R \rangle$ can be obtained at these particular times (Fig. 15.3b).
(4) Since I_0 is independent of the polydispersity g in the inflection points (Fig. 15.2a), the evolution of ε_2^* is obtained through several steps. (i) Determination of the time and I_0 values at the inflection points. (ii) Corresponding values of $\langle R \rangle$ by the interpolation of $\langle R(t) \rangle$ obtained in step (3) above. (iii) Looking into Eq. 15.29, which can be rewritten as:

$$\langle \tau \rangle = 1 - \varepsilon_2^* + \varepsilon_2^*(A + iD) \tag{15.32}$$

A and D are independent of ε_2^*. Then $I_0 = |\langle \tau \rangle|^2$ (Eq. 15.22) leads to:

$$\varepsilon_2^* = \frac{\{1 - A(\{I_0[D^2 + (1 - A)^2 - D^2]\}^{1/2}}{D^2 + (1 - A)^2} \tag{15.33}$$

Since $\langle \tau \rangle$ is nearly independent of g, A and D can be calculated for any value of g and $\langle R(t) \rangle$ as obtained in step (ii) above. With the inflexion point values from step (i), surface coverage can be extracted as seen in Fig. 15.3b.

15.2 Scattered Light

Drops efficiently scatter light due to their small size, which compares during the early time with the light wavelength. The light scattered in the forward direction at angle θ corresponding to the scattering wavevector $k = 2k_0\sin\theta$ forms an annular luminous ring at $k = k_M$ (Fig. 15.4).

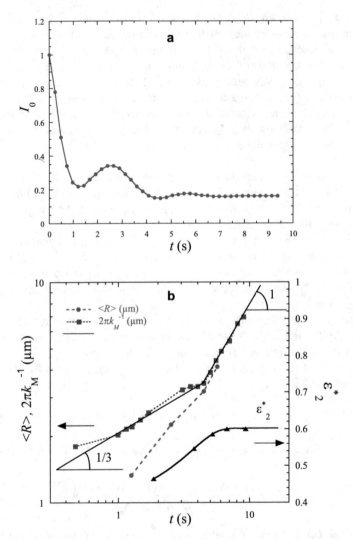

Fig. 15.3 Condensation on silanized glass (contact angle $\approx 90°$). (**a**) Evolution of transmitted intensity I_0, normalized by its initial value. (**b**) Evolution of surface coverage ε_2^*, (right ordinate, semi-log plot) and mean radius $\langle R \rangle$ (log–log plot) from I_0 or from the ring wavevector k_M (Eq. 15.34). The different growth law exponents, 1/3 for low surface coverage and 1 for high surface coverage, are evidenced, as the saturation of surface coverage at late times (see Sects. 6.3.1 and 6.3.2.2). (Adapted from Nikolayev et al., 1998, with permission)

The scattered intensity I is composed of the intensity scattered by individual particles, I_p, and an interference term, I_f, from the interference between particles (Hosemann and Bagchi, 1962). This last term is all the more pronounced as the particles are close together or their surface coverage ϵ_2 is large.

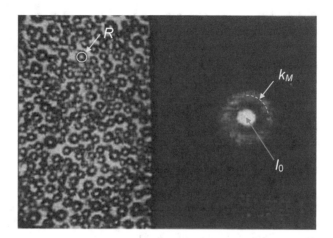

Fig. 15.4 Simultaneous observations of drop radius in direct space (left, from microscopy) and in the Fourier space (right, from light scattering). (Adapted from Beysens and Knobler, 1986, with permission)

The formation of a well-defined ring as in Fig. 15.4 is due to the contribution of the $I_p + I_f$ terms. Figure 15.5 contains typical scattered intensities for different surface coverages ϵ_2 and polydispersities g. The calculation was made assuming a

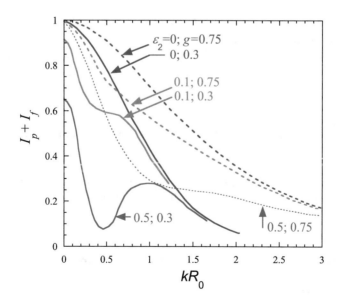

Fig. 15.5 Intensity $I_p + I_f$ versus kR_0 (notations: see text and Eq. 15.27) for different surface coverage ϵ_2 ($= 0, 0.1, 0.5$) and for polydispersities $g = 0.3, 0.75$. A ring is observed only for a large ϵ_2 and a small g. (Data from Hosemann and Bagchi, 1962)

Maxwellian distribution of sizes where polydispersity g and mean droplet radius $\langle R \rangle$ depend on parameters m and R_0 (see Eq. 15.27).

Concerning dropwise condensation, the shape of the scattered intensity $I(K)$ can be fitted by adjusting ϵ_2 and g. When I_p dominates and the particle size polydispersity is weak, assimilating the illuminated drops to be N black disks with a random position but the same radius $\langle R \rangle$, the diffracted light intensity will be the same as for one single drop (the Airy function), but with intensity N times larger. However, because of polydispersity, even if weak, only the first maximum of the Airy function is visible. It occurs at:

$$k_M \langle R \rangle \approx 1.63\pi \tag{15.34}$$

The measurement of ring radius, k_m in the scattered light intensity enables $\langle R \rangle$ to be determined. Figure 15.3b reports such measurements performed together with the analysis of the transmission data of Sect. 15.1.

15.3   Condensed Volume

An important quantity to be characterized is the total condensed volume V_T, which can be done by optical means as long as the distribution of the drop radius remains narrow enough such that a mean radius $\langle R \rangle$ can be characterized. It is thus limited to the first growth stages before new drop families nucleate between drops (see Sects. 6.7 and 6.8).

As noticed in Sect. 6.3.2.1, the condensed volume per unit surface area S_c, $h = V_T / S_c$, can be inferred from $\langle R \rangle$ and ε_2, both quantities that can be measured by optical means (Sects. 15.1 and 15.2). Equation 6.40 indeed relates h to $\langle R \rangle$ and ε_2 through:

$$h = \varepsilon_2 G(\theta_c) \langle R \rangle \tag{15.35}$$

The function $G(\theta_c)$ is involved in the volume evaluation of the drop (Appendix D, Eq. D.7).

Note that when the incident light is sent but not perpendicular to the condensing substrate, the transmitted intensity should become sensitive to the drop volume. While the theory of such a process is still lacking, some experiments (Flura et al., 2009) indeed show the effect.

15.4   Surface Emissivity Change During Condensation

15.4.1   General

Natural dew formation involves cooling by radiative exchange between the substrate and the surrounding atmosphere. With ε_s the atmosphere emissivity and T_a the atmosphere absolute temperature, the radiative power P_i received on ground from the atmosphere, integrated over solid angle and wavelength (Stefan-Boltzmann law, see e.g. Born and Wolf 1999 and Appendix 0), is given by:

$$P_i = \varepsilon_s \sigma T_a^4 \tag{15.36}$$

Here $\sigma = \frac{2k_B^4 \pi^5}{15c^2 h^3} = 5.670 \; 10^{-8}$ W.m^{-2}.K^{-4} is the Stefan–Boltzmann constant. With ε_c the substrate emissivity, which is also the substrate absorptivity according to the Kirchhoff's law of thermal radiation, the power absorbed by the substrate from the atmosphere radiation, P_{ci}, is:

$$P_{ci} = \varepsilon_c P_i = \varepsilon_c \varepsilon_s \sigma T_a^4 \tag{15.37}$$

With T_c the substrate temperature, the power emitted by the substrate, P_r, is given by:

$$P_r = \varepsilon_c \sigma T_c^4 \tag{15.38}$$

Because of the convective heat losses with surrounding air, $T_a - T_c$ never exceeds a few K, which allows the approximation $T_a \approx T_c$ to be made. With the help of Eqs. 15.36 and 15.37, the radiation power balance of the substrate can thus be evaluated as $R_i = P_{ci} - P_r$:

$$R_i = \varepsilon_c \varepsilon_s \sigma T_a^4 - \varepsilon_c \sigma T_c^4 \approx -\varepsilon_c (1 - \varepsilon_s) \sigma T_a^4 \tag{15.38}$$

The substrate emissivity thus plays a key role in the cooling process. However, during condensation the presence of water droplets, with emissivity ε_w, locally modifies the emissivity of the substrate. A new emissivity results, ε_c^*, corresponding to the average of the dry and wet fractions of the substrate. ε_c^* is then a function of the apparent droplet surface coverage seen from above, ε_2^* as evaluated in Eq. 15.28.

15.4.2   Wet Substrate Emissivity

Trosseille et al. (2021b) studied the effect of condensation on the substrate emissivity of two surfaces, with low and high emissivity Fig. 15.6 presents the radiance (corresponding to P_c plus reflected light from the environment, see Eq. 15.39) as received by a thermal camera working in a room in the atmospheric window (7.5–14 μm wavelength). Figure 15.6a corresponds to a smooth surface, S^-, of low emissivity (aluminium-like, $\varepsilon_c = 0.05$) and (Fig. 15.6b) a smooth surface, S^+, of high emissivity (PVC-like, $\varepsilon_c = 0.88$) maintained at temperature $T_c < T_a$. Both substrates are at the same temperature and are treated to exhibit the same wetting properties for water (same θ_c), thus ensuring the same droplet pattern evolution.

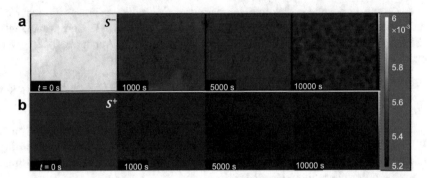

Fig. 15.6 Dropwise condensation observed with a thermal camera in the (7.5–14) μm wavelength range (atmospheric window) on (**a**) a low emissivity substrate S^- and (**b**) a high emissivity substrate S^+ with same conditions of wetting (85° contact angle) and thus dropwise evolution. Experimental conditions are (air) $T_a = 26$ °C, RH = 48%, $T_d = 13.9$ °C and (substrate) $T_c = 7.6$ °C. The mean substrate emissivity varies from (S^-) 0.05 at $t = 0$ s to 0.87 at $t = 10,000$ s and (S^+) from 0.88 at $t = 0$ s to 0.95 $t = 10,000$ s (see Fig. 15.9). The substrate area is ≈ 500 mm², and the image dimension is 12.5 mm × 15 mm. (Adapted from Trosseille et al., 2021b)

Energy conservation means that absorbed radiation (absorptivity ε_c) and reflected radiation (reflectivity r_c) obey:

$$\varepsilon_c + r_c = 1 \tag{15.39}$$

The radiation power P'_c received by the camera is the sum of the radiation emitted by the substrate and reflected from its environment, constituted by the atmosphere plus walls, obstacles, etc. (mean emissivity ε_{ao}), at air temperature T_a, that is:

$$P'_c = \varepsilon_c \sigma T_c^4 + (1 - \varepsilon_c)\varepsilon_{ao}\sigma T_a^4 \tag{15.40}$$

For the dry low emissivity substrate S^- with $\varepsilon_c = 0.05$, the reflectivity is large $r_c^- = 0.95$ (Eq. 15.39) and the radiance received by the camera is large (Fig. 15.6a, $t = 0$ s), corresponding to Eq. 15.40:

$$P_{S^-} \approx \varepsilon_{ao}\sigma T_a^4 \tag{15.41}$$

Concerning the high emissivity substrate S^+ where $\varepsilon_c = 0.88$ and $r_c^+ = 0.12$, the radiance is mainly driven by the radiative power at temperature $T_c < T_a$, corresponding to Eq. 15.40:

$$P_{S^+} \approx \varepsilon_c \sigma T_c^4 \tag{15.42}$$

Because $T_c < T_a$ and $\varepsilon_{ao} \approx \varepsilon_c \approx 1$, the initial image of S^+ (Fig. 15.6b) received by the camera will be darker than the initial image of S^- (Fig. 15.6a).

When condensation proceeds, water drops with emissivity $\varepsilon_w = 0.98$ in the atmospheric window (Downing and Williams, 1975 and Appendix R) modify the substrate emissivity according to (i) their thickness and (ii) the water fraction on the surface as measured by the surface coverage, ε_2. At $t = 1000$ s, the S^- radiance is lowered, corresponding to the nucleation of water drops on the substrate increasing the local emissivity. At $t \geq 5000$ s, circular spots of lower radiance appear, corresponding to the growth of drops as displayed in Chap. 6, with larger (water) emissivities. In contrast, for high emissivity substrates S^+, only a few radiance modifications are observed from $t = 0$ s to 10,000 s, in spite of the presence of condensed droplets.

It is possible to extract the mean substrate emissivity ε_c^* from the images by performing a surface average of the radiance P_c'. According to Eq. 15.40, it comes, assuming $\varepsilon_{ao} \approx 1$:

$$\varepsilon_c^* = \langle \varepsilon_c \rangle = \frac{\langle P_c' \rangle - \sigma T_a^4}{\sigma T_c^4 - \sigma T_a^4} \tag{15.43}$$

Figure 15.9a shows the evolution of ε_c^* during condensation calculated from Eq. 15.43 for both substrate S^- and S^+. They are compared with the emissivity of water in the same wavelength range $\varepsilon_w = 0.98$ (from the imaginary part of the refractive index in Downing and Williams, 1975, see Appendix R). Initially, both substrates exhibit values corresponding to the dry substrates (0.05 for S^- and 0.88 for S^+). Emissivities then increase due to water condensation until reaching values close to pure water emissivity. The slight S^+ decreases within a very short time due to a decrease in substrate temperature, soon compensated by the growing emissivity due to condensing water. As condensation proceeds, the temperature becomes stable, and the mean emissivity of the substrate is only subjected to an increase due to water condensation on the substrate. The dynamics of S^- emissivity obviously shows a much larger amplitude than for curve S^+.

15.4.3 Wet Surface Emissivity Calculation

The dynamics of emissivity can be explained by considering that in the drop pattern evolution, the dry fraction of the substrate exhibits the emissivity ε_c while the wet fraction shows an emissivity which, depending on the thickness of the droplet, goes from ε_c to ε_w. In order to analyze the light transmission and absorption (emission) through a drop, one can schematize (Trosseille et al., 2021b) the drops as cylinders (see Fig. 15.7) with height h the distance substrate-top of drop. For drop contact angles $\theta_c < 90°$, the cylinder radius $r = R$ and for $\theta_c > 90°$, $r = \rho$. From geometry, the height h is written as:

$$h = \rho(1 - \cos\theta_c) = \frac{R}{\sin\theta_c}(1 - \cos\theta_c) \tag{15.44}$$

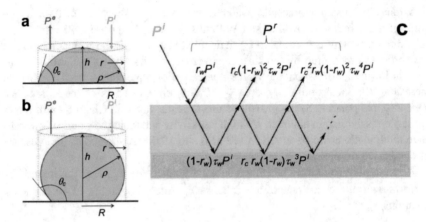

Fig. 15.7 (**a–b**) Drop schematization for (**a**) contact angle $\theta_c < 90°$ and (**b**), $\theta_c > 90°$. (**c**) Radiation propagation through a layer of liquid water (non-zero incidence angle was considered for the sake of clarity). Notations: see text. (Adapted from Trosseille et al., 2021b).

The global radiative properties of emission and reflection of the assembly (drop + substrate) are calculated by considering the trajectory of radiation through a layer of water with thickness h (Fig. 15.7c). With P^i (respectively P^r) the incident (respectively reflected) power, r_w (respectively r_c) the reflection factor of water (respectively substrate), $\tau_w(h)$ the transmission factor through the water layer, one obtains for the reflection coefficient $R = P^r/P^i$ the following geometric series, where p is an integer number:

$$R = r_w + r_c(1 - r_w)^2\tau_w^2 \sum_{p=0}^{\infty} \left(r_c r_w \tau_w^2\right)^p = r_w + \frac{r_c(1 - r_w)^2\tau_w^2}{1 - r_c r_{w\tau_w^2}} \quad (15.45)$$

From Kirchoff's law, one obtains the emissivity of the water layer:

$$\varepsilon_{wc}(h) = 1 - R \quad (15.46)$$

Using the real and imaginary (absorption) parts of the spectral values of the refractive index from Downing and Williams (1975) (see Appendix R) and integrating in the wavelength range [7.5–14] μm and in a larger range [2–50] μm, one can deduce the variation of ε_{wc} with the layer thickness h (Fig. 15.8a) for the S^- substrate. It appears that absorption and thus emissivity becomes is not thickness dependent when typically $h > h_{1,2} \approx 10$ μm. An important result is thus that small drops with $h < 10$ μm cannot be efficiently cooled by radiation deficit.

The effective surface emissivity derives from the substrate wet fraction ε_2 and dry fraction $1 - \varepsilon_2$ as:

$$\varepsilon_{wc} = \varepsilon_2\varepsilon_w + (1 - \varepsilon_2)\varepsilon_c \quad (15.47)$$

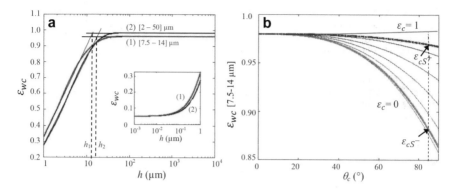

Fig. 15.8 **a** Effective emissivity ε_{wc}(substrate + drop) for two wavelength ranges (1) and (2) with respect to the thickness h of the drop (semi-log plot) for substrate S^- ($\varepsilon_c = 0.05$). The vertical lines determine the characteristic lengths h_1 and h_2 above which the drop exhibits a constant effective emissivity. Inset: effective emissivities for both wavelength ranges (1) and (2) at small drop thickness. **b** Mean effective emissivity at long times (Eq. 15.50) versus drop contact angle θ_c for substrate emissivities ranging from $\varepsilon_c = 1$ to $\varepsilon_c = 0$. Interrupted line: Values corresponding to Figs. 15.6 and 15.7. (From Trosseille et al., 2021b)

The wet fraction in the self-similar regime, $\varepsilon_{2,0}$, is dependent on the drop contact angle (see Sect. 6.5, Eq. 6.78). With θ_c in deg.:

$$\varepsilon_{2,0} = 1 - \frac{\theta_c}{180} \qquad (15.48)$$

At longer times, m self-similar families, which have nucleated at different times, coexist on the surface thus increasing the surface coverage to $\varepsilon_{2,m}$. From Eqs. 6.82 and 15.48, one obtains:

$$\varepsilon_{2,m} = 1 - \left(1 - \varepsilon_{2,0}\right)^m = 1 - \left(\frac{\theta_c}{180}\right)^m \qquad (15.49)$$

Equation 15.47 thus becomes, with the use of Eqs. 15.48, 15.49:

$$\varepsilon_{wc} = \varepsilon_c + (\varepsilon_w - \varepsilon_c)\left[1 - \left(\frac{\theta_c}{180}\right)^m\right] \qquad (15.50)$$

Figure 15.8b represents the variation of ε_{wc} with θ_c for different substrate emissivities and $m = 3$. The values for $\theta_c = 85°$ corresponding to Figs. 15.6 and 15.9 agree well with the reported data at long times. Note that when $\theta_c = 0$, the effective emissivity $\varepsilon_{wc} = \varepsilon_w$ since the substrate is completely wet by water. When $\varepsilon_c > \varepsilon_w$, the effective emissivity increases with angle as the dry fraction increases according to Eq. 15.50. In contrast, when $\varepsilon_c < \varepsilon_w$, the effective emissivity decreases with angle.

Fig. 15.9 a Mean emissivities $\varepsilon_c^* = \langle \varepsilon_c \rangle$ during condensation on S^+ and S^- substrates in the wavelength range 7.5–14 μm (atmospheric window). The vertical lines correspond to the pattern at 1000 s, 5000 s and 10,000 s in Fig. 15.6, and the horizontal lines $(S^-)_0$ and $(S^-)_0$ to emissivities measured on the dry substrates. W is the emissivity of an infinite layer of water. Experimental conditions are the same as in Fig. 15.6. **b** Evolution of the volume per unit area h condensed on S^+ and S^- substrates cooled by a radiation deficit on order 50 W.m^{-2} typical of natural dew formation. The slopes $\dot{h}_{S+} \approx \dot{h}_{S-} \approx 70$ nm.s^{-1}). Other notations: see text. Experimental conditions: T_a = 25.1 °C, RH = 95%. (From Trosseille et al., 2021b)

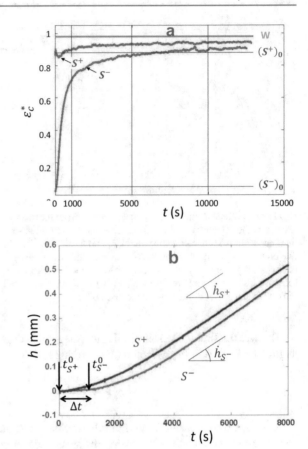

15.4.4 Emissivity and Condensed Mass Evolution

Emissivity evolution clearly has a strong influence on the condensed mass cooled by radiation deficit as encountered in natural dew formation. Trosseille et al. (2021b) performed measurements in a radiative chamber reproducing outdoor conditions (see Trosseille et al., 2021a). Figure 15.9b presents the evolution of condensed volume per surface area h for the S^+ and S^- substrates cooled by a radiation deficit of about 50 W.m^{-2}, a value typical of natural dew formation (see Fig. O.6 in Appendix O). The different volume evolution curves exhibit a similar shape and present two growth regimes: a transient regime characterized by a non-linear increase of the condensed volume, followed by a permanent regime with constant condensation rate. Although the condensation rates in the permanent regime are the same, condensation does not start at the same time for both substrates. S^+ substrate nucleation occurs nearly instantaneously (time t_{S+}^0) due to its large emissivity. In contrast, a longer time is needed to observe nucleation on the S^- substrate and corresponds to a significant delay $\Delta t = t_{S-}^0 - t_{S+}^0$, (~ 1000 s). This delay is the time for water to condense on substrate defects with mean emissivity to reach values (~0.75 in Fig. 15.9a) large enough to ensure efficient cooling.

For low substrate emissivity, the transient regime limits the condensation yield because, for the same time of condensation, the condensed mass is smaller than on high emissivity substrates (Fig. 15.9b). Moreover, by initially limiting the cooling radiative heat transfer, a low emissive substrate can even prohibit condensation if the threshold in drop thickness $h_{1,2}$ is never reached.

Correction to: The Physics of Dew, Breath Figures and Dropwise Condensation

Correction to:
D. Beysens, *The Physics of Dew, Breath Figures and Dropwise Condensation,* **Lecture Notes in Physics 994,**
https://doi.org/10.1007/978-3-030-90442-5

In the original version of the book, belated corrections have been incorporated in Chaps. 3, 4 and 13. The book has been updated with the changes.

The updated orginal version of these chapters can be found at
https://doi.org/10.1007/978-3-030-90442-5_3,
https://doi.org/10.1007/978-3-030-90442-5_4 and
https://doi.org/10.1007/978-3-030-90442-5_13

D. Beysens, *The Physics of Dew, Breath Figures and Dropwise Condensation*,
Lecture Notes in Physics 994, https://doi.org/10.1007/978-3-030-90442-5_16

Glossary—Acronyms

Latin symbols	Units	Definition
a	W m^{-2} K^{-1}	Convective heat transfer coefficient; heat transfer coefficient
a		Prefactor or constant
a	m	Substrate modulation spatial period; plateau width or top size of a microstructure; molecular size
a_a	W m^{-2} K^{-1}	Air heat transfer coefficient
a_d	W m^{-2}K^{-1}	Drop heat transfer coefficient by conduction
a_f	W m^{-2} K^{-1}	Film heat transfer coefficient (pure vapor)
a_i	W m^{-2}K^{-1}	Interface heat transfer coefficient
a_w	W m^{-2} K^{-1}	(Water) vapor mass transfer coefficient
a_{cc}	W m^{-2} K^{-1}	Coating heat transfer coefficient
a_{phob}	m	Hydrophobic stripe width
a_{cc}^{CB}	W m^{-2} K^{-1}	Cassie-Baxter heat transfer coefficient
a_{cc}^{W}	W m^{-2} K^{-1}	Wenzel heat transfer coefficient
a_f^*	W m^{-2} K^{-1}	Film heat transfer coefficient (vapor with non-condensable gas)
$a_{i,d2}^*$	W m^{-2} K^{-1}	Diffusive interface heat transfer coefficient for vapor with non-condensable gases (late stage of growth)
$a_{i,d1}^*$	W m^{-2} K^{-1}	Diffusive interface heat transfer coefficient for vapor with non-condensable gases (early stage of growth)
$a_{i,f}^*$	W m^{-2} K^{-1}	Diffusive interface heat transfer coefficient for vapor with non-condensable gases (film)
A, A_γ		Prefactor or constant
A	m^{-3} s^{-1}	Nucleation rate prefactor

© Springer Nature Switzerland AG 2022

D. Beysens, *The Physics of Dew, Breath Figures and Dropwise Condensation*,
Lecture Notes in Physics 994, https://doi.org/10.1007/978-3-030-90442-5

Latin symbols	Units	Definition
A	m^2	Surface area fraction covered by drops having a radius greater than R
A_c	m^2	Cross-sectional area
A_i		parameters (i=1, 2, 3)
A_i	m^2	i—Voronoi polygon surface area
A_N	s^2	Drop number-drop radius prefactor
A_s	$m^2 \, s^{-1}$	Growth law prefactor
A_s	m^2	Isothermal surface area
A_{pi}	$m^3 \, s^{-1}$	Growth law prefactor
b		Exponent
b	m	Channel microstructure width
b_{phil}	m	Hydrophilic stripe width
B	$mm \, K^{-1}$	Condensation rate—temperature correlation parameter
B		Prefactor or constant
B_i		Parameters ($i = 1, 2$)
B_s	$m^2 \, s^{-1}$	Rate at which the substrate is renewed by drop sliding
B_λ^G	$W \, m^{-2} \, \mu m^{-1} \, sr^{-1}$	Grey body thermal spectral radiance or intensity
B_λ	$W \, m^{-2} \, \mu m^{-1} \, sr^{-1}$	Black body thermal spectral radiance or intensity
Bo		Bond number
$B(m)$		Normalization constant
c		Defects concentration in a structure (fraction of polygons that do not have six sides)
c	m	Channel microstructure depth
c	$m \, s^{-1}$	Light velocity in a medium
c	$m \, s^{-1}$	Velocity of light
c	$kg \, m^{-3}$	Vapor concentration in mass per volume
c_m		(Mass/mass) salt concentration
c_s	$kg \, m^{-3}$	Water vapor concentration at saturation with liquid water in mass per volume
c_s		Maximum volume of vapor absorbed per unit volume of liquid at temperature T_v
c_0	m^{-3}	Volumic concentration of nucleation sites
c_0	m^{-3}	Volumic concentration of nucleation sites
c_0	$kg \, m^{-3}$	Vapor concentration at distance r_0 from the substrate
c_{sc}		Maximum volume of vapor absorbed per unit volume of liquid at temperature T_c
c_{si}	$kg \, m^{-3}$	Water vapor concentration at saturation with ice in mass per volume

Latin symbols	Units	Definition
c_∞	kg m^{-3}	Vapor concentration far from substrate
c^s	kg m^{-2}	Vapor surface concentration in mass per surface area
c_0^s	kg m^{-2}	Vapor surface concentration far from the drop in mass per surface area
c_s^s	kg m^{-2}	Vapor surface concentration at saturation in mass per surface area
c_∞^s	kg m^{-2}	Vapor surface concentration far from the substrate in mass per surface area
$\left(\vec{\nabla} c\right)_d$	kg m^{-4}	Dendrite concentration gradient
C	m^{-1}	Curvature
C		Dimensionless drop velocity
C_{pa}	J Kg^{-1} K^{-1}	specific heat of dry air
C_{riv}		Dimensionless drop velocity at pearling transition
CAH	deg., rd	Contact angle hysteresis
C	m	Oscillation amplitude
Ca		Capillary number
Ca$_{cr}$		Critical capillary number
Ca$_{riv}$		Pearling transition capillary number
C_{pl}	J Kg^{-1} K^{-1}	Liquid water specific heat
C_{ps}	J Kg^{-1} K^{-1}	Solid water specific heat; solid PCM specific heat
C_{pm}	J Kg^{-1} K^{-1}	Humid air specific heat
C_{pv}	J Kg^{-1} K^{-1}	Water vapor specific heat
$C(\psi)$		ψ—angle dependence of light transmission
d	m	Projectile horizontal range
d	m	Interdrop distance
d	m	Molecule diameter
d	m	Fiber diameter
d	m	Microstructure spatial period
d_c	m	Thread length
d_f		Drop pattern fractal exponent
d_f		Fractal exponent of the dry surface contours
d_i	m	$i + 1, i$ interdrop distance
d_m	m	Minimum distance between drops to coalesce
d_M	m	Projectile maximum attainable range
d_0	m	Distance between nucleation sites
D		Prefactor
D	m^2 s^{-1}	Water-air mutual diffusion coefficient
D	m	Projectile displacement

Latin symbols	Units	Definition
D_d		Drop dimensionality
D_h	$\mathrm{m^2\,s^{-1}}$	Hydrodynamic diffusion coefficient
D_s		Substrate dimensionality
D_T	$\mathrm{m^2\,s^{-1}}$	Thermal diffusivity
D_{ps}	$\mathrm{m^2\,s^{-1}}$	Solid PCM thermal diffusivity
D_{Tl}	$\mathrm{m^2\,s^{-1}}$	Liquid thermal diffusivity
e	$\mathrm{J\,m^{-3}}$	Free energy per unit volume
e	$\mathrm{J\,kg^{-1}}$	Internal energy per unit mass
e	m	Thickness; liquid PCM thickness
$\vec{e}_i,\ \vec{e},\ e$	m	Unity vector
E		Prefactor or constant
E	J	Energy
E	$\mathrm{V\,m^{-1}}$	Electric field
E_c	J	Kinetic energy
E_i		Incident light wave amplitude after hitting the substrate-water interface (dimensionless)
E_s		Incident light wave amplitude (dimensionless)
E_s	J	Surface energy
E_v	J	Film viscous dissipation
E_0	$\mathrm{V\,m^{-1}}$	Electric field amplitude
E_{Xi}	$\mathrm{J\,m^{-2}}$	Interface energy per unit surface area for configurations $X = A$, W and $i = 1, 2, 3$
$E_\parallel^i$		Component of the incident field parallel to the plane of incidence (dimensionless)
$E_\perp^i$		Component of the incident field perpendicular to the plane of incidence (dimensionless)
$E_\parallel^t$		Component of the transmitted field parallel to the plane of incidence (dimensionless)
$E_\perp^t$		Component of the transmitted field perpendicular to the plane of incidence (dimensionless)
E_ξ^t		Transmitted field component in direction ξ (dimensionless)
E_η^t		Transmitted field in direction η (dimensionless)
$E\left(\vec{k}\right)$		Diffracted field amplitude in direction $\vec{k}$ (dimensionless)
$E_1(f)$		Fourier transform of 1D surface coverage $\varepsilon_1(\tau)$
f	N	Force
f	$\mathrm{s^{-1}}$	Frequency
f		Surface energy fraction; dry substrate fraction; prefactor
f^*	$\mathrm{s^{-1}}$	Monomers collision rate with critical cluster

Latin symbols	Units	Definition
f_n		Total surface area fractions of interfaces under a drop
$f_0(\theta_c)$		Correction function for non-hemispherical drop growth
F	N	Force; vertical force on a water lens
F	J	Free energy
F	$m^3 \, s^{-1}$	Volume flow rate; exit flux
F_a	N	Attractive force between two droplets; air friction force
F_c	N	Capillary force
F_b	N	Buoyancy force
F_d	N	Driving force
F_g	N	Gravity force
F_s	N	Pinning or retention force
F_v	N	Viscous force; viscous drag
F_M	N	Thermocapillary Marangoni force
F_0	$W \, s^{-n/2} \, m^{-2}$	Heat flux prefactor
F_{01}	$W \, s^{1/2} \, m^{-2}$	Heat flux prefactor
$F_{s,\xi}$	N	Pinning or retention forces in the direction ξ
$F(x)$		Drop jump velocity dependence on radius ratio x of coalescing drops
$F(\theta_c) = \psi(\theta_c)/3$		Form factor relating the drop curvature ρ and volume V
$F(\theta, \theta_e)$		Non-dimensional function of the contact angle for driving force
F_v	N	Viscous force
F_d^*	N	Resulting force acting on a thin band of liquid
g		Polydispersity
g	$m \, s^{-2}$	Earth acceleration constant
g_x	$m \, s^{-2}$	X-component of earth acceleration
g_z	$m \, s^{-2}$	Z-component of earth acceleration
g_∞		Long time asymptotic polydispersity
$g(r)$		Pair function
$g(\theta)$		Groove filling function
$g(x)$		Modified Gaussian function for correction to $F(x)$
$\vec{g}$	$m \, s^{-2}$	Earth's gravitational acceleration
G		Prefactor or constant
G	J	Gibbs free energy
G	Pa	Shear modulus

Latin symbols	Units	Definition
$G(\theta_c)$		Form factor relating the drop cap radius and volume V
$G_6(r)$		Spatial sixfold orientational correlation function
h	m	Free standing drop bridge radius; sessile drop bridge height; drop cap height
h	m	Equivalent film thickness; film thickness; solid PCM layer
h	m	Interdroplet distance
$h = 6.626 \times 10^{-34}$	J s	Planck's constant
h	kJ kg^{-1}	Specific enthalpy of humid air
h	m; mm	Condensed volume per surface area; free standing drop bridge radius; sessile drop bridge height; film height
h_a	kJ kg^{-1}	Specific enthalpy of dry air
h_c	m	Coating thickness
h_i	m	Distance between points $i+1$ and i
h_l	kJ kg^{-1}	Specific enthalpy of liquid water
h_m	m	Minimum distance between points
h_p	m	Pillar height
h_s	kJ kg^{-1}	Condensation volume per unit surface at critical sliding onset
h_v	kJ kg^{-1}	Specific enthalpy of water vapor
h_w	kJ kg^{-1}	Specific enthalpy of water
h_L	m	Film thickness at lower end
h_0	m	Drop cap height
h_3	m	Amplitude of bending of the oil-air interface
$h_{1,2}$	m	Lens heights above (1) and below (2) the contact line; water layer thickness above which emissivity is not thickness dependent
$h.$	hour	Time unit
h_∞	m	Bridge height limit dimension
h^*		Scaled bridge height
$\dot{h} = dh/dt$	m; mm; L m^{-2}	Rate of condensation volume per surface area
$\dot{h}_{d,1}$	m s^{-1}	Drop volume growth rate per surface area (early stage of growth)
$\dot{h}_{d,2}$	m s^{-1}	Drop volume growth rate per surface area (late stage of growth)
$\dot{h}_{\text{phil}}$	m s^{-1}	Condensation rate in volume per surface area
$\dot{h}_{S+}$	m s^{-1}	S^+ condensed volume rate per surface area
$\dot{h}_{S-}$	m s^{-1}	S^- condensed volume rate per surface area
$h_s(V_0)$	m	Specific enthalpy of solid water (ice)

Latin symbols	Units	Definition
H	kJ	Enthalpy of humid air; interface amplitude oscillation
H_c	Pa^{-1}	Henry's constant of solubility of vapor in liquid at temperature T_c
H_v	Pa^{-1}	Henry's constant of solubility of vapor in liquid at temperature T_v
$H(R)$		Drop Maxwellian distribution function
$H(\tau)$		Heaviside-like function of τ
i		Integer number
$\mathbf{I}$	Pa	Unity tensor
I		Scattered light intensity (dimensionless)
I_f		Interference term (dimensionless)
I_p		Individual particles scattered intensity (dimensionless)
I_λ	W m^{-2} sr^{-1}	Spectral radiance or intensity
I_0		Zero diffraction order intensity (dimensionless)
I_λ^A	W m^{-2} sr^{-1}	Absorbed spectral radiance or intensity
I_λ^E	W m^{-2} sr^{-1}	Emitted spectral radiance or intensity
I_λ^R	W m^{-2} sr^{-1}	Reflected radiance or intensity
I_λ^T	W m^{-2} sr^{-1}	Transmitted radiance or intensity
I_i^*	W m^{-2} sr^{-1}	Transmitted radiance or intensity emitted by layer z_i
j		Integer number; $\sqrt{-1}$
j_c	kg m^2 s^{-1}	Condensation diffusive flux
j_e	kg m^2 s^{-1}	Evaporation diffusive flux
$\vec{j}$	kg m^2 s^{-1}	Monomer diffusive flux
J		Light transmission factor
k		Viscosity corrective factor for plot micropatterned substrate
k	m^{-1}	Scattering wavevector
k, k_P	m s^{-1}	Drop radius growth rate
k_a	N m^{-1} s	Air friction coefficient
k_i	m s^{-1}	Drop radius growth rate near geometry of i-type (p: plane; c: corner; e: edge; G: inside groove)
k_c	m^{-1}	Plateau-Rayleigh critical wavenumber
k_m	m^{-1}	Plateau-Rayleigh fastest growing wavenumber
$k_B = 1.38 \times 10^{-23}$	J K^{-1}	Boltzmann constant
k_C	m s^{-1}	Drop radius growth rate near a corner; Plateau-Rayleigh critical wavenumber
k_E	m s^{-1}	Drop radius growth rate near an edge

Latin symbols	Units	Definition
k_G	m s^{-1}	Drop radius growth rate inside a groove
k_M	m^{-1}	Upper cutoff wavenumber; maximum scattered intensity wavevector
k_0	m^{-1}	Incident light wavenumber
$k_{x,y}$	m s^{-1}	Growth rate of elliptical drop x-major and y-minor axis
k^*		Drop shape numerical constant in the pinning force expression
$\tilde{k}$	m^{-1}	Complex wave number
$\vec{k}$	m^{-1}	Scattering wavevector
K		Angle factor
K		Constant of contact line friction
l	m	Sessile drop bridge width; domain mean length; drop contact line length; ray path length in a drop; slab length; light penetration depth
l_c	m	Capillary length
l_m	m	Dendrite maximum length
l_v	m	Viscous length
l_{mfp}	m	Mean free path length
l_∞	m	Bridge width limit dimension
l^*	m	Scaled bridge width; reduced dendrite length
L	m	Length; thickness; system length; fiber length; slab thickness
L	m	Free standing drop bridge diameter
L	m	Sessile drop bridge length
L	J kg^{-1}	Condensation/evaporation latent heat
L_c	m	Capillary rise in a channel
L_c, L_v, L_w	J kg^{-1}	(Water) condensation/evaporation latent heat
L_i	m	Interface length between hydrophilic and hydrophobic regions
L_p	J kg^{-1}	PCM melting latent heat
L_s	J kg^{-1}	(Water) solidification latent heat
L_∞	m	Bridge length limit dimension
L^*	m	Mean interdrop distance at the sliding onset
L^*		Scaled bridge length
m		Integer number; $(1 - A)$ power law exponent; parameter of the Maxwellian distribution $H(R)$
m	kg	Mass; condensed mass; projectile mass
m_a	kg	Air mass
m_d	kg	Droplet mass
m_l	kg	Liquid mass

Latin symbols	Units	Definition
m_s	kg m^{-2}	Mass per unit surface area
m_v	kg	Vapor mass
m_w	kg m^{-2}	Vapor mass per unit surface (density length, precipitable water)
m_{phil}	kg	Condensed mass on hydrophilic regions
m_{phob}	kg	Condensed mass on hydrophobic regions
$\dot{m}$	kg s^{-1}	Condensed mass per unit time
$\dot{m}_s$	kg m^{-2} s^{-1}	Rate of condensation per unit surface area
$\dot{m}_{phil}$	kg s^{-1}	Condensation rate on hydrophilic regions
M	kg	Mass
M	g	Molar mass
M		Radius mismatch
$M_a = 29$	g	Dry air molar mass
$M_{NaCl} = 58.44$	g	Nacl salt molar mass
$M_v, M_{H_2O} = 18.02$	g	Water molar mass
Ma		Marangoni number
n		Integer number; number of moles or molecules; number of drop coalescences; real refractive index
n_a		Number of air moles or molecules; air refractive index
n_s	m^{-2}	Surface density of nucleation sites
n_s		Substrate refractive index
n_w		Water refractive index
n_λ		Spectral real refractive index
n_0	m$^{-(Ds + Dd)}$	Amplitude of the distribution number of drops per drop volume and substrate surface area
n_v, n_{H_2O}		Number of water moles or molecules
n_{NaCl}		Number of salt moles
$n_{H_2O}^s$		Number of water moles to saturate the solution for given n_{Nacl} At given temperature
n^*		Reduced drop number; critical cluster molecule number
n_h^*		Critical cluster molecule number (heterogeneous case)
$\tilde{n}$		Complex refractive index
$\tilde{n}_\lambda$		Spectral complex refractive index
$\vec{n}$		Unit vector normal to drop surface
$n(R), n_c(R)$	m$^{-(Ds + 1)}$	Distribution number of drops per drop radius and substrate surface area
$n(V)$	m$^{-(Ds + Dd)}$	Distribution number of drops per drop volume and substrate surface area

Latin symbols	Units	Definition
$n_c(R)$	m^{-Ds+1}	Line distribution number of drops per drop volume and substrate surface area at the sliding onset
$n_s^1(V_0)$	$1 - D_s$	Line distribution number of drops per drop volume and substrate surface area at the sliding onset
$n_s^1(V_0)$	$1 - D_s$	Line distribution number of drops per drop volume and substrate surface area at the sliding onset
N		Integer value; number of drops; number of drops whose radius is larger than R; number of coalescences; number of points; number of holes
N	m^{-3}	Number of drops of critical size per volume
N_s	m^{-2}	Surface density of nucleation sites
$N_A = 6.022 \times 10^{23}$	$mole^{-1}$	Avogadro constant
N_0		Number of drops; number of plateaus covered by a drop
$\dot{N} = dN/dt$	$m^{-3} \, s^{-1}$	Nucleation rate of drops of critical size per volume
$\dot{N}_0$	$m^{-3} \, s^{-1}$	Threshold number of nucleation rate of drops of critical size per volume
$N(R_g)$		Number of holes whose radius of gyration is less than R_g
Oh		Ohnesorge number
p	Pa	Pressure
p		Integer number; number of time steps; ratio base radius of a water lens / radius of curvature of the lower section of lens; length of filled grooves in units of substrate length
p_a	Pa	Dry air partial pressure
p_c, p_L	Pa	Laplace capillary pressure
p_g	Pa	Gas pressure
p_h	Pa	Hydrostatic pressure
p_i	Pa	Pressure at interface; pressure at distance r_i
p_l	Pa	Liquid pressure
p_m	Pa	Atmospheric pressure
p_s	Pa	(Water) saturation vapor pressure
p_t		Probability that a site is not visited by a random walker
p_v	Pa	(Water) vapor pressure
p_η	Pa	Viscous pressure
p_{si}	Pa	Saturation water vapor pressure for ice

Latin symbols	Units	Definition
p_{s0}	Pa	Saturation water vapor pressure for saturated salty water
p_{ss}	Pa	Saturation water vapor pressure for unsaturated salty water
p_{ext}	Pa	Exterior pressure
p_{iner}	Pa	Dynamical inertial pressure
p_{int}	Pa	Interior pressure
p_0	Pa	Nucleation pressure
p_∞, $p_{v\infty}$	Pa	(Water) vapor pressure far from substrate
P	W	Light intensity
$\mathbf{P}$	Pa	Stress tensor
P_a	W m^{-2}	Total radiative heat flux from atmosphere
P_a	W	Heat flux from air
P_c	W	Cooling heat flux
P_r	W m^{-2}	Condenser radiative power per surface area
P_i	W m^{-2}	Atmosphere radiative power per surface area
P_B	W m^{-2}	Black body radiative power per surface area
P_{ci}	W m^{-2}	Atmosphere radiative power absorbed by the substrate
P_{Li}	W	Latent heat flux for water ($i = w$) or PCM ($i = p$)
P_0	W	Light intensity amplitude
P^i	W	Incident power
P^r	W	Reflected power
P_B^G	W m^{-2}	Grey body radiative power per surface area
$P_c{}'$	W	Radiation power received by camera
P_i^*	W m^{-2}	Apparent atmosphere radiative power per surface area
Pe$_M$		Mass diffusion Peclet number
Pe$_T$		Thermal Peclet number
Pr		Prandtl number
$P(\tau)$		Stepwise function
$\vec{q}$, q	W m^{-2}	Heat flux per unit surface
q	W m^{-2}	Heat power per unit surface
q	m^2 s^{-1}	Volumic flux per unit length
q_a	W m^{-2}	Air heat flux per unit surface
q_c	W m^{-2}	Substrate heat flux per unit surface
q_i	W m^{-2}	Interface heat flux per unit surface; latent heat flux per unit surface
q_l	W m^{-2}	Liquid film heat flux per unit surface
q_m	m^{-1}	Minimum wave number

Latin symbols	Units	Definition
q_M	m^{-1}	Maximum wave number
Q	W	Heat flux; total heat flux
Q_d	W	Drop heat flux
Q_i	W	Interfacial heat flux
Q_p	J	Solid PCM sensitive heat
Q_L	J	Heat amount
r, r_i	m	Distance; distance from drop center; bridge negative curvature radius; cylinder radius; droplet radius
r	$J\,kg^{-1}\,K^{-1}$	Humid air specific constant
r		Ratio of the total surface area to the projected area of the solid; Wenzel roughness factor
$r_a = 287$	$J\,kg^{-1}\,K^{-1}$	Dry air specific constant
r_c		Condensing surface reflectivity
r_m		Reflection coefficient at the vacuum/medium interface
r_p	m	Pore radius
r_s	$s\,m^{-1}$	Bulk surface resistance
r_t		Fraction of spins that never flipped
$r_v = 462$	$J\,kg^{-1}\,K^{-1}$	Water vapor specific constant
r_w		Water reflectivity
$r_{m\lambda}$		Spectral reflection coefficient at the vacuum/medium interface
r^*		Reduced distance between drop centers
r_c^+		S^+ high emissivity substrate reflectivity
r_c^-		S^- low emissivity substrate reflectivity
$r(\phi)$	m	Ellipse equation versus angle ϕ
$\vec{r_i}$	m	Position of drop (i)
R	m	Drop cap radius; arithmetic average of roughness absolute values
R		Reflection coefficient
$R = 8.314$	$J\,mole^{-1}\,K^{-1}$	Molar gas constant
R, R_{cd}	$W^{-1}\,K\,m^2$	Conductive thermal resistance
R_a	m	Roughness arithmetic average of absolute values
R_c	m	Critical drop radius
R_g	m	Radius of gyration
R_i	m	Individual drop radius in a pattern
R_i	$W^{-1}\,m^2\,K$	Interface thermal resistance
R_i	$W\,m^{-2}$	Radiation deficit or cooling energy per unit surface

Latin symbols	Units	Definition
R_p	m	Maximum roughness peak
R_q	m	Roughness root mean squared value
R_t	m	Maximum roughness height
R_v	m	Maximum roughness depth
R_λ		Spectral reflectivity
R_x	m	Elliptical drop major axis
R_y	m	Elliptical drop minor axis
R_0	m^3	Critical drop cap radius at sliding onset; initial drop cap radius; typical drop cap radius
$R_{0,1,2}$	m	0, 1, 2 drop radius
R_{cv}	$K\ W^{-1}$	Convective thermal resistance
R_{no}	m	Typical minimum filling height between pillars
$R_{0,1x}$	m	Elliptical drop major axis amplitudes
$R_{0,1y}$	m	Elliptical drop minor axis amplitudes
R'	m	Composite drop radius; drop inner transparent region
R_h^*	m	Critical drop cap radius (heterogeneous case)
$R^{(p)}$		P-weighted moment of R
Ra		Rayleigh number
Re		Reynolds number
RH	%	Relative humidity
Re		Reynolds number
RH	%	Relative humidity
$\vec{R}$	m	Composite drop position
$\vec{R_i}$	m	Ith drop position; radiative heat flux
$\vec{R_{1,2}}$	m	1, 2 drop position
s_c	m^2	Puddle cross-sectional surface area
s_j	m	Drop sliding distance
s_v	m^{-3}	Number of target particles per unit volume
s_1	m^2	Illuminated area per drop
S	m^2	Surface area; hole surface
S_c	m^2	Condensation surface area
S_d	m^2	Drop interface area; dendrite condensing surface area
S_e	m^2	Evaporation surface area
S_E	m^2	Edge condensation surface
S_T	m^2	Total surface wetted by drops
S_1	m^2	Illuminated area
S_{LG}	m^2	Liquid surface in contact with gas

Latin symbols	Units	Definition
S_{LS}	m^2	Liquid surface in contact with solid
S_{phil}	m^2	Hydrophilic regions surface area
S_{phob}	m^2	Hydrophobic regions surface area
$S_{os/w}$	$J\ m^{-2}$	Oil on solid spreading coefficient in the presence of water
$S_{ow/a}$	$J\ m^{-2}$	Oil on water spreading coefficient in the presence of air
$S_{wo/a}$	m^2	Water on oil spreading coefficient in the presence of air
S_{oa}^l	m	Oil-air interface area of a lens of water on oil in presence of air
S_{wo}^l	m	Oil-water interface area of a lens of water on oil in presence of air
S_{wa}^l	m	Air-water interface area of a lens of water on oil in presence of air
$S_{LS/G}$	$J\ m^{-2}$	Liquid on solid spreading coefficient in the presence of gas
S'	m^2	Composite drop wetted surface
S^*		Superhydrophilic surface area ratio
S^+		High emissivity surface
S^-		Low emissivity surface
S_M^*		Highest efficiency superhydrophilic surface area ratio
SR		Supersaturation ratio
SR^*		Critical supersaturation ratio
SR_i^*		Critical supersaturation ratio for substance i
SR_0^*		Critical supersaturation ratio
Sc		Schmidt number
$S(R_g)$	m^2	Variation of holes surfaces with respect to hole gyration radius R_g
$S_N(R_g)$	m^2	Total surface of N holes with a radius of gyration R_g Smaller than a given value
t, t_i	s	Time
t_c	s	Typical time; lag time; molecular velocity scale time
t_j	s	Lag time for flow to occur ($j = 0, 1, 2$)
t_m	s	Melting time
t_M	s	Time of maximum projectile height
t_{c0}	s	Lag time
$t_\parallel$		Light transmission factor parallel to the plane of incidence

Latin symbols	Units	Definition
$t_\perp$		Light transmission factor perpendicular to the plane of incidence
t^*	s	Typical evolution time; reduced time
t_f	s	Lag or final time; projectile time of flight
t_0	s	Initial time
t_{S+}^0		S^+ lag time
t_{S-}^0		S^- lag time
T	°C or K	Temperature
T_a	°C or K	Air temperature; atmosphere temperature
T_b	°C or K	Fiber base temperature
T_c	°C or K	Temperature of the condensing surface; liquid temperature
T_d	°C or K	Dew point temperature near ground
T_i	°C or K	Interface temperature
T_m	°C or K	Melting temperature
T_s	K	Saturation temperature; sky temperature
T_t	°C or K	Fiber tip temperature
T_v	°C or K	Liquid temperature
T_λ		Spectral transmissivity or transmittance
T_0	°C or K	PCM lower boundary temperature
T_∞	°C or K	Gas or vapor temperature far from the substrate
T_{cc}	°C or K	Coating temperature
T_{cc}	°C or K	substrate coating temperature
u	m s^{-1}	Flow velocity; drop velocity; object velocity
u_d	m s^{-1}	Drop growth velocity
u_x	m s^{-1}	x-component of projectile velocity
u_z	m s^{-1}	z-component of projectile velocity
u_0	m s^{-1}	Drop jump velocity; initial projectile velocity
u_1	m s^{-1}	Film flow velocity inside posts
u_2	m s^{-1}	Film flow velocity outside posts
u_{cl}	m s^{-1}	Contact line velocity
$u_{s,m}$	m s^{-1}	Monolayer spreading front velocity
$u_{s,n}$	m s^{-1}	Nanofilm spreading front velocity
u_{0m}	m s^{-1}	Minimum drop jump velocity
U		Scaling velocity
$\vec{U}, U$	m s^{-1}	Velocity; film flow velocity
U_m	m s^{-1}	Maximum air velocity
U_x	m s^{-1}	x-component of flow velocity
U_z	m s^{-1}	z-component of flow velocity

Latin symbols	Units	Definition
U_0	m s^{-1}	Flow velocity far from the substrate; drop sliding velocity
v	m s^{-1}	Moving particle velocity
v_c	m^3	Volume of dripping drop
v_m	m^3	Molecule volume
v_n	m s^{-1}	Contact line velocity normal to line
v_p	m^3	PCM melted volume
v'	m^3 kg^{-1}	Humid air specific volume
v^*	m s^{-1}	Reduced dendrite velocity
V	m^3	Volume; drop volume
V_d	m^3	Drop volume
V_i	m^3	Water ($i = w$) or PCM ($i = p$) volume
V_l	m^3	Liquid volume
V_v	m^3	Vapor volume
V_w	m^3	Water condensed volume
V_T	m^3	Total condensed volume
V_0	m^3	Critical drop volume at sliding onset; lens volume
V_1	m^3	Stored water volume; stationary film volume above posts
$V_{1,2}$	m^3	Air-water (1) and oil-water (2) fraction volumes in a water lens
V_{ml}	m^3	Liquid molar volume
V_{mv}	m^3	Vapor molar volume
V_{0c}	m^3	Sliding drop maximum volume
$V_{T,E}$	m^3	Condensed volume near an edge
V^*	m^3	Critical volume; reduced volume
V^l	m^3	Water lens volume
$\dot{V} = dV/dt$	m^3.s^{-1}	Drop volume growth rate
w	m	Rivulet width
w	g kg^{-1}; kg kg^{-1}	Humidity ratio; mass mixing ratio; absolute humidity; specific humidity; moisture content
w_s	g kg^{-1}; kg kg^{-1}	Humidity ratio at saturation
W	J	Energy; mechanical work; nucleation total energy; liquid-vapor potential energy difference
W_h	J	Heterogeneous nucleation total energy
W^V	J	Energy gain
W^S	J	Surface energy
W^*	J	Critical cluster work of formation
W_h^S	J	Heterogeneous case surface energy
W_h^V	J	Heterogeneous case volume energy

Latin symbols	Units	Definition
W^S	J	Surface energy
W_i^S	J	Surface energy for water in contact with air and substance i
W_h^*	J	Critical heterogeneous nucleation energy
$W_{w/o}^S$	J	Surface energy for water drop engulfed by oil
$W_{wo/a}^S$	J	Surface energy for water drop oil-air contact
$W_{ws/a}^S$	J	Surface energy for water drop solid-air contact
$W_{ws/o}^S$	J	Surface energy for water drop solid-oil contact
x	m	Spatial coordinate parallel to substrate and/or gravity
x		Flux power law exponent; ratio of coalescing drops radii
x_g	m	Center-of-mass abscissa
x_i	m	Abscissa of element i
x_s	m^{-3}	Water molecule concentration in air at saturation
X_a		Mole fraction of air
X_v		Mole fraction of water vapor
y	m	Spatial coordinate parallel to substrate
y_g	m	Center-of-mass ordinate
y_i	m	Ordinate of element i
Y_a		Mass fraction of dry air
Y_v		Mass fraction of water in humid air
z	m	Spatial coordinate perpendicular to substrate
z_i	m	Layer elevation
z_M	m	Projectile maximum height
z_{MM}	m	Projectile highest achievable height
Z		Zeldovich factor

Greek symbols	Units	Definition
α	deg. or rd	Drop jump direction with horizontal; substrate tilt angle with horizontal; light angle of incidence; direction with horizontal of projectile initial velocity
α		Proportionality constant; prefactor; $N\left(R_g\right)$ power law exponent
α	m^{-1}	Attenuation coefficient or absorptivity
α_λ	m^{-1}	Spectral absorptivity
α_m		Monomer sticking coefficient
α_{cr}		Incidence critical angle for full internal reflection
β	deg. or rd	Light angle of refraction
β	K^{-1}	Air volumetric thermal expansion coefficient

Greek symbols	Units	Definition
β		Drop growth law exponent; plot dissipation factor; ellipse shape factor
β_p	K^{-1}	Volumetric thermal expansion at constant pressure p
χ		Ratio of transparent-opaque drop regions
χ_s	m^{-3}	Water molecule concentration at saturation
δ	m	Radius of inhibited condensation region; contact line width; drop-dendrite spacing
δ		$S(R_g)$ power law exponent
δ_c	m	Coating thickness
δ_F	m	Free convection hydrodynamic boundary layer
δ_H	m	Hydrodynamic boundary layer
δ_N	m	Radius of nucleation inhibited condensation region
δ_T	m	Thermal boundary layer
δ_0	m	Typical interdrop distance
δ_{FC}	m	Free convection diffuse boundary layer
δ_{FF}	m	Forced flow diffuse boundary layer
$\delta X, \Delta X$		Difference of variable X
δ^*	m	Thickness of inhibited condensation region
δ_{is}^*	m	Thickness of ice- salty water condensation region
δ_{iw}^*	m	Thickness of ice-liquid water inhibited condensation region
δ_{ws}^*	m	Thickness of liquid water-salty water inhibited condensation region
Δ	m	Drop center mean displacement
$\overrightarrow{\Delta_n}, \Delta_n$	m	Drop center mean displacement after n coalescences
ε		Grey body emissivity
ε	m s	Drop radius growth rate
$d\varepsilon$	m	Distance between two isothermal surfaces
ε_c		Condensing surface emissivity
ε_s		Atmosphere emissivity
ε_s		Sky emissivity
ε_w		Water emissivity
ε_λ		Spectral emissivity
ε_E		Border drop surface coverage
ε_1		1D surface coverage fraction
ε_2		2D surface coverage fraction
ε_{ao}		Environment emissivity
ε_{wc}		Water layer emissivity
$\varepsilon_{2,m}$		m self-similar families, 2D surface coverage fraction
$\varepsilon_{2,n}$		Long time asymptotic 2D surface coverage fraction for n coexisting patterns

Greek symbols	Units	Definition
$\varepsilon_{2,0}$		1 self-similar family, 2D surface coverage fraction
$\varepsilon_{2,\infty}$		Long time asymptotic 2D surface coverage fraction
ε_c^*		Wet condensing surface emissivity
ε_i^*		Emissivity of layer z_i
ε_s^*		Apparent sky emissivity
ε_2^*		Apparent 2D surface coverage fraction
ϕ		Angle; azimuthal angle; inclination angle
ϕ_s, ϕ		Surface fraction of liquid-solid contact
ϕ_0	kg s^{-1}	Monomer flux
ϕ^*		Corrected inclination angle
γ	t.b.d.	Γ density
γ	deg., rd	Angle
γ		Drop growth law exponent; ratio $\sin\alpha_{cr} = n_a/n_w$
γ	Pa K^{-1}	Psychrometric constant
Γ		Gamma function
Γ	t.b.d.	Extensive variable
η	Pa s	Dynamic or shear viscosity
η	m	Axis perpendicular to the direction of polarization ξ of the incoming light wave
η		Power law exponent of the orientational order correlation function $G_6(r)$
η_0	Pa s	Oil dynamic or shear viscosity
η^*	Pa s	Enhanced dynamic or shear viscosity
φ	deg. or rd	Azimuthal angle
φ		Pillar density
κ	m^2	Coefficient of permeability
κ		Imaginary part of the complex refractive index
κ_λ		Spectral imaginary part of the complex refractive index
λ	W m^{-1} K^{-1}	Thermal conductivity
λ		Homomorphic factor
λ		Prefactor or constant
λ	W m^{-1} K^{-1}	Thermal conductivity
λ	m	Wavelength
λ_a	W m^{-1} K^{-1}	Air thermal conductivity
λ_c	W m^{-1} K^{-1}	Coating thermal conductivity
λ_l	W m^{-1} K^{-1}	Liquid thermal conductivity
λ_m	W m^{-1} K^{-1}	Humid air thermal conductivity
λ_m	m	Plateau-rayleigh fastest growing wavelength
λ_p	W m^{-1} K^{-1}	Pillar thermal conductivity

Greek symbols	Units	Definition
λ_w	W m^{-1} K^{-1}	Water thermal conductivity
λ_0	m	Incident light wavelength
λ_{cs}	W m^{-1} K^{-1}	Solid PCM thermal conductivity
λ_{sp}	W m^{-1} K^{-1}	Solid PCM thermal conductivity
$\lambda(\theta)$		Form factor relating the lens radius r and volume v
μ	m^{-1}	Coefficient of light absorption
μ_l	J mole^{-1}	Chemical potential of liquid
μ_v	J mole^{-1}	Chemical potential of v
ν	m^2 s^{-1}	(Water) kinematic viscosity
ν		Radius asymmetry
θ	deg., rd	Angle; polar angle
θ	°C	Temperature
θ		Surface dry fraction exponent; volumic drop distribution exponent
θ_a	deg., rd	Advancing contact angle
θ_a	°C	Ambient air temperature
θ_b	deg., rd	Oil-air interface angle with horizontal
θ_c	deg., rd	Drop contact angle
θ_d	deg., rd	Drop mean dynamic contact angle
θ_d	°C	Dew point temperature
θ_e	deg., rd	Drop cap internal angle
θ_i	deg., rd	Drop contact angle on substance i
θ_n	deg., rd	(n) interface contact angle
θ_r	deg., rd	Receding contact angle
θ_s	°C	Dew point temperature
θ_w	°C	Wet bulb temperature
θ_A	deg., rd	Drop contact angle on substance A
θ_B	deg., rd	Drop contact angle on substance B
θ_W	deg., rd	Wenzel contact angle
θ_{wa}	deg., rd	Water-air angle with horizontal
θ_{wo}	deg., rd	Water-oil angle with horizontal
θ_{aB}	deg., rd	Drop advancing contact angle on substance B
θ_{CB}	deg., rd	Cassie-Baxter contact angle
θ_{Sa}	deg., rd	Advancing contact angle
θ_{Sr}	deg., rd	Receding contact angle
θ_{ij}	deg., rd	Contact angle of substance i in contact with j
θ_{rA}	deg., rd	Drop receding contact angle on substance A
$\theta_{os/a}$	deg., rd	Oil-solid-air contact angle
$\theta_{os/w}$	deg., rd	Oil-solid-water contact angle

Greek symbols	Units	Definition
$\theta_{ws/a}$	deg., rd	Water-solid-air contact angle
$\theta_{1,2,3,4}$	deg., rd	Water-solid-liquid cyclohexane ridge contact angles
θ^*	deg., rd	apparent angle
$\theta_s{}'$	deg., rd	Non equilibrium contact angle
θ_c^*	deg., rd	Cassie-Baxter / Wenzel critical contact angle
ρ	Kg m^{-3}	Density or mass per unit volume
ρ	m	Drop curvature radius; lens upper segment radius of curvature; equivalent spherical drop radius of a lens; object typical size
ρ_0	m	Initial drop curvature radius
ρ_a	kg m^{-3}	Air density
ρ_c	m	Critical drop curvature radius when coalescences start
ρ_g	kg m^{-3}	Gas density
ρ_i	kg m^{-3}	Ice density
ρ_l	kg m^{-3}	Liquid density
ρ_m	kg m^{-3}	Humid air density
ρ_s	m	Smaller drop radius
ρ_s	kg m^{-3}	Salt density
ρ_v	kg m^{-3}	(Water) vapor density
ρ_w	kg m^{-3}	(Water) liquid density
ρ_0	kg m^{-3}	Oil density
ρ_0	m	Curvature radius of the departing drop by gravity
ρ_1	m	Curvature radius of drop 1
ρ_2	m	Curvature radius of drop 2
ρ_{lp}	kg m^{-3}	Liquid PCM density
ρ_{ps}	kg m^{-3}	Solid PCM density
ρ_{wa}	m	Lens radius of curvature of the upper section
ρ_{wo}	m	Lens radius of curvature of the lower section
$\rho_{1,2}$	m	Surface radii of curvature
ρ_w^s	kg m^{-2}	(Water) liquid density
ρ^*	m	Drop critical radius of curvature; reduced radius
ρ_h^*	m	Critical radius (heterogeneous case)
$\rho_s{}' = d\rho_s/dc_m$	kg m^{-3}	Salt density ρ_s derivative w.r.t salt concentration c_m
$\dot{\rho}$	m s^{-1}	Drop growth rate; drop radius growth velocity
$\dot{\rho}_1$	s^{-1}	Reduced drop overall growth rate
$\sigma = 5.670\times10^{-8}$	W m^{-2} K^{-4}	Stefan-Boltzmann constant
σ	J m^{-2}	Surface tension
σ	m^2	Cross-sectional surface area
σ	m	Drop radius standard deviation
σ_c		Condensation coefficient

Greek symbols	Units	Definition
σ_e		Evaporation coefficient
σ_{al}	J m^{-2}	Surface tension liquid-air
σ_{oa}	J m^{-2}	Surface tension oil-air
σ_{wa}	J m^{-2}	Surface tension water-air
σ_{wo}	J m^{-2}	Surface tension water-oil
σ_{AW}	J m^{-2}	Substance A-vapor surface tension
σ_{BW}	J m^{-2}	Substance B-vapor surface tension
σ_{LG}	J m^{-2}	Liquid-gas surface tension
σ_{LS}	J m^{-2}	Liquid-solid surface tension
σ_{LV}	J m^{-2}	Liquid-vapor surface tension
σ_{SG}	J m^{-2}	Solid-gas surface tension
Σ	Pa	Viscous stress tensor
τ	s	Damping time; typical time; reduced or dimensionless time; sweeping period
τ		Power law exponent of the volume distribution of sliding drops; light transmittancy
τ_w		Water layer transmission factor
τ_R		Drop complex transmittancy
τ_λ		Spectral transmittancy
τ_{ic}	s	Drop inertial-capillary timescale
τ^*	s	Reduced time; reduced damping time
$\widetilde{\tau}_R$		Angle dependent drop complex transmittancy
ω	rd s^{-1}	(Light) angular frequency
$\omega_{i/j}$		Ratio of surface energies i / j
Ω	sr	Solid angle
ξ	m	Spatial coordinate parallel to substrate; surface boundary layer; cross-over length; capillary waves amplitude; axis in the direction of polarization of the incoming light wave
$\delta\xi$	m	Roughness parallel to surface
$\xi(\theta)$		Form factor
ψ		Degree of saturation
ψ	deg., rd	Angle
$\psi(\theta_c) = 3F(\theta_c)$		Form factor relating the drop curvature radius ρ and volume V
$\psi_i(\theta_c)$		Form factor corresponding to state i = I, II, III
$\psi_{cm}(v)$		Inertial center-of-mass function of radius asymmetry
$\psi_{ss}(v)$		Contact line dissipation center-of-mass function of radius asymmetry
ζ		Relation drop cap radius-drop radius of curvature-contact angle

Greek symbols	Units	Definition
ζ, ζ^*	m	Spatial coordinate perpendicular to substrate; diffuse boundary layer thickness
$\delta\zeta$	m	Roughness perpendicular to surface

The Clausius-Clapeyron Relation

A

The Clausius-Clapeyron relation gives the slope of the saturation vapor pressure curve. One can derive it according to the elegant demonstration from MIT (2016) based on the Carnot cycle (Fig. A.1).

One considers a Carnot cycle ABCD in the Clapeyron phase diagram. The phase change A-B (condensation) gives a mass $m = \rho_l V_l$ of liquid. The following heat amount must be removed from the vapor:

$$Q_L = mL_v \tag{A.1}$$

where L_v ($\approx 2.5 \times 10^6$ J kg^{-1} at 0 °C) is the latent heat of vaporization. The thermal efficiency of the ABCD Carnot cycle is

$$\frac{dW}{Q_L} = \frac{dT}{T_d} \tag{A.2}$$

where dW is the area enclosed by the rectangle ABCD:

$$dW \approx dp(V_v - V_l) \tag{A.3}$$

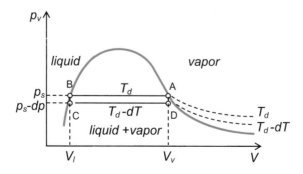

Fig. A.1 Carnot cycle

© Springer Nature Switzerland AG 2022
D. Beysens, *The Physics of Dew, Breath Figures and Dropwise Condensation*,
Lecture Notes in Physics 994, https://doi.org/10.1007/978-3-030-90442-5

It follows, frm Eqs. A.1 to A.2, that

$$\frac{dp}{dT} = \frac{mL_v}{(V_v - V_l)T_d} \tag{A.4}$$

As $V_v \gg V_l$ and using the equation of ideal gases for water to express V_v as a function of p_s and T (see Chap. 1, Eq. 1.3), one readily obtains the Clausius-Clapeyron relation:

$$\frac{dp}{dT} = \frac{\rho_v L_v}{T_d} = \frac{p_v L_v}{r_w T_d^2} \tag{A.5}$$

The second term is obtained from the equation of ideal gases (Chap. 1, Eq. 1.3), with ρ_v the vapor or humid air density.

Relation Between Vapor and Heat Transfer Coefficients

B.1   Vapor Transfer

Vapor condensation is the result of water molecules diffusing in the air to the condensing surface in a concentration gradient of extent ζ, the boundary layer thickness where the Peclet number is smaller than unity (see Sect. 2.2, Eq. 2.12). The condensation rate dm/dt (where m is condensed mass) in the steady-state where the thin-film approximation holds (see Sect. 6.1.2) can be written following Eqs. 6.13 and 6.14 as

$$\frac{dm}{dt} = \rho_w S_c \frac{dh}{dt} = S_c \frac{D(c_\infty - c_s)}{\zeta} \tag{B.1}$$

Here h is the equivalent water film thickness, S_c the condenser surface area, ρ_w the liquid water density, ζ the diffuse boundary later thickness and $(c_\infty - c_s)$ the supersaturation concentration counted in water vapor mass per volume. When expressed as a function of water vapor pressure $p_s(T_c)$ (saturation water vapor pressure at condenser temperature T_c) and $p_v(T_a)$ (water pressure in the humid air above the condenser), it becomes from Eq. 1.45 in Sect. 1.2.10, with ρ_a the air density:

$$\frac{dm}{dt} \approx \frac{r_a}{r_v} \rho_a S_c \frac{D}{P_m \zeta}(p_v - p_s) \tag{B.2}$$

The water vapor transfer coefficient, a_w, enters the transfer equation Eq. 1.42:

$$\frac{dm}{dt} = S_c a_w (p_v - p_s) \tag{B.3}$$

From Eqs. B.2 to B.3 it readily comes

$$a_w = \frac{r_a}{r_v} \frac{\rho_a D}{P_m \zeta} \left(= \frac{0.745 D}{P_m \zeta}; \text{S.I. units} \right) \tag{B.4}$$

© Springer Nature Switzerland AG 2022
D. Beysens, *The Physics of Dew, Breath Figures and Dropwise Condensation*,
Lecture Notes in Physics 994, https://doi.org/10.1007/978-3-030-90442-5

B.2   Heat Transfer

Heat losses at a condensing surface thermally isolated from below comes from heat conduction in a thermal boundary layer of extent δ_T, where the thermal Peclet number is smaller than unity (see Sect. 2.2, Eqs. 2.13 and 2.16). Heat flux R_{he} can be thus written as

$$R_{he} = \lambda_m S_c \frac{T_a - T_c}{\delta_T} \tag{B.5}$$

where λ_m is the humid air thermal conductivity. It corresponds to the heat transfer equation, with heat transfer coefficient a:

$$R_{he} = a S_c (T_a - T_c) \tag{B.6}$$

Expressing in Eq. B.5 λ_m as a function of ρ_m, humid air specific heat C_{pm} and thermal diffusivity $D_T = \frac{\lambda_m}{\rho_m C_{pm}}$, it comes

$$R_{he} = \frac{\rho_m C_{pm} D_T}{\delta_T} S_c (T_a - T_c) \tag{B.7}$$

From Eqs. B.6 to B.7, one readily deduces

$$a = \frac{\rho_m C_{pm} D_T}{\delta_T} \tag{B.8}$$

B.3   Ratio of Transfer Coefficients

It is now straightforward to express the vapor and heat transfer coefficients from Eqs. B.4 to B.8

$$a_w = \frac{r_a}{r_v} \frac{1}{C_{pm} p_m} \frac{D}{D_T} \frac{\delta_T}{\zeta} a \tag{B.9}$$

Following Sect. 2.2 the ratio $\delta_T D / \zeta D_T \approx 1$. Equation B.9 thus becomes

$$a_w \approx \frac{r_a}{r_v} \frac{1}{C_{pm} p_m} a = \frac{0.6212}{C_{pm} p_m} a \tag{B.10}$$

This relationship is quite similar to Eq. 1.43 in Sect. 1.2.8 concerning the wet-bulb temperature and the evaporation mass transfer coefficient. With the psychrometer "constant" (Eq. 1.39)

$$\gamma = \frac{C_{\text{pm}} p_m}{0.6212 L_v} \left(\approx 65 \text{Pa.K}^{-1}\right) \tag{B.11}$$

one finds

$$a_w = \frac{a}{\gamma L_v} \tag{B.12}$$

Navier-Stokes Equations

C

The Navier-Stokes differential describes the motion of viscous fluids substances. The equations are named after the French engineer and physicist Claude-Louis Navier and Anglo-Irish physicist and mathematician George Gabriel Stokes. They basically express the conservation of mass and momentum for Newtonian fluids. These equations stem from Newton's second law and the description of the stress in the fluid as the sum of a diffusing viscous term and a pressure term.

The conservation law for a given extensive variable Γ, whose density is γ submitted to velocity U and containing a volumic term of production S can be expressed as

$$\frac{\partial \gamma}{\partial t} + \vec{\nabla}.\left(\gamma \vec{U}\right) = S \tag{C.1}$$

The most used equations are those concerning the Eulerian formulation, where the problem is stationary and the frame of reference is fixed. The above equation is used for the conservation of volumic mass or density ρ, momentum ρU and total energy ρE:

Mass conservation

$$\frac{\partial \rho}{\partial t} + \vec{\nabla}.\left(\rho \vec{U}\right) = 0 \tag{C.2}$$

Momentum conservation

$$\frac{\partial \rho \vec{U}}{\partial t} + \vec{\nabla}.\left(\rho \vec{U}\vec{U}\right) = \vec{\nabla}.\mathbf{P} + \rho\vec{g} = -\vec{\nabla}p + \vec{\nabla}.\mathbf{\Sigma} + \rho\vec{g} \tag{C.3}$$

Energy balance

$$\frac{\partial \rho E}{\partial t} + \vec{\nabla}\cdot\left(\rho E\vec{U}\right) = \vec{\nabla}\cdot\left(\mathbf{P}\cdot\vec{U}\right) + \rho\vec{g}\cdot\vec{U} + \vec{\nabla}\cdot\vec{q} + \vec{\nabla}\cdot\vec{R}_l \tag{C.4}$$

© Springer Nature Switzerland AG 2022
D. Beysens, *The Physics of Dew, Breath Figures and Dropwise Condensation*,
Lecture Notes in Physics 994, https://doi.org/10.1007/978-3-030-90442-5

In the above equations, pressure is p, $\vec{g}$ (x, y, z, t) is gravity and **P** stands for stress tensor (pressure tensor). The quantity $\vec{q}$ is the heat flux per surface area and $\overrightarrow{R_i}$ is the radiative heat flux per surface area. With **Σ** the viscous stress tensor and **I** the tensor unity

$$\mathbf{P} = \mathbf{\Sigma} - \mathbf{I} \tag{C.5}$$

E holds for the total energy per unit mass. Expressing the internal energy per unit mass by e, one can write E as

$$E = e + \frac{1}{2}\left|\vec{U}\right|^2 \tag{C.6}$$

To the above equations the particular boundary conditions must be added.

Volume of a Spherical Cap

D

The volume of a spherical cap of sphere radius ρ, contact radius R and contact angle θ_c can be calculated from geometry (Fig. D.1). One considers a horizontal slab of length l, thickness dz, at ordinate z leading to the following equations:

$$R = \rho \sin \theta$$
$$z = \rho \cos \theta$$
$$l = \rho \sin \theta$$
$$dz = -\rho \sin \theta d\theta \qquad \text{(D.1)}$$

The volume of the slab is

$$dV = \pi l^2 dz \qquad \text{(D.2)}$$

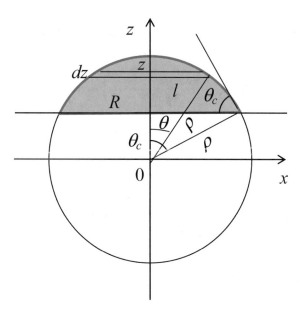

Fig. D.1 Spherical cap

© Springer Nature Switzerland AG 2022
D. Beysens, *The Physics of Dew, Breath Figures and Dropwise Condensation*,
Lecture Notes in Physics 994, https://doi.org/10.1007/978-3-030-90442-5

Integration on the cap gives the cap volume

$$V = \int_{\theta_c}^{0} \pi \rho^2 \sin \theta^2 (-\rho \sin \theta) d\theta \tag{D.3}$$

or

$$V = \pi \rho^3 \int_{0}^{\theta_c} \sin \theta^3 d\theta = \pi \rho^3 \left[\int_{0}^{\theta_c} \sin \theta \left(1 - \cos \theta^2 \right) d\theta \right] \tag{D.4}$$

The integration gives $V = \pi \rho^3 [-\cos \theta]_0^{\theta_c} + \frac{1}{3} [\cos \theta^3]_0^{\theta_c}$ or

$$V = \pi \rho^3 \left(\frac{2 - 3\cos\theta_c + \cos\theta_c^3}{3} \right) = \pi \rho^3 F(\theta_c) = \frac{\pi \rho^3}{3} \psi(\theta_c) \tag{D.5}$$

with

$$F(\theta_c) = \left(\frac{2 - 3\cos\theta_c + \cos\theta_c^3}{3} \right) = \frac{(2 + \cos \theta_c)(1 - \cos \theta_c)^2}{3} \tag{D.6}$$

$$\psi(\theta_c) = 2 - 3\cos\theta_c + \cos\theta_c^3 = (2 + \cos \theta_c)(1 - \cos \theta_c)^2$$

When expressed as a function of R, the volume eventually becomes, using Eqs. D.1 and D.5

$$V = \pi R^3 \left(\frac{2 - 3\cos\theta_c + \cos\theta_c^3}{3\sin\theta_c^3} \right) = \pi R^3 G(\theta_c) \tag{D.7}$$

where

$$G(\theta_c) = \frac{2 - 3\cos\theta_c + \cos\theta_c^3}{3\sin\theta_c^3} \tag{D.8}$$

Figure D.2 reports the variation of V/R^3 and V/ρ^3 with respect to contact angle θ_c ranging from $0°$ to $180°$. They both exhibit the same value for $\theta_c = 90°$ Fig. D.2.

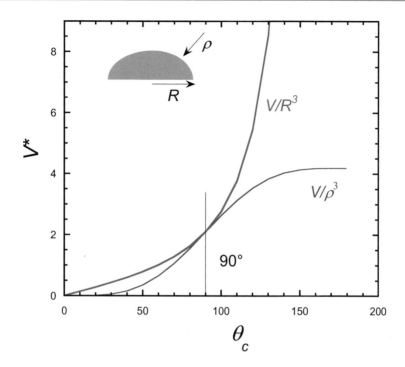

Fig. D.2 Reduced volume $V^* = V/R^3$ and V/ρ^3 with respect to contact angle θ_C

Wetting and Super Wetting Properties

E

E.1 Ideal Surface

E.1.1 Young-Dupré Relation

The shape of a liquid drop plunged in gas and which rests on a solid surfais governed by three interactions, namely liquid (L), solid (S) and gas (G). These interactions are found at the microscopic level between the molecules of the three media. They can be modeled at the macroscopic level by means of the capillary forces related to different surface energy or surface tension σ_{LS}, σ_{LG} and σ_{SG}, which act at the interfaces L-S, L-G and S-G.

At equilibrium (Fig. E.1), the forces acting per unit length along the three phases contact line should be zero. The components of net force in the direction along each of the interfaces are given by the Young-Dupré relation, where θ is the liquid contact angle:

$$\sigma_{SG} = \sigma_{LS} + \sigma_{LG} \cos \theta_c \tag{E.1}$$

High energy solid surfaces (strong σ_{LS}) correspond to hydrophilic substrates ($\theta < 90°$) while low energy surfaces (weak σ_{LG}) correspond to hydrophobic substrates ($\theta_c > 90°$), as shown in Fig. E.2. In order to describe the tendency of the

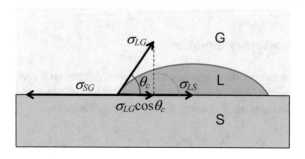

Fig. E.1 Sessile drop and forces acting at the three phases contact line

© Springer Nature Switzerland AG 2022
D. Beysens, *The Physics of Dew, Breath Figures and Dropwise Condensation*,
Lecture Notes in Physics 994, https://doi.org/10.1007/978-3-030-90442-5

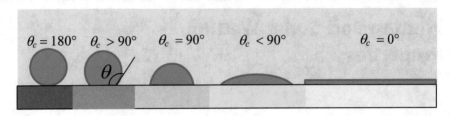

Fig. E.2 Water on different solids. From $\theta_c = 180°$ (perfect non-wetting) to $\theta_c = 90°$, the substrate is hydrophilic, corresponding to weak solid/liquid energy. From $\theta_c = 90°$ to $\theta_c = 0$ (perfect wetting = film), the substrate is hydrophilic, corresponding to strong solid/liquid energy

liquid phase to spread (complete wetting case) on a substrate phase, a spreading coefficient S can be defined. It distinguishes the two different regimes of wetting by measuring the difference between the surface energy (per unit area) of the substrate when dry and wet, that is

$$S_{LS/G} = \sigma_{SG} - (\sigma_{LS} + \sigma_{LG}) = \sigma_{LG}(\cos \theta_c - 1) \qquad (E.2)$$

The second term makes use of Eq. E.1. When $S > 0$, the liquid perfectly wets the surface and the second part of Eq. E.2 has no physical solution for θ_c ($\cos \theta_c > 1$). When $S < 0$, wetting is partial with the contact angle θ_c defined by Eq. E.1.

E.1.2   Energy of Adhesion

The Gibbs free energy of adhesion ΔG can be defined as the reversible work to separate solid and liquid initially at contact and to bring them at a distance where they no longer interact (Fig. E.3):

$$\Delta G = -\pi R^2(\sigma_{SG} + \sigma_{LG} - \sigma_{LS}) = -\pi R^2 \sigma_{LG}(1 + \cos \theta_c) \qquad (E.3)$$

The second equality makes use of the Young-Dupré equation, Eq. E.1.

E.1.3   Laplace-Young Equation

The Laplace-Young equation connects the pressure difference across the gas–liquid interface to its shape. It is merely the account of the normal stress balance at an interface that is treated as a surface, that is without thickness:

$$\Delta p = p_{ext} - p_{int} = -\sigma_{LG}\vec{\nabla}.\vec{n} = -\sigma_{LG}C \qquad (E.4)$$

where Δp is the pressure difference (exterior pressure p_{ext} minus interior pressure p_{int}) across the fluid interface (Laplace pressure), $\vec{n}$ is the unit vector normal to

Fig. E.3 Work ΔG per unit surface area A to separate solid and liquid initially at contact

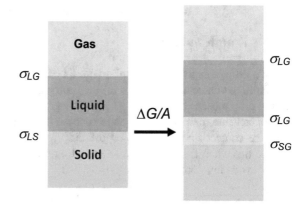

the surface, pointing out of it, and C is the mean surface curvature. With ρ_1 and ρ_2, the surface radii of curvature, $C = \rho_1^{-1} + \rho_2^{-1}$ and Eq. E.4 becomes

$$\Delta p = -\sigma_{\text{LG}}\left(\frac{1}{\rho_1} + \frac{1}{\rho_2}\right) \tag{E.5}$$

E.1.4 Capillary Length

On horizontal substrates, the shape of the drop can be altered such that a spherical cap becomes a pancake. This is due to the non-negligible effect of hydrostatic pressure p_h on height ~ρ as compared to capillary pressure p_c.

Let us consider for simplicity a spherical drop. With ρ_l the liquid water density and g the earth acceleration constant, the hydrostatic pressure corresponding to a column of liquid of 2ρ height is $p_h = 2\rho_l g\rho$. The Laplace-Young equation Eq. E.5 relates the capillary pressure p_c to the radius of curvature of the drop such as $p_c = 2\sigma_{\text{LG}}/\rho$. The hydrostatic pressure will begin to overcome the capillary pressure, and then gravity dominates over capillary effects, when $p_h = 2\rho_l g\rho > p_c = 2\sigma_{\text{LG}}/\rho$, or

$$\rho > l_c = \sqrt{\frac{\sigma}{\rho_l g}} \tag{E.6}$$

The typical height l_c where capillary and hydrostatic effects balance is the capillary length. It corresponds for water in the air to ≈ 2.7 mm with $\sigma \approx 73$ mN m^{-1}, $g = 9.81$ m s^{-2} and $\rho_l = \rho_w = 10^3$ kg m^{-3}.

E.2   Rough and Micro-patterned Surfaces

E.2.1   Rough Substrate

When a drop is deposited on a solid with a syringe (Fig. E.4), one can distinguish two limiting contact angles when (a) the drop expands due to the syringe push, or (b) the drop recedes due to syringe pull. In each case, the drop volume varies with the contact line, remaining immobile because it is pinned on the substrate defects until the contact line moves when the drop reaches the limiting contact angles. In case (a), when the drop starts to spread on the solid, its limiting contact angle is the advancing contact angle θ_a. In case (b), when the drop starts to recede on the solid, its contact angle is the receding contact angle θ_r. When the syringe is removed, the equilibrium contact angle θ_c is non-unique and there is an infinite number of equilibrium states between θ_a and θ_r:

$$\theta_r < \theta_c < \theta_a \qquad (E.7)$$

Note, therefore, that the current definition of contact angle taken as

$$\theta_c = \frac{\theta_a + \theta_r}{2} \qquad (E.8)$$

is arbitrary.

The dynamics of spreading or receding have been the subject of many studies (see, e.g., Bonn et al. 2009). Here only the work related to dew formation (see Chap. 6) is considered. During condensation, a drop grows most of the time with an advancing contact angle θ_a, then coalesces with a neighboring drop and forms a composite drop with non-equilibrium contact angle θ. Relaxation to a circular

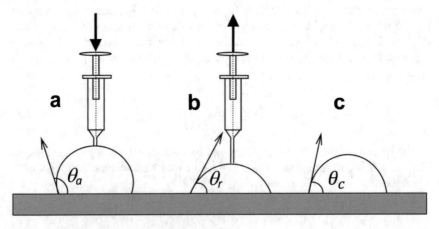

Fig. E.4 Examples of advancing (**a**) and receding (**b**) contact angles on a non-ideally smooth solid substrate. The equilibrium contact angle (**c**) is between θ_r and θ_a

drop is triggered by the force related to the difference in the receding angle θ_r and occurs at times 10^5-10^6 larger than the typical viscous dissipation time in the internal liquid flow (see Sect. 5.3.5). This spectacular slowing down is due to the high dissipation during the contact line motion (see Andrieu et al. 2002; Narhe et al. 2004b, 2008).

E.2.2 Micro-patterned Substrate. Cassie-Baxter and Wenzel States

A patterned substrate can be considered as being a substrate with well-controlled defects (Fig. E.5). A drop can wet the substrate materials and fill the microstructures (e.g. the drop is firmly pushed onto the substrate) and in that case it is said to be in the Wenzel (W) or penetration state. In this state, the drop is highly pinned. When it does not wet the substrate materials (e.g. when the drop is gently pushed towards the substrate), it can sit at the top of the microstructure with composite contact and is said to be in a Cassie-Baxter (CB) or air pocket state. Wenzel state amplifies the hydrophilic properties of the materials while Cassie-Baxter increases its hydrophobic features. In both cases, the apparent drop contact angle is thus changed from the materials smooth surface angle θ_c into a larger (CB) or smaller (W) apparent angle θ_W or θ_{CB}.

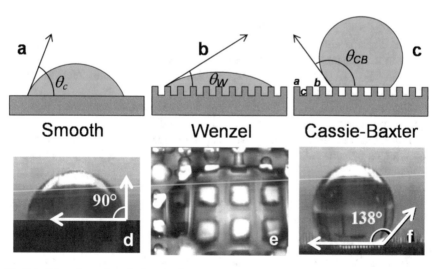

Fig. E.5 Water drops deposited on (**a, d**) a smooth surface with contact angle θ_C ($= 90°$ in (**d**)) and on (**b, c, e, f**) a patterned surface (square pillars with side $a = 32° \mu m$, spacing $b = 32° \mu m$, thickness $c = 62 \mu m$) with same materials as in (**a, d**). (**b**—side view) and (**e**—top view) correspond to the Wenzel-penetration state with contact angle θ_W 0°. (**c, f**) Cassie-Baxter state with composite contact angle $\theta_{CB} = 138°$. The Wenzel state is the minimum energy state as the critical contact angle $\theta_c^* = 106° > \theta_c = 90°$ (see text). Adapted from Narhe and Beysens (2007)

The apparent angle is classically calculated from energy arguments (see, e.g., Patankar 2003; Erbil and Cansoy 2009; Milne and Amirfazli 2012) where what only matters is the wetting contact area between the drop and the microstructures.

In Wenzel's approach, where the liquid fills the microstructures, the apparent contact angle, θ_W, is given by

$$\cos \theta_W = r \cos \theta_c \qquad (E.9)$$

where r is the surface roughness defined as the ratio of the actual area to the projected area and θ_c is the equilibrium contact angle of the liquid drop on the flat surface with the same materials.

The apparent contact angle in the CB state, θ_{CB}, derives from the relation

$$\cos \theta_{CB} = \phi_s \cos \theta_c + \phi_s - 1 \qquad (E.10)$$

Here ϕ_s is the area fraction of the liquid–solid contact.

The equilibrium state depends on whether, for a given θ_c, r, ϕ_s, the minimum energy is in a W or CB state. A critical contact angle θ_c^* corresponding to equaling Cassie-Baxter and Wenzel angles, Eqs. E.9 and E.10, can be defined as

$$\cos \theta_c^* = \frac{\phi_s - 1}{r - \phi_s} \qquad (E.11)$$

When $\theta_c > \theta_c^*$, the most stable state is CB, whereas when $\theta < \theta_c^*$, it is the W state (Lafuma and Quéré 2003). In the latter situation, if $\theta_c > 90°$, the drop can be in a metastable state, with the existence of an energy barrier to make the transition CB to W (Ishino et al. 2004; Patankar 2004; Dupuis and Yeomans 2005; Porcheron and Monson 2006).

Milne and Amirfazli (2012) define f_n as the total areas of each solid–liquid or liquid–vapor interface under the drop, with the condition that $\sum f_n = 1$ and each of the n interfaces under the drop with contact angle θ_n can have a total area in excess of its planar area. In that case, a full form of the Cassie equation can be written as

$$\cos \theta_c = \sum_n f_n \cos \theta_n \qquad (E.12)$$

It applies to all forms of wetting (e.g. Cassie–Baxter, Wenzel and Young). Equation E.12 can thus predict the contact angle for an arbitrary substrate (rough or smooth, chemically homogeneous or heterogeneous, wet with air remaining or completely wetted). A heterogeneous surface can thus be rough and can be made wet by a liquid with or without air remaining under the drop. The presence of air enhances the heterogeneous interface under the drop.

As an example, let us consider the pattern of Fig. E.5 composed of square pillars with side $a = 32$ μm, spacing $b = 32$ μm and thickness $c = 62$ μm. The roughness factor is $r = 1 + \left[4ac/(a+b)^2\right] = 2.93$, giving $\theta_W = 0$. The solid–liquid interface area fraction $\phi_s = a^2/(a+b)^2 = 0.25$, giving $\theta_{CB} = 138°$. The critical angle (Eq. E.11) is $\theta_c^* = 106°$, thus larger than $\theta_c = 90°$, which means that the W state corresponds to the minimum energy state. It is not guaranteed, however, that a drop will always exist in this lower-energy state because the state in which the drop will settle depends on how the drop is formed. In general, the transition from a higher-energy CB state to a lower-energy W state is possible only if the required energy barrier is overcome by the drop (e.g. by slightly pressing the drop, releasing the drop from some height etc.).

Patterning can be at different scales such as the Lotus leaf. This leaf is well-known for self-cleaning because water drops exhibit a very large contact angle and are only very weakly pinned, thus rolling off easily from the rough surface. This surface exhibits two typical patterning length scales. One is at the nano-scale ("nanograss") and the other is at the micro-scale (Fig. E.6). The apparent contact angle is enhanced and the minimum energy state is always CB (Patankar 2004). Thus if by accident (or during a condensation process, see Chap. 10 and Chen et al. 2007) a drop comes in the W state, it will soon reach the CB state. Double roughness structure pillars help amplify the apparent contact angle and, more importantly, it also helps in making the composite drop energetically much more favorable, ensuring that a composite drop is always formed.

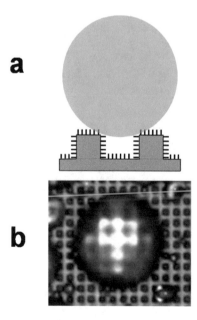

Fig. E.6 Double-scale roughness geometry. **a** Schematic side view. **b** Top view. Adapted from Chen et al. (2007)

Coalescence Bridge Geometry

The bridge which appears during the coalescence between two spherical cap of contact radius R_0 and contact angle θ_c obeys a number of relations that can be determined from geometry (Fig. F.1).

F.1   Radius of Curvature r in the X, Z Plane

The sides of the triangle OIO′ obey the relation:

$$\overline{OO'}^2 = \overline{OI}^2 + \overline{IO'}^2 \tag{F.1}$$

With $\overline{OO'} = r + \rho$, $\overline{OI} = \rho \sin \theta_c$ and $\overline{IO'} = \overline{IB} + \overline{BC} + \overline{CO'} = \rho \cos \theta_c + h + r$, one obtains

$$r = \frac{h}{2} \frac{2\rho \cos \theta_c + h}{\rho(1 - \cos \theta_c) - h} \tag{F.2}$$

F.2   Large Contact Angle

When θ_c is near 90°, simplifications can be made (Eddi et al. 2013) by neglecting the curvature and writing $h = \overline{BC} \approx \overline{BD}$. It corresponds to confound bow and circle rope to lead to the simplification $\overline{DG} = r \sin \theta_c \approx r$. In the triangle OGJ, one can thus write

$$\overline{OJ}^2 + \overline{JG}^2 = \overline{GO}^2 \tag{F.3}$$

or $(\rho \sin \theta_c - r)^2 + (\rho \cos \theta_c + h)^2 = \rho^2$, which gives, after development

$$r \approx \rho \sin \theta_c - \left[\rho^2 - (h + \rho \cos \theta_c)^2 \right]^{1/2} \tag{F.4}$$

© Springer Nature Switzerland AG 2022
D. Beysens, *The Physics of Dew, Breath Figures and Dropwise Condensation*,
Lecture Notes in Physics 994, https://doi.org/10.1007/978-3-030-90442-5

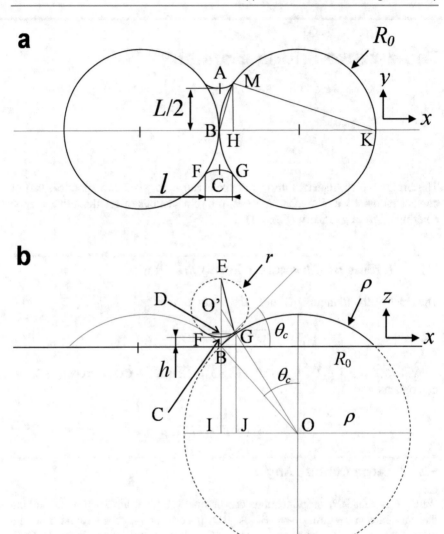

Fig. F.1 Coalescence geometry of two sessile drops of the same radius. **a** View from above, **b** side view, **c** enlargement of the central part of (**b**)

F.3   Small Contact Angle

Equation F.2 can be developed for small θ_c, using $h/\rho \ll 1$:

$$r \approx \frac{h}{2} \frac{2\rho + h}{\rho \theta_c^2 / 2 - h} \approx 2 \frac{h}{\theta_c^2} \tag{F.5}$$

The parameters of interest for the bridge geometry are the length $L = \text{CA}$, width $l = \text{FG}$ and height $h = \text{BD}$. A relation can be drawn between L and l from the right-angle triangle BAK (front view (a) in Fig. F.1) where $\overline{\text{BA}}^2 = \overline{\text{HM}}^2 = \text{BH.HK}$, giving $\left(\frac{L}{2}\right)^2 = \frac{l}{2}\left(2R_0 - \frac{l}{2}\right) \approx R_0 l$ or

$$L \approx 2(R_0 l)^{1/2} \tag{F.6}$$

A relation between h and l can be found (from side view (b) in Fig. F.1) from $\text{BC} \approx \text{BD} \approx (\text{FG}/2) \tan \theta_c$, that is

$$h \approx \frac{l}{2} \tan \theta_c \approx \frac{l}{2} \theta_c \tag{F.7}$$

Mean Values

The definition of the arithmetic mean values of variable R_i is classically written as

$$\langle R \rangle = \frac{\sum_{i=1}^{N} R_i^1}{\sum_{i=1}^{N} R_i^0} = \frac{\sum_{i=1}^{N} R_i}{N} \tag{G.1}$$

$$\langle R^2 \rangle = \frac{\sum_{i=1}^{N} R_i^2}{N} \tag{G.2}$$

$$\dots$$

$$\langle R^p \rangle = \frac{\sum_{i=1}^{N} R_i^p}{N} \tag{G.3}$$

with N is the number of data.

In some cases, it is advantageous to generalize the writing of the first term of Eq. G.1 by calculating other weighted averages, such as

$$\text{length-weighted } \left\langle R^{(2)} \right\rangle = \frac{\langle R^2 \rangle}{\langle R \rangle} \tag{G.4}$$

$$\text{area-weighted } \left\langle R^{(3)} \right\rangle = \frac{\langle R^3 \rangle}{\langle R^2 \rangle} \tag{G.5}$$

etc.

$$\dots$$

$$\left\langle R^{(p)} \right\rangle = \frac{\langle R^p \rangle}{\langle R^{p-1} \rangle} \tag{G.6}$$

© Springer Nature Switzerland AG 2022
D. Beysens, *The Physics of Dew, Breath Figures and Dropwise Condensation*,
Lecture Notes in Physics 994, https://doi.org/10.1007/978-3-030-90442-5

Table G.1 Different expressions used for mean values by assuming a standard normal distribution of variables R_i. The approximation corresponds to a Taylor expansion. The parameter $g = \sigma/R$ is polydispersity (from the standard normal distribution 2019)

order p	Mean value $\langle R^p \rangle$	$\frac{\langle R^{(p)} \rangle}{\langle R \rangle} = \frac{1}{\langle R \rangle}\frac{\langle R^p \rangle}{\langle R^{p-1} \rangle}$	$\frac{\langle R^{2(p)} \rangle}{\langle R \rangle^2} = \frac{1}{\langle R \rangle^2}\frac{\langle R^p \rangle}{\langle R^{p-2} \rangle}$
1	$\langle R \rangle$	1	–
2	$\langle R \rangle^2 + \sigma^2$	$1 + g^2$	–
3	$\langle R \rangle^3 + 3\langle R \rangle \sigma^2$	$1 + 2\frac{g^2}{1+g^2}$	$1 + 3g^2$
		$\approx 1 + 2g^2$	
4	$\langle R \rangle^4 + 6\langle R \rangle^2 \sigma^2 + 3\sigma^4$	$1 + 3g^2\frac{(1+g^2)}{1+3g^2}$	$1 + g^2\frac{5+3g^2}{1+g^2}$
		$\approx 1 + 3g^2$	$\approx 1 + 5g^2$

It can be also interesting to calculate the mean values of R^2 by using the same kind of averages, e.g.,

$$\left\langle R^{2(p)} \right\rangle = \frac{\langle R^p \rangle}{\left\langle R^{p-2} \right\rangle} \tag{G.7}$$

In particular, the area-weighted average can be written as

$$\left\langle R^{2(4)} \right\rangle = \frac{\langle R^4 \rangle}{\langle R^2 \rangle} \tag{G.8}$$

The relationship between the $\left\langle R^{(p)} \right\rangle$ and $\langle R \rangle$ are given in Table G.1 assuming a Gaussian, normal distribution of variables R_i

$$f\big((R|\langle R \rangle, \sigma^2\big) = \frac{1}{\sqrt{2\pi\sigma^2}}\exp\left[-\frac{(R - \langle R \rangle)^2}{2\sigma^2}\right] \tag{G.9}$$

with standard deviation σ and variance

$$\sigma^2 = \left\langle (R - \langle R \rangle)^2 \right\rangle = \left\langle R^2 \right\rangle - \langle R \rangle^2 \tag{G.10}$$

When $\frac{\sigma^2}{\langle R \rangle^2} \ll 1$, which is the case for a droplet pattern in the coalescence stage (see Sect. 7.2.2) where the square of polydispersity $g^2 = \sigma^2/\langle R \rangle^2 \approx 0.04$ (see Eqs. 6.36 and 6.37):

$$\left\langle R^{(p)} \right\rangle \approx \langle R \rangle \tag{G.11}$$

and

$$\left\langle R^{2(p)} \right\rangle \approx \left\langle R^2 \right\rangle \tag{G.12}$$

However, p must remain small (Table G.1) for Eqs. G.6 and G.7 to remain valid.

Note that the averages in Table G.1 can be obtained using the expression of variance.

Concerning $\left\langle R^{(2)} \right\rangle$, from Eq. G.10 it readily comes

$$\left\langle R^{(2)} \right\rangle = \frac{\langle R^2 \rangle}{\langle R \rangle} = \langle R \rangle \left[1 + \frac{\sigma^2}{\langle R \rangle^2} \right] \tag{G.13}$$

The same kind of relation can be found with $\left\langle R^{(3)} \right\rangle$, developing $(\langle R \rangle - \langle R \rangle)^3$ as $R^3 - 3R^2 \langle R \rangle + 3R \langle R \rangle^2 - \langle R \rangle^3$ and taking the average:

$$\left\langle (R - \langle R \rangle)^3 \right\rangle = \left\langle R^3 \right\rangle - 3 \langle R \rangle^2 \langle R \rangle + 2 \langle R \rangle^3 \tag{G.14}$$

Since the power of $(R - \langle R \rangle)^3$ is odd, its average is zero. Therefore, Eq. G.14 can be written as

$$.\left\langle R^3 \right\rangle - 3 \langle R \rangle^2 \langle R \rangle + 2 \langle R \rangle^3 = 0 \tag{G.15}$$

or

$$\left\langle R^{(3)} \right\rangle = \frac{\langle R^3 \rangle}{\langle R^2 \rangle} = \langle R \rangle \left[3 - 2 \frac{\langle R \rangle^2}{\langle R^2 \rangle} \right] = \langle R \rangle \left[1 + \frac{2}{\langle R \rangle^2 + \sigma^2} \right] \tag{G.16}$$

At order p, one obtains (Standard normal distribution 2019):

$$\left\langle (R - \langle R \rangle)^p \right\rangle = f(x) = \begin{cases} 0 \text{ if } p \text{ is odd} \\ \sigma^p (p-1)!! \text{ if } p \text{ is even} \end{cases} \tag{G.17}$$

Here !! denotes the double factorial, i.e., the product of all numbers from p to 1 that have the same parity as p. This double factorial increases rapidly with p, thus limiting the range of applicability of Eq. G.11.

Drop Pearling Transition

A drop of liquid sliding on an inclined plane can exhibit different shapes according to its velocity U. The latter can be tuned by varying for instance the plane tilt angle with horizontal. In this process, the corresponding non-dimensional number is the capillary number Ca, which represents the relative importance of viscous drag forces versus surface tension forces:

$$Ca = \frac{\eta U}{\sigma} \tag{H.1}$$

Here σ is the liquid–air surface tension and η is the liquid dynamic viscosity. The velocity of a drop of volume V results from the balance between the effective drop weight $\rho g V \sin \alpha$, the viscous drag on order of $\eta V^{1/3} U$ and the pinning force on order $\sigma \Delta(\cos \theta_c) V^{1/3}$. The quantity $\Delta(\cos \theta_c) = \cos \theta_r - \cos \theta_a$ is the difference between advancing and receding cosines angles (see Sect. 5.2.3.1, Eqs. 5.24–5.26 and Sect. 13.1.2, Eq. 14.14). In non-dimensional numbers, making use of the Bond number

$$Bo = \left(\frac{V^{1/3}}{l_c} \right)^2 \tag{H.2}$$

Here $l_c = \sqrt{\sigma/\rho g}$ is the capillary length (see Appendix E, Sect. E.1.4), one obtains

$$ca = \eta U / \sigma \sim Bo \sin\alpha - \Delta(\cos \theta_c) \tag{H.3}$$

From Eq. H.3 one can deduce the drop velocity U. At low velocity, the drop departs slightly from a spherical cap with a rounded base (Fig. H.1a, d). Upon increasing velocity, the drops deform because the contact line cannot follow the imposed speed, thus causing dynamic wetting (see Voinov 1976; Cox 1986). The contact line therefore develops a corner along the trailing edge (Fig. H.1b), associated with a conical structure of the interface (Fig. H.1e).

A first interpretation for the formation of corners was given by Blake and Ruschak (1979) who postulated that there exists a maximum speed or capillary number Ca_{cr} at which a contact line can move. When $Ca > Ca_{cr}$, the contact line

© Springer Nature Switzerland AG 2022
D. Beysens, *The Physics of Dew, Breath Figures and Dropwise Condensation*,
Lecture Notes in Physics 994, https://doi.org/10.1007/978-3-030-90442-5

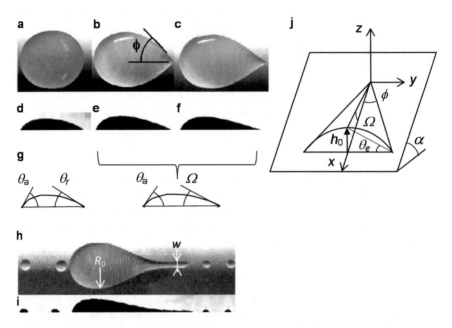

Fig. H.1 Drop shapes observed for increasing drop velocity U or increasing capillary number $Ca = \eta U / \sigma$. The motion is from right to left. As the velocity increases, a corner first forms, which then becomes unstable leading to the ejection of drops ("pearling" transition). (**a**), (**b**) and (**c**): Respectively rounded drops, cornered drops and cusped drops of silicon oil viewed from above; (**d**), (**e**) and (**f**): the same drops viewed from the side; (**g**) definition of side views angles; (**h**) and (**i**): pearling drop. (**j**) Corner (schematics). Adapted from Limat et al. (2001), with permission

inclines so as to keep the speed in the direction of its normal constant (see Rio et al. 2005), such as

$$Ca = \frac{Ca_{cr}}{\sin \phi} \qquad (H.4)$$

Here ϕ is the inclination angle, measured from the direction of the motion of the contact line (Fig. H.1b). Experimental data from Blake and Ruschak (1979) and Petrov and Sedev (1985) are in reasonable agreement with this expression.

As the velocity still increases, the corner becomes sharper, up to ϕ close to 30°. At this value, a cusp forms at the corner tip (Fig. H.1c, f), which eventually becomes a rivulet where the tail releases tiny droplets (Fig. H.1h, i). This is the so-called "pearling" transition (see, e.g., Duineveld 2003). More information on the transition rounded/corner can be found in Limat et al. (2001) and Snoeijer et al. (2007).

Equation H.4 does not define any maximum speed above which the assumption of inclination can fail and rivulet/pearls can form. Snoeijer et al. (2007) showed that Eq. H.4 is only the low-speed limit of a more complete law, which gives solutions above a critical speed and thus explains the breakdown of the corner solution.

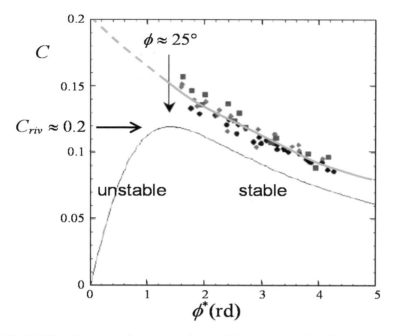

Fig. H.2 Relationship between the corner angle $\phi \approx 3.3\phi$ (see text) and the dimensionless drop velocity C (Eq. H.5; solid line), compared to experiments for different values of the viscosity (red diamonds: 10 cP; blue squares: 104 cP; black dots: 1040 cP). The solid line displays a maximum for $\phi = 25°$, which corresponds to the onset of the pearling instability. Adapted from Snoeijer et al. (2007)

This law stems from simplified lubrication equations and further assuming that the drop shape is slender at the rear. In the low opening angle ϕ limit, the following selection of ϕ with Ca is found (Snoeijer et al. 2007) through the variation of dimensionless drop velocity C with ϕ:

$$C = \frac{3\text{Ca}}{\varepsilon \theta_e^3} = \frac{6\phi^*}{35 + 18\phi^{*2}} \tag{H.5}$$

Here $\theta_e = 2h_0/R_0$ is defined in Fig. H.1j and $\phi^* = \phi/\varepsilon$ with $\varepsilon \approx 0.3$ comes from a logarithmic correction in the Cox-Voinov relation (see Voinov 1976; Cox 1986). Equation H.5 possesses a maximum ($C_{riv} \approx 0.12$, $\phi \approx 25°$) above which no solution exists (Fig. H.2). The region $C > C_{riv}$ thus corresponds to having a drop leaving behind it a rivulet that transforms into droplets, thanks to an instability reminiscent of the Plateau-Rayleigh instability (see below).

The condition $C = C_{riv}$ then corresponds to the pearling transition. For typical value $\theta_e \approx 41°$ (see Snoeijer et al. 2011), the value $C_{riv} = 0.12$ gives the capillary number $\text{Ca}_{riv} \approx 0.007$ where rivulets form.

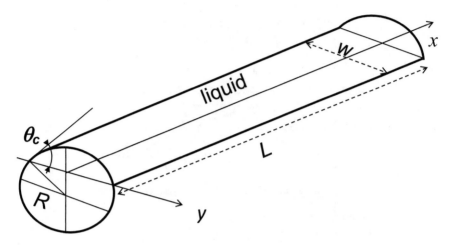

Fig. H.3 Fluid rivulet on a solid surface. Adapted from Diez et al. (2009)

The rivulet width, w, can be expressed (Snoeijer et al. 2007) near the pearling transition as

$$w \approx 2R_0\exp\left(-\frac{1}{\sqrt{C/C_{\mathrm{riv}} - 1}}\right) \tag{H.6}$$

The instability where the rivulet breaks up into droplets resembles the instability that a falling free jet exhibits when it breaks up into smaller drops. The drops have the same volume but less surface area than the jet, that is, less surface energy thanks to surface tension. This instability, called Plateau-Rayleigh, is well-known (Plateau 1873; Rayleigh 1879; see, e.g., de Gennes et al. 2003). In this instability, a sinusoidal perturbation of the interface reduces the surface energy for wavenumbers below the critical value k_c. For cylindrical jets of radius R, the critical wavenumber $k_c = 1/R$. This calculation has been extended by Davis (1980) and Diez et al. (2009) to infinite cylindrical rivulets with contact angle θ_C, as shown in Fig. H.3. The critical wavenumber depends on the contact angle and is written following Diez et al. (2009) as

$$k_c^2 R^2 = k_c^2\frac{w^2}{2\sin\theta_c^2} = 2\left[1 + \sin\theta_c^2 - \theta_c\cot\theta_c\right]^{-1} \tag{H.7}$$

This relation is expressed as a function of R or rivulet width $w = 2R\sin\theta_c$. For hemispherical rivulet ($\theta_c = \pi/2$) the critical wavenumber $k_c = 1/R$, the same as for a free jet of radius R. In Fig. H.4 is shown the details of a rivulet break up with interface periodic undulation that leads to droplets with radius r.

Winkels (2013) studied the breakup of rivulets in 20 cP viscosity silicon oil and compared the results to the Rayleigh-Plateau instability. The fastest growing wavenumber $k_m = 2\pi/\lambda_m$ is determined from the dispersion relation resulting

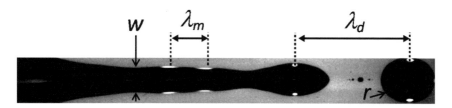

Fig. H.4 Rivulet break up on a solid surface. Adapted from Winkels (2013)

from a stability analysis. The latter has been numerically performed by Diez et al. (2009) for various contact angles and Winkels (2013) found a mean estimate $k_m \approx 0.7k_c$. For $\theta_c = 45°$, the latter value gives the prediction $k_m w = 2\pi w/\lambda_m \approx 1.7$. However, the interface modulation wavelength λ_m is experimentally found much larger than expected. In addition, in contrast to what is found with free jets, the droplet spacing λ_d (see Fig. H.4) is larger than λ_m (with free jets $\lambda_m \approx \lambda_c$). As outlined by Winkels (2013), there are indeed some differences between a rivulet and the free jet in the Plateau-Rayleigh geometry. Due to the transition region from "cone" to "rivulet" on the sliding drop, the jet does not exhibit a well-defined cylindrical shape. The break-up then should not be determined by a Plateau-Rayleigh instability but rather by the contact line dynamics at the tip of the rivulet leading to the final pinching of the droplets.

In any case, the diameters of the droplets emitted by the rivulet still remain in the order of the rivulet width

$$2r \sim w \qquad\qquad (H.8)$$

whose value can be calculated by using Eq. H.7.

Frost Propagation

At the same temperature, the saturation vapor pressure of ice, p_{sI}, is always lower than that of liquid water, p_{sw} (Lide 1998). For example, at $-12\,°C$, $p_{sI} = 220$ Pa and $p_{sW} = 245$ Pa. This difference is also encountered with salty drops, which makes the condensation of supercooled (SC) droplets in the presence of ice a similar problem (see Chap. 7). Although weak, the difference in saturation vapor pressures has nevertheless important consequences. Two situations can be usually encountered: (i) ice crystals on a dry substrate on which condensation starts, and (ii) the most common, condensation of supercooled (SC) droplets where ice forms locally due to substrate heterogeneity that favors ice nucleation.

In situation (i), the vapor concentration profile around ice prevents water or ice nucleation in regions of inhibited condensation (RIC) with thickness δ_I (ice) or δ_W (SC water). The evaluation of the nucleation RIC can be performed by evaluating the vapor concentration profile as discussed in Chap. 7.

In situation (ii), the ice crystal grows at the expense of the surrounding SC droplets that evaporate, thus also creating a RIC around the ice crystal. The RIC thickness can be evaluated by balancing the evaporating flux from the SC droplets towards the ice crystal with the droplet condensation rate (Chap. 7). Depending on ice growth rate and SC droplet decline rate, two situations can be met.

The first situation corresponds to a weak vapor concentration gradient around the ice crystal and then weak SC droplet evaporative flux on ice. Ice undergoes faceted growth and the RIC is steady (Guadarrama-Cetina et al. 2013a). In contrast, when the concentration gradient near the ice crystal and then the SC droplet evaporative flux on ice is strong, ice develops by fast-growing dendrites that have enough time to hit a SC droplet before it fully evaporates. The droplet then freezes at the ice contact and in turn emits a dendrite, which will hit another SC droplet and so on, resulting in a very fast percolating ice front.

Another situation corresponds to the simultaneous freezing of all SC droplets. In this case, there are no vapor concentration gradients between the nucleated ice crystals, which grow above the substrate by desublimation.

The condition for observing fast ice front propagation corresponds to the condition where the ice dendrite can touch the SC droplet just before it has completely evaporated. It thus corresponds (Nath and Boreyko 2016) to having a droplet mass, m_d, equal to the dendrite mass, m_i. With two initial hemispherical droplets of

© Springer Nature Switzerland AG 2022

D. Beysens, *The Physics of Dew, Breath Figures and Dropwise Condensation*,
Lecture Notes in Physics 994, https://doi.org/10.1007/978-3-030-90442-5

Fig. I.1 Schematics of dendrite growth from the left frozen drop

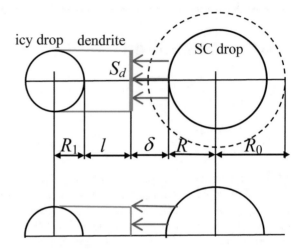

about the same radius R and dendrite maximum length l_m, the equality scales as $\frac{1}{2}\rho_i \pi R^2 l_m \approx \frac{2}{3}\pi \rho_w R^3$ or

$$l_m \approx \frac{4}{3}\frac{\rho_w}{\rho_i} R \tag{I.1}$$

Here ρ_w is water density and ρ_i is ice density.

A more refined treatment needs to establish both the dendrite length evolution and SC droplet evaporation to determine the time when the dendrite hits the drop.

Evaporation of the SC water droplet follows an inverse process of condensation (see Sect. 4.2, Eq. 4.21):

$$\frac{dV}{dt} = -\frac{1}{\rho_w}\int_S \vec{j}(r=R)\cdot \vec{n}\, dS = -\frac{1}{\rho_w}D\int_S \left(-\vec{\nabla}c\right)_d \cdot \vec{n}\, dS \tag{I.2}$$

Here the gradient $\left(-\vec{\nabla}c\right)_d$ corresponds to evaporation on dendrite and can be written, with $\delta(t)$ the drop-dendrite spacing (Fig. I.1), as

$$\left(-\vec{\nabla}c\right)_d \sim \frac{c_s - c_{si}}{\delta} = \frac{\Delta c}{\delta} \tag{I.3}$$

In the above equation, c_s, and c_{si} are, respectively, water and ice vapor concentration at saturation and $\Delta c = c_s - c_{si}$.

Let us denote S_d the dendrite surface area issued from the icy drop where condensation takes place. It is also the water drop surface where evaporation occurs (Fig. I.1). The drop evaporation mass flux reads as (drop is assumed hemispherical for simplification)

$$2\pi \rho_w R^2 \frac{dR}{dt} \sim -S_d \frac{D\Delta c}{\delta(t)} \tag{I.4}$$

Since $R \leq R_1$ and $S_d \sim \pi R_1^2 / 2$, Eq. I.4 becomes

$$4\rho_w \frac{R^2}{R_1^2} \frac{dR}{dt} \sim -\frac{D\Delta c}{\delta(t)} \tag{I.5}$$

This mass flux feeds the dendrite whose growth is ensured by the same gradient as found in evaporation $\nabla c \sim \Delta c / \delta(t)$. Dendrite volume evolution can be written (Eq. I.2) as

$$S_d \rho_i \frac{dl}{dt} \sim S_d \frac{D\Delta c}{\delta(t)} \tag{I.6}$$

Comparing Eqs. I.5 and I.6 leads to

$$\frac{dl}{dt} = -4 \frac{\rho_w}{\rho_i} \frac{R^2}{R_1^2} \frac{dR}{dt} \tag{I.7}$$

Integrating Eq. I.7 gives with the boundary conditions $t = 0$, $l = 0$, $R = R_0$; $t = t_1$, $*\delta = 0$, $R = R_1$

$$l = \frac{4}{3} \frac{\rho_w}{\rho_i R_1^2} \left(R_0^3 - R^3 \right) \tag{I.8}$$

or equivalently

$$R = R_0 \left(1 - \frac{3}{4} \frac{\rho_i}{\rho_w} \frac{R_1^2}{R_0^3} l \right)^{1/3} \tag{I.9}$$

A simple solution for l evolution in Eq. I.6 can be found by noting that the drop radius evolution is weak when compared to l evolution (exponent 1/3 in Eq. I.8). Making then $R \approx R_0 \approx R_1$, and noting (Fig. I.1) that

$$\delta = \langle d \rangle - l - R - R_1 \tag{I.10}$$

Equation I.6 becomes

$$\frac{dl}{dt} (l - \delta_0) = -\frac{D\Delta c}{\rho_i} \tag{I.11}$$

with the typical interdrop distance

$$\delta_0 = \langle d \rangle - 2R_0 \tag{I.12}$$

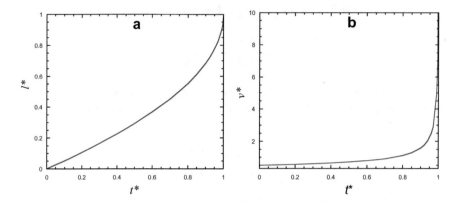

Fig. I.2 Schematics of dendrite growth. Evolution in reduced units. **a** Dendrite length $l^* = l/\delta_0$ and **b** velocity in reduced units $v^* = dl^*/dt^*$. Reduced time $t^* = \frac{2D\Delta c}{\rho_i \delta_0^2} t$

Integration of Eq. I.11 gives, with boundary condition $t = 0$, $l = 0$,

$$l = \delta_0 - \left(2\frac{D\Delta c}{\rho_i}\right)^{1/2}\left(\frac{\delta_0^2}{2D\Delta c/\rho_i} - t\right)^{1/2} \tag{I.13}$$

The maximum length corresponds to $l = \delta_0$ at time $t_0 = \frac{\delta_0^2}{2D\Delta c/\rho_i}$. Defining the reduced units

$$l^* = \frac{l}{\delta_0} \; ; t^* = \frac{2D\Delta c}{\rho_i \delta_0^2} t \tag{I.14}$$

Equation I.13 can be thus written as

$$l^* = 1 - \left(1 - t^*\right)^{1/2} \tag{I.15}$$

Dendrite velocity, in reduced units $v^* = dl^*/dt^*$, can be written as

$$v^* = \frac{1}{2}\left(1 - t^*\right)^{-1/2} \tag{I.16}$$

Figure I.2 reports such dendrite length and velocity evolution. Initial evolution is made at near-constant velocity $v^* \approx \frac{1}{2}$ or $v \approx \frac{D\Delta c}{\rho_i \delta_0}$, corresponding to a near-constant concentration gradient since $\delta_0 \gg R$. Late evolution exhibits a diverging velocity when $\delta \to 0$.

In the definition of reduced quantities (Eq. I.14), note that one can relate δ_0 to the mean distance between droplets centers $\langle d \rangle$ or the mean radius R_0 by means of apparent drop surface coverage ε_2^*. The latter is the ratio of drop surface observed from above the substrate to the substrate surface (see Sect. 6.5 and Fig. 6.18). The drop surface coverage indeed corresponds to the process of coalescence, which occurs at the contact line for contact angle $\theta_c \leq 90°$ and at the drop interface for larger contact angles. For $\theta_c \leq 90°$, this surface corresponds to the surface effectively wetted by the drops and $\varepsilon_2^* = \varepsilon_2$. For $\theta_c > 90°$, it is the projection of the drop surface on the substrate and $\varepsilon_2^* \approx 0.55$ (Fig. 6.18). The relation between the mean droplet radius R_0, $\langle d \rangle$ and ε_2 is

$$\varepsilon_2 = \alpha^2 \frac{R_0^2}{\langle d \rangle^2} \tag{I.17}$$

Drops on a square lattice gives $\alpha = \sqrt{\pi} \approx 1.8$; drops on a triangular lattice gives $\alpha = 2\sqrt{\frac{\pi}{3\sqrt{3}}} \approx 1.6$. From Eq. I.17 one deduces the value of $\delta_0 = \langle d \rangle - 2R_0$ as a function of either $\langle d \rangle$ or R_0:

$$\delta_0 = R_0\left(\frac{\alpha}{\sqrt{\varepsilon_2}} - 2\right) = \langle d \rangle \left(1 - 2\frac{\sqrt{\varepsilon_2}}{\alpha}\right) \tag{I.18}$$

When ε_2 is large, drops are close together and δ_0 is small. The time $t_0 = \frac{\delta_0^2}{2D\Delta c/\rho_i}$ for a dendrite to hit a neighboring drop is thus small and frost propagation is fast. The threshold in Eq. I.1 to observe fast frost propagation, $\delta_0 = l$, corresponds, using Eq. I.18, to the threshold in surface coverage:

$$\varepsilon_{2,\min} = \left(\frac{\alpha}{2 + \frac{4}{3}\frac{\rho_w}{\rho_i}}\right)^2 \tag{I.19}$$

Using a value $\alpha \approx 1.7$, one deduces $\varepsilon_{2,\min} \approx 0.26$ for fast frost propagation.

Raoult Law and Salty Drops

The following is the detailed calculation made by Guadarrama-Cetina et al. (2014). In order to evaluate the variation of the saturation pressure with the salty drop radius in Sect. 7.4.2, one makes use of the Raoult law. In salty water, it can be written as

$$p_{ss} = p_s \frac{n_{H_2O}}{n_{H_2O} + n_{NaCl}} \tag{J.1}$$

where n_{NaCl} is the number of moles of salt, which is constant in an experiment. n_{H_2O} is the number of moles of water and depends on the vapor condensation history and the initial size of the drop, R_0. Therefore,

$$p_s - p_{ss} = p_s \frac{n_{NaCl}}{n_{H_2O} + n_{NaCl}} \tag{J.2}$$

Under saturated conditions corresponding to salt in water at T_a (=10 °C)

$$p_s - p_{s0} = p_s \frac{n_{NaCl}}{n_{H_2O}^s + n_{NaCl}} \tag{J.3}$$

where $n_{H_2O}^s$ is the number of water moles that are needed to saturate the solution with the specified n_{NaCl} at the substrate temperature, T_c.

Dividing Eq. J.2 by Eq. J.3 one obtains

$$p_s - p_{ss} = (p_s - p_{s0}) \frac{n_{H_2O}^s + n_{NaCl}}{n_{H_2O} + n_{NaCl}} \tag{J.4}$$

With m_d the total mass of the drop (which increases with time) and the molecular masses of water and salt, M_{H_2O} (=18.02 g mole^{-1}) and M_{NaCl} (= 58.44 g mole^{-1}), it becomes

$$m_d = n_{H_2O} M_{H_2O} + n_{NaCl} M_{NaCl} \tag{J.5}$$

© Springer Nature Switzerland AG 2022
D. Beysens, *The Physics of Dew, Breath Figures and Dropwise Condensation*,
Lecture Notes in Physics 994, https://doi.org/10.1007/978-3-030-90442-5

Then

$$n_{H_2O} = \frac{m_d - n_{NaCl} M_{NaCl}}{M_{H_2O}} \tag{J.6}$$

When considering a hemispherical drop (a good approximation for the conditions of Sect. 7.4.2)

$$m_d = \frac{2\pi}{3} \rho_s R^3 \tag{J.7}$$

Here ρ_s is salty drop density. It depends on temperature and salt concentration, thus on the drop radius R. However, the weak evolution of R (in $\sim t^{1/5}$, see Eq. 7.19) makes the density variation small and ρ_s can be taken as the salty drop density at saturation. In Sect. 7.4, temperature is constant. Using Eqs. J.5 and J.6, one can eventually write Eq. J.4 as

$$p_s - p_{ss} = (p_s - p_{s0}) \frac{\frac{2\pi}{3} \rho_s R_0^3 - n_{NaCl}(M_{H_2O} - M_{NaCl})}{\frac{2\pi}{3} \rho_s R^3 - n_{NaCl}(M_{H_2O} - M_{NaCl})} \tag{J.7}$$

An approximation of Eq. J.7 can be obtained in the following way. Taking the logarithmic derivative of Eq. J.2, one obtains, after some algebra

$$\frac{d(p_s - p_{ss})}{p_s - p_{ss}} = -\frac{dn_{H_2O}}{n_{H_2O} + n_{NaCl}} \tag{J.8}$$

One defines the (mass/mass) salt concentration

$$c_m = \frac{n_{NaCl} M_{NaCl}}{m_t} \tag{J.9}$$

whose derivative is

$$dc_m = -\frac{n_{NaCl} M_{NaCl}}{m_t^2} dm_d \tag{J.10}$$

One differentiates Eq. J.6

$$dn_{H_2O} = \frac{dm_d}{M_{H_2O}} \tag{J.11}$$

and Eq. J.7

$$dm_d = 2\pi R^2 \rho_s dR + \frac{2\pi}{3} R^3 \rho' dc_m \tag{J.12}$$

One uses $\rho_s = \rho_s(c_m)$ and noted $\rho'_s = d\rho_s/dc_m$. The variation $\rho_s(c_m)$ and its derivative $\rho'_s(c_m)$ can be found in the tables of Nikolskij et al. (1964).

From Eqs. J.10, J.11 and J.12, one eventually finds

$$dn_{H_2O} = \frac{1}{M_{H_2O}} \frac{2\pi R^2 \rho_s}{1 + c_m \frac{\rho_s'}{\rho_s}} \tag{J.13}$$

Inserting the expression Eq. J.13 in Eq. J.8, one obtains

$$\frac{d(p_s - p_{ss})}{p_s - p_{ss}} = -\left(\frac{1}{n_{NaCl} + \frac{m_t - n_{NaCl}M_{NaCl}}{M_{H_2O}}}\right) \frac{1}{M_{H_2O}} \frac{2\pi R^2 \rho}{1 + c_m \frac{\rho_s'}{\rho_s}} dR \tag{J.14}$$

which develops into the following expression:

$$\frac{d(p_s - p_{ss})}{p_s - p_{ss}} = -3\left(\frac{dR}{R}\right)\left(\frac{1}{1 + c_m \frac{\rho_s'}{\rho_s}}\right)\left(\frac{1}{1 + \frac{n_{NaCl}(M_{H_2O} - M_{NaCl})}{\frac{2\pi}{3}\rho_s R^3}}\right) \tag{J.15}$$

Equation J.15 can be exactly integrated and obviously leads to Eq. J.7. As ρ_s and R vary very smoothly, $1 + c_m \frac{\rho_s'}{\rho_s}$ can be considered in the salty drop evolution as a constant (≈ 1.002 from the values in Nikolskij et al. (1964)) as well as $1 + \frac{n_{NaCl}(M_{H_2O} - M_{NaCl})}{\frac{2\pi}{3}\rho_s R^3}$ (≈ 0.911 from the values in Nikolskij et al. (1964)). Equation J.15 then becomes

$$\frac{d(p_s - p_{ss})}{p_s - p_{ss}} \approx -b\left(\frac{dR}{R}\right) \tag{J.16}$$

where $b = 3.3$. Equation J.16 can be integrated to give

$$p_s - p_{ss} \approx (p_s - p_{s0})\left(\frac{R_0}{R}\right)^b \tag{J.17}$$

Error and Gamma Functions

K.1 Error and Complementary Error Functions

The error function, denoted erf(z), is obtained by integrating the normal distribution (the normalized form of the Gaussian function). It is thus defined by

$$\text{erf}(z) = \frac{2}{\sqrt{\pi}} \int_0^z e^{-t^2} dt \tag{K.1}$$

The complementary error function is $\text{erfc}(z) = 1 - \text{erf}(z)$. From Eq. K.1, it can be written as

$$\text{erfc}(z) = \frac{2}{\sqrt{\pi}} \int_z^\infty e^{-t^2} dt \tag{K.2}$$

whose special values are $\text{erfc}(-\infty) = 2$, $\text{erfc}(0) = 1$, $\text{erfc}(\infty) = 0$.

Its derivative is written as

$$\frac{d}{dz}(\text{erfc}(z)) = \frac{2}{\sqrt{\pi}} e^{-t^2} dt \tag{K.3}$$

Its indefinite integral gives

$$\int \text{erfc}(z) dz = z\text{erfc}(z) - \frac{1}{\sqrt{\pi}} e^{-z^2} + \text{const.} \tag{K.4}$$

The function satisfies the identity

$$\text{erfc}(-z) = 2 - \text{erfc}(z) \tag{K.5}$$

© Springer Nature Switzerland AG 2022   403
D. Beysens, *The Physics of Dew, Breath Figures and Dropwise Condensation*,
Lecture Notes in Physics 994, https://doi.org/10.1007/978-3-030-90442-5

Iterated integrals of the erfc function are written as

$$i^{-1}\text{erfc}(z) == \frac{2}{\sqrt{\pi}}e^{-z^2} \tag{K.6}$$

$$i^0\text{erfc}(z) = \text{erfc}(z) \tag{K.7}$$

For positive integer arguments $n = 0, 1, 2, \ldots$

$$i^n\text{erfc}(z) = \int_z^\infty i^{n-1}\text{erfc}(t)dt = \frac{2}{\sqrt{\pi}}\int_z^\infty i^n\frac{(t-z)^n}{n!}e^{-z^2}dt \tag{K.8}$$

A general recurrence formula is

$$2ni^n\text{erfc}(z) = i^{n-2}\text{erfc}(z) - 2zi^{n-1}\text{erfc}(z) \tag{K.9}$$

It follows for erfc(0)

$$i^n\text{erfc}(0) = \frac{1}{2^n\Gamma\left(1+\frac{n}{2}\right)} \tag{K.10}$$

where Γ is the gamma function (see Sect. K.2).

K.2   The Gamma Function

(i) For the positive integer arguments $n = 0, 1, 2, \ldots$ the gamma function is defined as (Legendre's form)

$$\Gamma(n) = (n-1)! \tag{K.11}$$

where ! denotes the factorial [$n! = n(n-1)(n-2)\ldots1$]. The Γ-values are then $\Gamma(1) = 1$, $\Gamma(2) = 1$, $\Gamma(3) = 1$, $\Gamma(4) = 6$, $\Gamma(1) = 24$, $\ldots$

(ii) For positive half-integers, it becomes

$$\Gamma\left(\frac{n}{2}\right) = \sqrt{\pi}\frac{(n-2)!!}{2^{\frac{n-1}{2}}} \tag{K.12}$$

Here !! is the double factorial [$n! = n(n-2)(n-4)\ldots$]. Equation K.13 can be written in the equivalent forms:

$$\Gamma\left(\frac{1}{2}+n\right) = \sqrt{\pi}\frac{(2n-1)!!}{2^n} = \sqrt{\pi}\frac{(2n)!}{4^n n!} \tag{K.13}$$

$$\Gamma\left(\frac{1}{2} - n\right) = \sqrt{\pi}\,\frac{(-2)^n}{(2n-1)!!} = \sqrt{\pi}\,\frac{(-4)^n n!}{2^n!} \qquad \text{(K.14)}$$

Particular values are $\Gamma\left(\frac{1}{2}\right) = \sqrt{\pi}$, $\Gamma\left(\frac{3}{2}\right) = \frac{1}{2}\sqrt{\pi}$, $\Gamma\left(\frac{5}{2}\right) = \frac{3}{4}\sqrt{\pi}$, $\Gamma\left(\frac{7}{2}\right) = \frac{15}{8}\sqrt{\pi}$.

(iii) For positive real numbers, the gamma function can be defined as a definite integral (Euler's integral):

$$\Gamma(z) = \int_0^\infty t^{z-1} e^{-t} dt = 2\int_0^\infty e^{-t^2} t^{2z-1} dt \qquad \text{(K.15)}$$

When integrating by part, it comes

$$\Gamma(z) = (z-1)\Gamma(z-1) \qquad \text{(K.16)}$$

Then, when z is an integer, one obtains $\Gamma(z) = (z-1)(z-2)(z-3)\ldots 1 = (z-1)!$, which is the definition of the gamma function, Eq. K.12.

Sand Blasting Roughness

L.1 Roughness Amplitudes

During sand blasting, hard particles (e.g. silica spheres) of radius ρ hit a softer surface (e.g. Duralumin). Metal can be removed depending on the angle the particle hits the surface. In most cases, however, a sand jet is generally directed perpendicular to the surface. Spheres explode at the impact, making individual craters with spherical lens shapes according to the sphere size (Figs. L.1 and L.2). The depth of the impact, $\delta\zeta$, is related to the kinetic energy of the bead, itself proportional to the jet pressure p, according to the Bernoulli principle (see, e.g., Batchelor 2000). The bead has the same velocity u of the jet, then

$$\delta\zeta \sim u^2 \sim p \tag{L.1}$$

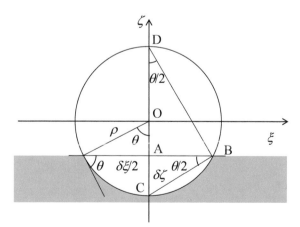

Fig. L.1 Crater made by a sphere hitting a surface

© Springer Nature Switzerland AG 2022
D. Beysens, *The Physics of Dew, Breath Figures and Dropwise Condensation*,
Lecture Notes in Physics 994, https://doi.org/10.1007/978-3-030-90442-5

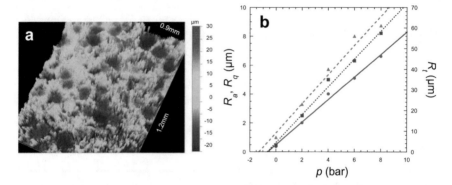

Fig. L.2 **a**. Scanned portion (0.9 mm × 1.2 mm) of the surface of a 175 mm × 175 mm × 5 mm Duralumin plate hit perpendicularly by silica beads of 25 μm diameter under air pressure of 8 bars during approximately 3 min, with an impact density of about 45 mm^{-2}. The jet was scanned parallel and perpendicular to one side of the square. Roughness due to the bead impacts is on the order $R_a = \overline{\delta\zeta} = 6.6$ μm, corresponding to $\overline{\delta\xi} = 36.3$ μm from Eq. L.2. **b** Variation of R_a, R_q (left ordinate) and R_t (right ordinate) with pressure showing a linear dependence as expected from Eq. L.1 (notations: see text). R_a: Circles, full line; R_q: Squares, dotted line; R_t: Triangles, interrupted line. Adapted from Verbrugghe (2016)

The roughness perpendicular to the surface, $\delta\zeta$, can be related to the roughness parallel to it, $\delta\xi$, by expressing $tg(\theta/2) = AB/AD = CA/AB$

$$\delta\xi = 2(\delta\zeta(2\rho - \delta\zeta))^{1/2} \approx 2\sqrt{2}(\rho\delta\zeta)^{1/2} \qquad (L.2)$$

The approximation holds for the general case where $\delta\zeta/\rho << 1$.

Classically, the characteristics of surface roughness can be characterized by the following amplitude parameters (see, e.g., Whitehouse 2004):

Arithmetic average of absolute values:

$$R_a \equiv \overline{\delta\zeta} = \frac{1}{n}\sum_{i=1}^{n}|\delta\zeta_i| \qquad (L.3)$$

Root mean squared values:

$$R_q = \sqrt{\frac{1}{n}\sum_{i=1}^{n}\delta\zeta_i^2} \qquad (L.4)$$

Maximum height of the profile:

$$R_t = R_p + R_v \qquad (L.5)$$

with maximum valley depth $R_v = (\max)_i\ \delta\zeta_i$ and maximum peak height $R_p = (\max)_i\ \delta\zeta_i$.

L.2 Wenzel Roughness Factor

According to Eq. E.9 in Appendix E, the surface roughness factor for the determination of the Wenzel contact angle is the ratio of the actual area to the projecsted area. Making use of

$$AB = \delta\xi/2 = \rho \sin\theta \tag{L.6}$$

it follows

$$S = 2\pi\rho^2[1 - \cos\theta] \approx 2\pi\rho^2\left(\frac{\delta\xi^2}{8\rho^2} + \frac{\delta\xi^4}{384\rho^4}\right) \tag{L.7}$$

The approximation holds for the general case where $\delta\xi/\rho << 1$.
The roughness factor can be then estimated as

$$r = \frac{4S}{\pi\delta\xi^2} = \frac{2\pi\rho^2[1 - \cos\theta]}{\pi\rho^2\sin\theta^2} \approx 1 + \frac{\delta\xi^2}{16\rho^2} \tag{L.8}$$

Expressing $\delta\xi$ as a function of $\delta\zeta$ from Eq. L.2, the roughness factor can be then written under the simple form:

$$r \approx 1 + \frac{\delta\zeta}{2\varrho} \tag{L.9}$$

Drop Motion in a Wettability Gradient

A drop placed on a surface with a wettability gradient can undergo motion towards the region of highest wettability. Such an observation has been reported by Chaudhury and Whitesides (1992), Zhao and Beysens (1995) and further addressed by Daniel and Chaudhury (2002).

Following the latter authors, one can derive the force driving this motion from the free energy of adhesion (Eq. E.3 in Appendix E):

$$\Delta G = -\pi R^2 \sigma_{LG}(1 + \cos\theta_c) \tag{M.1}$$

Here R is the droplet base radius (Fig. M.1). The driving force F_d for the drop movement in the wettability gradient $\cos\theta_c(x)$ along axis x comes from the gradient of the free energy:

$$F_d = -\frac{d\Delta G}{dx} = \pi R^2 \sigma_{LG} \frac{d\cos\theta_c}{dx} \tag{M.2}$$

However, this relation ignores the hysteresis of the contact angle (advancing θ_a and receding θ_r contact angles, respectively). The force originating from the wettability gradient must indeed overcome a pinning force for the drop to move. In order to estimate the effect of hysteresis, one calculates the resulting force F_d^* acting on a thin band of liquid of thickness dy by integrating it over the drop perimeter (Fig. M.1):

$$F_d^* = \sigma_{LG} \int_{-R}^{+R} \int (\cos\theta_{aB} - \cos\theta_{rA}) dy \tag{M.3}$$

In this equation θ_{aB} and θ_{rA} are the advancing and receding angles at both ends of the thin strip in A and B, respectively (Fig. M.1). The integral can be calculated in the case of a smooth gradient by expanding the cosines around their values $\cos\theta_{a0}$ and $\cos\theta_{r0}$ at the central line of the drop. With r the chord at distance y (Fig. M.1)

© Springer Nature Switzerland AG 2022   411
D. Beysens, *The Physics of Dew, Breath Figures and Dropwise Condensation*,
Lecture Notes in Physics 994, https://doi.org/10.1007/978-3-030-90442-5

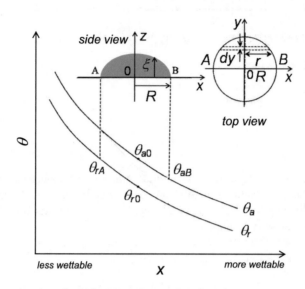

Fig. M.1 Advancing θ_a and receding θ_r contact angles of a drop on a wettability gradient surface. During motion towards the higher wettability region, the drop advancing edge B shows the advancing contact angle θ_{aB} and its receding edge experiences the receding angle θ_{rA}. Adapted from Daniel and Chaudhury (2002)

$$\cos \theta_{aB} = \cos \theta_{a0} + \left(\frac{d \cos \theta_a}{dx}\right) r \qquad (M.4)$$

$$\cos \theta_{rA} = \cos \theta_{r0} - \left(\frac{d \cos \theta_r}{dx}\right) r \qquad (M.5)$$

In weak gradients the drop keeps a near-circular shape, then Eq. M.3 can be evaluated by using Eqs. M.4 and M.5:

$$F_d^* = \sigma_{LG}\left(\frac{d \cos \theta_a}{dx} + \frac{d \cos \theta_r}{dx}\right) \int_{-R}^{+R} r \, dy$$

$$-\sigma_{LG}(\cos \theta_{r0} - \cos \theta_{a0}) \int_{-R}^{+R} dy \qquad (M.6)$$

Defining θ_d as the mean dynamic contact angle such that

$$\cos \theta_d = \frac{1}{2}(\cos \theta_a + \cos \theta_r) \tag{M.7}$$

it eventually comes

$$F_d^* = \pi R^2 \sigma_{LG} \left(\frac{d \cos \theta_d}{dx} \right) - 2\sigma_{LG} R(\cos \theta_{r0} - \cos \theta_{a0}) \tag{M.8}$$

One sees that there is a threshold in contact angle hysteresis above which motion cannot take place. It is for

$$(\cos \theta_{r0} - \cos \theta_{a0})_c = \frac{\pi R}{2} \left(\frac{d \cos \theta_d}{dx} \right) \tag{M.9}$$

As this threshold depends on the drop radius, for a given hysteresis it corresponds to a threshold in drop radius below which drop does not move.

$$R_c = \frac{2}{\pi} \frac{\cos \theta_{r0} - \cos \theta_{a0}}{(d \cos \theta_d / dx)} \tag{M.10}$$

This threshold can be reduced or even canceled at the favor of a coalescence event as observed by Zhao and Beysens (1995). When the contact line moves, the effect of pinning is much reduced (Gao et al. 2018). The substrate can also be vibrated (Daniel and Chaudhury 2002).

The velocity of drop motion can be evaluated by evaluating the viscous drag F_v generated in the liquid (Brochard 1989):

$$F_v = 3\pi \eta R u \int_A^B \frac{dx}{\xi(x)} \tag{M.11}$$

Here u is the drop velocity, η is liquid shear viscosity, $\xi(x)$ is drop thickness and A and B are the drop edges. As dissipation is mostly localized in the edges, Eq. M.10 can be approximated as

$$F_v \approx 6\pi \eta R u \int_A^0 \frac{dx}{(x - R) \sin \theta_d} \approx \frac{6\pi \eta R u}{\sin \theta_d} \ln \frac{R}{a} \tag{M.12}$$

Here a is in the order of molecular dimension, making $\ln \frac{R}{a} \sim 10 - 15$. Equaling $F_d^* = F_v$ from Eqs. M.8 and M.11, one deduces the drop velocity

$$u = \frac{\sigma_{LG} R \sin \theta_d}{6\eta \ln(R/a)} \left[\left(\frac{d \cos \theta_d}{dx} \right) - 2\frac{\cos \theta_{r0} - \cos \theta_{a0}}{\pi R} \right] \tag{M.13}$$

This relation simplifies for small contact angle hysteresis in

$$u \approx \frac{\sigma_{LG} R \sin \theta_d}{6 \eta \ln(R/a)} \left(\frac{d \cos \theta_d}{dx} \right) \tag{M.14}$$

It shows that velocity is proportional to drop radius, neglecting the weak log dependence.

Ballistic Motion

The ballistic motion of a projectile is the problem of the trajectory that follows an object projected on earth. It is a question that has been addressed since Galileo, who showed that the projectile follows a parabola (see, e.g., Projectile Motion (2020) from which is drawn what follows below).

N.1 Trajectory Without Air Resistance

Neglecting the air resistance, the motion of a projectile launched with initial velocity u_0 making a tilt angle α with horizontal (Fig. N.1) can be described by solving Newton's second law. With axis z (vertical) and x (horizontal) as shown in Fig. N.1, the accelerations to which the projectile is subjected are those imposed by the earth's gravitational acceleration $\vec{g}$, that is the components

$$\frac{du_x}{dt} = g_x = 0; \quad \frac{du_z}{dt} = g_z = g \tag{N.1}$$

Here t is time. Velocity is obtained by integrating Eq. N.1. The velocity u along x is thus constant while it increases linearly with time along z

$$u_x = u_0 \cos\alpha; \quad u_z = u_0 \sin\alpha - gt \tag{N.2}$$

The magnitude of the velocity $u = \left(u_x^2 + u_y^2\right)^{1/2}$ thus follows

$$u = \left(g^2 t^2 - 2u_0 \sin\alpha + u_0^2\right)^{1/2} \tag{N.3}$$

The vertical and horizontal displacements are obtained by integrating Eq. N.2:

$$x = (u_0 \cos\alpha)t; \quad z = (u_0 \sin\alpha)t - \frac{1}{2}gt^2 \tag{N.4}$$

© Springer Nature Switzerland AG 2022
D. Beysens, *The Physics of Dew, Breath Figures and Dropwise Condensation*,
Lecture Notes in Physics 994, https://doi.org/10.1007/978-3-030-90442-5

Fig. N.1 Forces acting on an object with velocity u (other notations: see text)

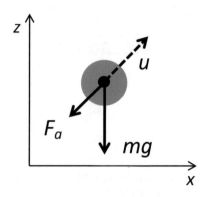

One deduces the *displacement* $D = \left(x^2 + z^2\right)^{1/2}$

$$D = t\left(\frac{1}{4}g^2t^2 - u_0 \sin \alpha gt + u_0^2\right)^{1/2} \tag{N.5}$$

From Eq. N.4, eliminating time, one finds the equation of a parabolic flight:

$$z = (\tan \alpha)x - \left(\frac{g}{2u_0^2\cos^2\alpha}\right)x^2 \tag{N.6}$$

The *time of flight* t_f, after which the projectile returns to a horizontal axis, corresponds in Eq. N.4 to $y = 0$ and is written as

$$t_f = \frac{2u_0 \sin \alpha}{g} \tag{N.7}$$

The *maximum height*, z_M, of the projectile comes from Eq. N.6 when $dz/dx = 0$:

$$z_M = \frac{u_0^2\sin^2\alpha}{2g} \tag{N.8}$$

The *highest achievable height*, z_{MM}, corresponds to $\alpha = 90°$:

$$z_{MM} = \frac{u_0^2}{2g} \tag{N.9}$$

The *horizontal range* d of the projectile is the horizontal distance it has traveled when it returns to its initial height ($z = 0$), that is at time $t = t_f$. From Eq. N.6, one finds

$$d = \frac{u_0^2}{g}\sin 2\alpha \tag{N.10}$$

The *maximum attainable range* is thus for angle $\alpha = 45°$, with value

$$d_M = \frac{u_0^2}{g} \tag{N.11}$$

N.2   Trajectory with Air Resistance

Air friction is, however, present during the motion of an object in the air (Fig. 14.2). Air friction, for an object with typical size ρ, velocity u in a fluid with kinematic viscosity v and low Reynolds number ensuring laminar flow is

$$\text{Re} = \frac{u\rho}{v} < 1000 \tag{N.12}$$

Air friction can be evaluated as a force that varies linearly with velocity:

$$F_a = -k_a u \tag{N.13}$$

The proportionality factor k_a is the air friction coefficient. With $v = 0.14 \times 10^{-2}$ m^{2}s^{-1} 4 (Table 1.2), the condition Eq. N.12 corresponds to the following condition:

$$\rho u < 1.4 \times 10^{-2} \tag{N.14}$$

With velocity ranging from 1 to 1000 m s^{-1}, Eq. N.14 corresponds to object sizes ranging from 1.4 cm to 14 μm. When the inequality Eq. N.12 is not fulfilled, the friction force becomes quadratic in u, F_a u^2. This more realistic assumption cannot be calculated analytically, but only by numerical simulations (Fig. N.2).

In the simplified linear conditions of Eq. N.12, the air friction coefficient can be derived from Stokes law. For a sphere with radius ρ in a fluid whose dynamic viscosity is η ($=v_{\text{air}}\rho_{\text{air}} = 1.7 \times 10^{-5}$ Pa s for air, see Table 2.2):

$$k_a = 6\pi \eta \rho \tag{N.15}$$

The mass of the projectile is noted m. Only the case $\alpha = [0 - 180°]$ is considered. The projectile starts from the (x, z) axes origin. Applying Newton's second law, it becomes

$$-k_a u_x = m\frac{du_x}{dt}; \quad -k_a u_z - mg = m\frac{du_z}{dt} \tag{N.16}$$

The solutions of the above differential equations, with boundary conditions at $t = 0$, $u_x = u_0 \cos\alpha$, $u_z = u_0 \sin\alpha$, and $x = 0$, $z = 0$, are the following:

$$u_x = u_{0x}\exp\left(-\frac{k_a}{m}t\right)$$

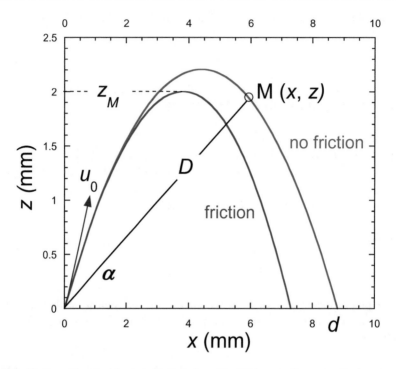

Fig. N.2 Motion with and without air friction of an object (0.1 mm radius water drop) sent at 45° with $u_0 = 0.3$ m s^{-1} initial velocity under earth's acceleration field (notations: see text)

$$x = \frac{m}{k_a} u_{0x} \left[1 - \exp\left(-\frac{k_a}{m} t \right) \right] \tag{N.17}$$

$$u_z = -\frac{mg}{k_a} + \left(u_{0z} + \frac{mg}{k_a} \right) \exp\left(-\frac{k_a}{m} t \right)$$

$$z = -\frac{mg}{k_a} t + \frac{m}{k_a} \left(u_{0z} + \frac{mg}{k_a} \right) \left[1 - \exp\left(-\frac{k_a}{m} t \right) \right] \tag{N.18}$$

The maximum height z_M corresponds to time t_M where $u_z = 0$, that is

$$t_M = \frac{m}{k_a} \ln\left(1 + \frac{k_a u_{0z}}{mg} \right); \; z_M = z(t_M) \tag{N.19}$$

One notes that for small air resistance $\frac{k_a}{m} t \ll 1$, Eq. N.18 can be expanded as

$$x = [u_{0x} t] - \frac{u_{0x} k_a}{2m} t^2 + \dots \tag{N.20}$$

$$z = \left[(u_{0z}) t - \frac{1}{2} g t^2 \right] - \frac{u_{0z} k_a}{2m} t^2 + \dots \tag{N.21}$$

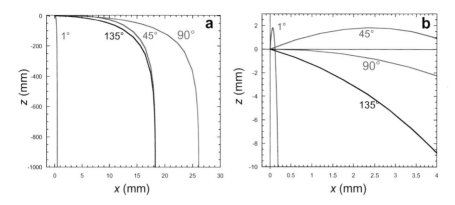

Fig. N.3 Trajectory of a 0.12 mm radius water drop sent at different angles from a vertical plane with initial velocity $u_0 = 0.2$ m s^{-1} corresponding to Fig. 10.18. **a** Down to $z = -1$ m. **b** Zoom down to $z = -10$ mm

The square bracket represents the evolution without air friction.

Figure N.1 reports the trajectories of a water drop assuming or not air friction, showing that air friction cannot be ignored. In Fig. N.3 are the trajectories calculated with air friction of a 0.12 mm water drop sent with initial velocity $u_0 = 0.2$ m s^{-1} at different angles from a vertical plane. It corresponds to Fig. 10.18 where the maximum height was ~1.5 mm. The air friction coefficient k_a is deduced from Eq. N.15.

Radiative Cooling

O

O.1 Radiative Properties of Materials

O.1.1 Definitions

Interaction of electromagnetic waves with matter involves absorption, transmission and reflection. Considering an incident monochromatic radiation with spectral intensity or radiance I_λ and wavelength λ, one can define spectral absorptivity, α_λ, reflectivity, R_λ and transmissivity or transmittance, T_λ, by

$$\alpha_\lambda = \frac{I_\lambda^A}{I_\lambda} \tag{O.1}$$

$$R_\lambda = \frac{I_\lambda^R}{I_\lambda} \tag{O.2}$$

$$T_\lambda = \frac{I_\lambda^T}{I_\lambda} \tag{O.3}$$

Here I_λ^A, I_λ^R and I_λ^T are, respectively, the spectral absorbed, reflected and transmitted intensities. Energy conservation implies that

$$I_\lambda = I_\lambda^A + I_\lambda^R + I_\lambda^T \tag{O.4}$$

O.1.2 Planck's Law and Black Body

Thermal excitation of matter atoms and molecules make them vibrate, rotate and results in emission of electromagnetic radiation. Let us consider a material with uniform temperature and composition that emits and thus absorbs all radiation (Kirchoff's law, see Sect. O.1.4), termed a "black body". Planck's law (Planck 1914) describes the electromagnetic radiation emitted by a black body in thermal

© Springer Nature Switzerland AG 2022
D. Beysens, *The Physics of Dew, Breath Figures and Dropwise Condensation*,
Lecture Notes in Physics 994, https://doi.org/10.1007/978-3-030-90442-5

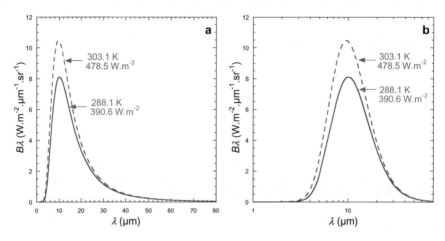

Fig. O.1 Black body spectral radiance at two temperatures 288.1 K (15 °C) and 300.1 K (30 °C) with integrated values

equilibrium at a definite temperature T. The radiation is homogeneous, isotropic and unpolarized. The spectral radiance, B_λ, describes the power radiated by a surface element at different wavelengths λ. It is measured per unit area of the body, per unit emission solid angle $d\Omega$ and per unit wavelength. It is given by

$$B_\lambda = \frac{2hc^2}{\lambda^5} \frac{1}{\exp\left(\frac{hc}{\lambda k_B T}\right) - 1} \tag{O.5}$$

Here k_B is the Boltzmann constant, h is the Planck's constant and c is the light velocity in the medium. Black body spectra at 288.1 K (15 °C) and 30.1 K (30 °C) are presented in Fig. O.1. In linear scales, they show a sharp cut-off at low wavelengths and a smooth decrease at long wavelengths. In a semi-log plot, the curves look nearly symmetrical.

O.1.3 Stefan-Boltzmann Law

The total radiative power P_B emitted by a surface element dS of a black body can be obtained by integrating Eq. O.5 over all wavelengths and over a hemispheric solid angle ($\Omega = 2\pi$) above the surface (Fig. O.2). Using spherical coordinates to define the radiation direction, with θ the polar angle (angle radiation direction – normal to the surface direction) and ϕ the azimuthal angle, the solid angle element can be written as $d\Omega = 2\pi \sin\theta d\theta d\phi$. According to Lambert's cosine law (Born and Wolf 1999), the emitting flux is connected to the apparent surface area $\cos\theta$

Fig. O.2 Radiation by surface element dA in solid angle $d\Omega$

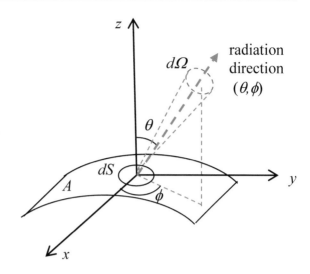

dA in the radiation direction. It thus comes as radiative power:

$$P_B = \int\limits_{0}^{\infty} d\lambda \int\limits_{\frac{1}{2}sph.} d\Omega B_\lambda \cos\theta = 2\pi \int\limits_{0}^{\infty} d\lambda \int\limits_{0}^{\pi/2} d\theta \int\limits_{0}^{2\pi} d\phi B_\lambda \sin\theta \cos\theta \qquad (O.6)$$

The integration over solid angle and wavelength gives

$$P_B = \sigma T^4 \qquad (O.7)$$

Here $\sigma = \frac{2k_B^4 \pi^5}{15c^2 h^3} = 5.670 \times 10^{-8}$ W m^{-2} K^{-4} is the so-called Stefan-Boltzmann constant.

O.1.4 Kirchhoff's Law of Thermal Radiation

The emissivity and absorptivity depend upon the distributions of states of molecular excitations. The Kirchhoff's law of thermal radiation states that absorptivity and emissivity of radiation are equal when a matter is at thermodynamic equilibrium at temperature T. Note that this equality is in general not respected when conditions of thermodynamic equilibrium are not met (the distributions of states of molecular excitations depend differently upon the distributions of states).

Considering the thermal radiation spectral intensity B_λ, the actual emitted spectral intensity I_λ^E or absorbed spectral intensity I_λ^A, one can define the spectral or monochromatic matter emissivity, ϵ_λ, as

$$\varepsilon_\lambda = \frac{I_\lambda^E}{B_\lambda} = \frac{I_\lambda^A}{I_\lambda} = \alpha_\lambda \qquad (O.8)$$

For the black body that emits and then absorbs all radiation, $\varepsilon_\lambda = 1$.

0.1.5 Grey Body

For materials other than black bodies, with discrete spectral bands, $\varepsilon_\lambda < 1$. Such materials are called *grey bodies*. This is especially the case of the atmosphere whose molecules absorb/radiate in specific spectral bands (see Section O.2 and Fig. O.3).

When going back to the total energy radiated by a grey body, Eq. O.5 becomes

$$B_\lambda^G = \varepsilon_\lambda \frac{2hc^2}{\lambda^5} \frac{1}{\exp\left(\frac{hc}{\lambda k_B T}\right) - 1} \tag{O.9}$$

Integration of Eq. O.9 over wavelength and solid angle give

$$P_B^G = \varepsilon \sigma T^4 \tag{O.10}$$

Here

$$\varepsilon = \frac{\int_0^\infty B_\lambda^G d\lambda}{\sigma T^4} \tag{O.11}$$

is the grey body emissivity. The grey body emissivity Eq. O.11 is thus the ratio between the energy the body radiates and the energy that a black body would radiate at the same temperature. It is a measurement of the capacity of a body to absorb and re-emit radiated energy. In the case of the black body, which absorbs all energy without re-emitting it, $\varepsilon = 1$. For any other body of uniform temperature, $\varepsilon < 1$. A material of weak emissivity, in particular a metal surface, therefore constitutes a good insulator of thermal radiation. Just as metals stop radio wave frequencies, good conductors stop infrared radiation. The largest emissivity (0.98) is for human skin, water, carbon graphite and plywood. The lowest is with metals (polished gold: 0.018).

0.2   Atmospheric Radiation

Atmospheric radiation is due to the atmosphere gases that absorb and emit (according to Kirchoff's law) radiation in the longwave part of the spectrum (3–100 μm). Oxygen and nitrogen, which comprise 99% of the atmosphere, do not absorb or emit radiation in the far-infrared as they are symmetrical molecules. Figure O.3 displays atmosphere absorption for sun radiations (shortwave) and earth IR radiations (longwave). High absorption in this spectral region corresponds to a black body at about 300 K temperature, except for a low absorption between 7 and 14 μm. The latter is known as the *atmospheric window* and does not contain water

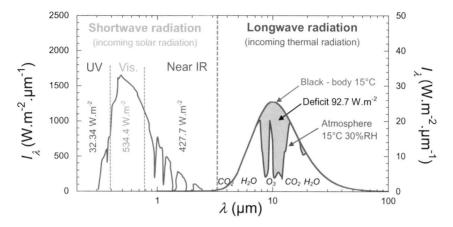

Fig. O.3 Typical spectral radiance received on earth from a 2π sr. hemisphere. Shortwave radiation (left ordinate) is of solar origin (ASTM G173-03, 2012) and longwave radiation (right ordinate) is from thermal excitation of mostly H_2O, CO_2 and O_3 molecules (MODTRAN®,[2] 1996))

contributions. Only a peak due to stratospheric O_3 is present, whose influence is relatively weak due to the stratosphere's low temperature.

Near the ground level, in the atmospheric boundary layer,[1] the contribution from water vapor (about 0.2–2% by volume) and carbon dioxide (about 0.03% in volume) is thus of great importance, with radiation from water vapor being by far the more important of the two.

Water vapor concentration usually decreases with altitude, which makes the boundary layer thickness the region where most IR radiation is emitted to the ground (Berger et al. 1984). Note that it is only the presence of small amounts of these two H_2O and CO_2 gases that prevent the atmosphere from being completely transparent in the far-infrared. As noted by Bliss (1961), if the air is completely dry (no water vapor) and completely free of carbon dioxide, the absence of a greenhouse effect would make the mean radiant temperature of the sky at night very near the absolute zero of outer space.

---

[1] The atmospheric boundary layer comprises the lowest part of the atmosphere extending from the ground (see, e.g., Wallace and Hobbs 2006). It is where ground and atmosphere exchange radiative, sensitive and latent heats. It extends until where cumulus clouds form, which marks the commencement of the free atmosphere. In this layer many physical quantities (air flow velocity, temperature, humidity…) display rapid and turbulent fluctuations and vertical mixing is strong. The boundary layer thickness can range from tens of meters to a few kilometers and varies with time.

[2] The MODTRAN® (MODerate resolution atmospheric TRANsmission) computer code is used worldwide by research scientists in government agencies, commercial organizations and educational institutions for the prediction and analysis of optical measurements through the atmosphere. MODTRAN was developed and continues to be maintained through a longstanding collaboration

O.2.1 Long-Wave Radiative Transfer in the Atmosphere

The problem to evaluate radiation from atmosphere or sky emissivity is estimating radiation from a gaseous mixture (water vapor, carbon dioxide etc.) in a hemisphere above ground level and relating it to ground parameters. Because of the strong influence of the water content in the boundary layer, parameters such as ground air temperature T_a and dew point temperature T_d will be particularly important to express the sky emissivity. Since atmosphere composition, temperature and pressure all vary with height above the ground, the task is quite complex and necessitates the use of a standard model for atmosphere temperature and density and concentration of IR emitting gases, together with IR radiometer measurements at different elevations above the ground. Below are described the solutions that eventually give rise to atmosphere emissivity.

O.2.2 Clear Sky Emissivity: Radiation Deficit

One first considers radiation from the atmosphere without clouds or aerosols (clear sky) and makes a plane-parallel approximation where the physical properties of the atmosphere components are assumed to be a function only of the height z above the ground (Fig. O.4). Sky radiation on the ground results from the balance between emission and absorption of thermal radiation issued by all atmosphere layers above the ground.

Bliss (1961) set up the basis for the study of atmospheric radiation by considering a simplified calculation where the contributions from each layer are added.

Fig. O.4 Sky emissivity at the ground level calculated from parallel slabs

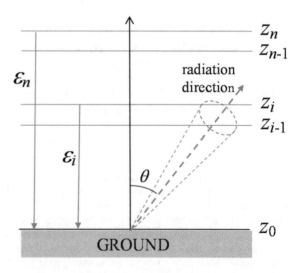

It can be indeed shown that the emissivity of a column and a hemisphere of gas (each having the same value of water content) are equal. Columnar emissivity and hemispherical emissivity are essentially merely two ways of defining the same thing.

Actual radiation from the atmosphere should then be considered neither as radiation from a column nor from a hemisphere, but rather as radiation from a series of horizontal layers, each of varying composition, temperature or pressure. For purposes of radiation calculations, those layers are considered infinite slabs.

Emissivity is assumed to be due to H_2O only (nevertheless, the small correction due to CO_2 traces can be accounted for). The main parameter is thus the water content of each layer, linearly proportional to the density-length product term $\rho_w L$, classically expressed in meteorology (see, e.g., Wallace and Hobbs 2006) by the density length (or "precipitable water"):

$$m_w = \rho_w L \tag{O.12}$$

Here ρ_w is the water vapor density and L is the slab thickness. m_w is usually expressed in g cm^{-2} and is numerically equal to the number of g of the radiating gas contained in a slab $L = 1$ cm long and 1 cm^2 in cross-section. Precipitable water and then emissivity varies with elevation depending on density length, atmosphere pressure, partial water vapor pressure and temperature (Fig. O.5). In general, the total emissivity increases with an increase of any of those variables.

However, over the range of pressure and temperature encountered, $\mathcal{E}$ is a weak function of partial water pressure and temperature. For the purpose of simplification, only the variation of $\mathcal{E}$ with density length and total atmosphere pressure is considered. The total spectral emissivity is obtained by integrating the emissivities over all θ angles, all wavelengths and slab elevations. Mean height dependence of density, temperature, concentration is considered following a standard dependence, the "U. S. Standard Atmosphere" (Haltiner and Martin 1957).

Below are the main steps of the calculation whose details are described in Bliss (1961). First, all layers (z_i) with emissivity ε_i^* are assumed to emit intensity:

$$I_i^*(z_i, T_a) = \varepsilon_i^*(z_i, T_a)\sigma T_a^4 \tag{O.13}$$

The layers temperature is first assumed to be the same as the ground temperature T_a, thus defining on the ground an apparent emissivity contribution:

$$\varepsilon_s^* = \frac{P_i^*}{\sigma T_a^4} = \sum_{i=1}^{n} \varepsilon_i^*(z_i, T_a) \tag{O.14}$$

where

$$P_i^* = \sum_{i=1}^{n} I_i^*(z_i, T_a) \tag{O.15}$$

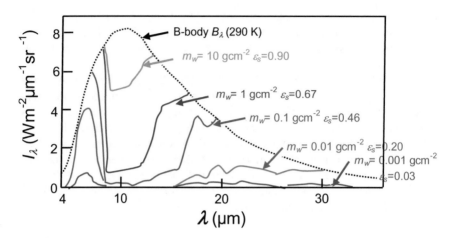

Fig. O.5 Spectral intensity of the water vapor radiation in the near IR for different condensable water thickness m_w as found at different elevation and corresponding sky emissivity ε_s. Temperature is 290 K. Adapted from Bliss (1961)

corresponds to the apparent clear sky total radiation intensity received on the ground. In a further step, a correction is given by considering each thermal emission σT_i^4 to account for the different layer temperatures $T_i(z_i)$, leading to the radiation intensity P_i received on the ground. A final emissivity ε_s follows

$$\varepsilon_s = \frac{P_i}{\sigma T_a^4} = \sum_{i=1}^{n} \varepsilon_i(z_i, T_i) \tag{O.16}$$

with

$$P_i = \left[\sum_{i=1}^{n} I_i^*(z_i, T_a) \left(\frac{T_i(z_i)}{T_a} \right)^4 \right] \tag{O.17}$$

The most important contribution to emissivity is the water content of the atmosphere in the boundary layer, itself a function of the dew point temperature. In most cases, the CO_2 contribution can be ignored. The variation of $\varepsilon_s^* \approx \varepsilon_s$ with respect to T_d indeed follows a linear variation in nearly all the studied range:

$$\varepsilon_s^* \approx \varepsilon_s \approx 0.8004 + 0.00396 T_d(°C) \tag{O.18}$$

From the above one can also define an apparent sky temperature T_s, which relates ground air temperature and emissivity to give the same sky radiance on the ground:

$$P_i = \sigma T_s^4 = \varepsilon_s \sigma T_a^4 \tag{O.19}$$

which gives the following expression for the sky temperature:

$$T_s = T_a \varepsilon_s^{1/4} \tag{O.20}$$

It is interesting to express the radiation deficit R_i (the available cooling energy for dew formation (see Chap. 7) of a black body on the ground when exposed to clear sky radiation:

$$R_i = \sigma T_a^4 - \varepsilon_s \sigma T_a^4 = \sigma (1 - \varepsilon_s) T_a^4 \tag{O.21}$$

Radiation deficit is thus a function of T_a and T_d (from ε_s) or, equivalently, of T_a and air relative humidity RH near the ground. Figure O.6 presents the deficit as a function of T_a for different air relative humidity RH and T_d. Not surprisingly, the radiation deficit increases when RH decreases at constant T_a (the air vapor content decreases). For typical nocturnal conditions where dew forms ($T_a = 15\ °C$ and $RH = 85\%$, $T_d = 12.5\ °C$), $R_i \approx 60\ W/m^2$.

Other studies have been concerned with measurements in different locations of earth-sky radiative fluxes using spectroradiometers and pyrgeometers. The ozone contribution and a few other small constituents that were forgotten in the above Bliss approach are taken into account. However, the formulations for emissivity somewhat differ between each other and to some extent depend on the place and elevation where the measurements were performed. The earlier determination was performed by Angstrom (1916). Table O.1 reports the main evaluations. Although the formulation by Berdahl and Fromberg (1982) is often cited for its simplicity,

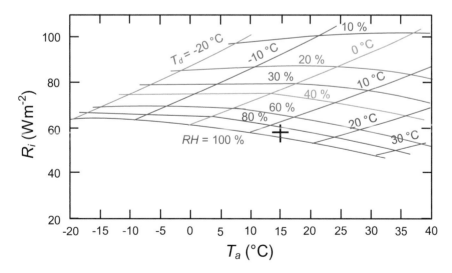

Fig. O.6 Radiation deficit as a function of ground air temperature T_a for various relative humidity RH and dew point temperature T_d. The large "+" represents typical diurnal conditions of dew formation ($T_a = 15\ °C$, $RH = 85\%$, $T_d = 12.5\ °C$, $R_i \approx 60\ W/m^2$). Adapted from Bliss (1961)

the most pertinent evaluation seems to be that from Berger et al. (1992) who included a new variable, the site elevation H.

Different formulations of clear sky emissivities are expressed in Table O.1 by using either the saturated vapor pressure p_s or the dew point temperature T_d. With p_s in mbar and T_d in °C, these quantities can be related by the following formulation derived from Antoine's Eq. 1.10:

$$p_s = \frac{1013}{760}\exp\left(20.519 - \frac{5179.25}{T_d + 273.15}\right) \qquad (O.22)$$

Table O.1 Different evaluations of nocturnal clear sky emissivity. T_a(°C): Air temperature near the ground; T_d(°C): Air dew point temperature near the ground; RH (%): relative humidity; p_w (mb): air water vapor pressure near the ground; H (km): site elevation a.s.l. Relation between p_s and T_d is given by Eq. O.22

Nocturnal emissivity ε_s	Remark	Ref
$0.25 - 0.32 \times 10^{-0.052 - p_w}$		Angstrom (1916)
$0.564 + 0.059\sqrt{p_w}$		Brunt (1932, 1940)
$0.8004 + 0.00396 T_d$		Bliss (1961)
$0.66 + 0.040\sqrt{p_w}$		Kondratyev (1969)
$0.67 p_w^{0.080}$		Staley and Jurica (1972)
$1 - 0.261 e^{-(7.77 \times 10^{-4} T_a^2)}$	No humidity dependence	Idso and Jackson (1969)
$1.24\left[\frac{p_w}{T_a + 27A11.15}\right]^{1/7}$		Brutsaert, 1975
$0.787 + 0.0028 T_d$		Clark and Allen (1978)
$0.741 + 0.0062 T_d$	Close to Tang et al. (2004)	Berdahl and Fromberg (1982)
$\begin{bmatrix} 5.7723 + 0.9555(0.6017)^H \end{bmatrix}$ $\times T_a^{1.893} \times RH^{0.065} \times 10^{-4}$	Elevation dependence is considered	Melchor Centeno (1982)
$0.770 + 0.0038 T_d$		Berger et al., 1984
$0.711 + 0.56(T_d/100) + 0.73(T_d/100)^2$	Similar to Brunt (1932, 1940)	Berdahl and Martin (1984)
$0.73223 + 0.006349 T_d$		Chen et al. (1991)
$0.75780 - 0.049487 H + 0.0057086 H^2$ $+ (4.3628 - 0.25422 H$ $+ 0.05302 H^2) \times 10^{-3} T_d$	Elevation dependence is considered	Berger et al. (1992)
$1 - \left[1 + 46.5\left(\frac{p_w}{T_a + 27A11.15}\right)\right]$ $\times e^{-\left[1.2 + 139.5\left(\frac{p_w}{T_a + 27A11.15}\right)\right]^{1/2}}$		Prata (1996)

The Kelvin Equation

One follows here the classical demonstration as clearly exposed in Anonymous (2020). One considers a liquid in contact with its own vapor whose interface is a spherical cap with the radius of curvature ρ, defined >0 if seen convex from the gas phase. The liquid–vapor interfacial tension is σ_{LV}. The Laplace-Young equation (see Appendix E, Eq. E.5) indicates that gas and liquid pressures, p_g and p_v, respectively, are connected:

$$p_l - p_v = \frac{2\sigma_{LV}}{\rho} \tag{P.1}$$

At liquid–vapor equilibrium, with a flat interface, the chemical potentials of liquid, μ_l, and vapor, μ_v, are equal:

$$\mu_l = \mu_v \tag{P.2}$$

Let us assume that one slightly moves to a new equilibrium. This gives $\mu_l + d\mu_l = \mu_v + d\mu_v$ and therefore

$$d\mu_l = d\mu_v \tag{P.3}$$

From the definition of the chemical potential one can relate the above transformation when performed at constant temperature T to the pressures and molar volumes V_{ml} and V_{mv} in the liquid and vapor phase, respectively. Equation P.3 becomes

$$V_{ml}dp_l = V_{mv}dp_v \tag{P.4}$$

After derivation, the Laplace-Young equation Eq. P.1 gives

$$dp_l - dp_v = d\left(\frac{2\sigma_{LV}}{\rho}\right) \tag{P.5}$$

From Eqs. P.4 and P.5, one obtains

© Springer Nature Switzerland AG 2022
D. Beysens, *The Physics of Dew, Breath Figures and Dropwise Condensation*,
Lecture Notes in Physics 994, https://doi.org/10.1007/978-3-030-90442-5

$$\frac{V_{mv} - V_{ml}}{V_{ml}} dp_v \approx \frac{V_{mv}}{V_{ml}} dp_v = d\left(\frac{2\sigma_{LV}}{\rho}\right)$$ (P.6)

The approximation accounts for the fact that $V_{ml} \ll V_{mg}$. Assuming that the vapor can be considered as an ideal gas and making use of the specific vapor constant $r_v = R/M_v$ (see Sect. 1.2), one deduces from $V_{ml}/V_{mv} = \rho_v/\rho_l$ and $\rho_v = p_v/r_v T$, with ρ_l and ρ_v the liquid and vapor density, respectively:

$$\rho_l r_v T \frac{dp_v}{p_v} = d\left(\frac{2\sigma_{LV}}{\rho}\right)$$ (P.7)

Integrating now from the state flat interface—saturation pressure p_s and curved interface—vapor pressure p_v, one obtains

$$\ln\left(\frac{p_v}{p_s}\right) = \frac{2\sigma_{LV}}{\rho_l r_v T \rho}$$ (P.8)

For small differences in pressure, one can write

$$\frac{\Delta p_v}{p_s} = \frac{2\sigma_{LV}}{\rho_l r_v T_d \rho}$$ (P.9)

In wetted capillaries, where the meniscus is concave as seen from the vapor phase, $\rho < 0$ and $p_v < p_s$. It is thus possible to condense the liquid at a pressure lower than the saturation pressure (the so-called capillary condensation). In contrast, for convex interfaces such as sessile drops, $\rho > 0$, $p_v > p_s$ and condensation needs a higher pressure than the saturation pressure.

Meniscus in a Groove

Q

Consider a two-dimensional section of a groove (Fig. Q.1) with the objective to estimate the area of the liquid in the groove when the meniscus reaches the edge. The liquid contact angle is θ_a (the advancing contact angle during condensation). The width of the groove is b and its depth is c. Groove width is assumed to be always less than the capillary length $b < l_c$. The meniscus is then circular with radius ρ.

The liquid area ABCDE to be determined can be considered as the sum of areas $ABCD = bc$ and ADE. The latter is itself the difference between areas OAED and OAD.

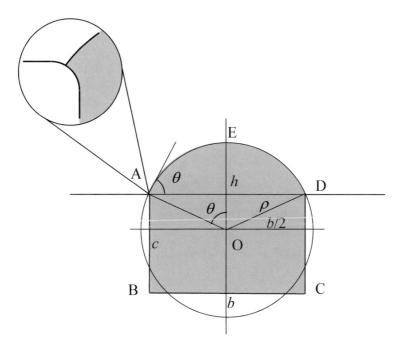

Fig. Q.1 Meniscus in a groove

© Springer Nature Switzerland AG 2022

433

D. Beysens, *The Physics of Dew, Breath Figures and Dropwise Condensation*,
Lecture Notes in Physics 994, https://doi.org/10.1007/978-3-030-90442-5

With

$$\rho = \frac{b}{2 \sin \theta} \qquad (Q.1)$$

And

$$h = \frac{b}{2 \tan \theta} \qquad (Q.2)$$

one can express the area OAED as

$$\text{OAED} = \theta \rho^2 = \frac{b^2}{4 \sin^2 \theta} \theta \qquad (Q.3)$$

The area OAD can be written as

$$\text{OAD} = \frac{bh}{2} = \frac{b^2}{4 \tan \theta} \qquad (Q.4)$$

Thus for the area ADE = OAED – OAD:

$$\text{ADE} = \frac{b^2}{8 \sin^2 \theta} (2\theta - \sin 2\theta) \qquad (Q.5)$$

Defining now the function

$$g(\theta) = \frac{2\theta - \sin 2\theta}{8 \sin^2 \theta} \qquad (Q.6)$$

one can write

$$\text{ADE} = b^2 g(\theta) \qquad (Q.7)$$

Eventually, the total area ABCDE = ABCD + ADE can be written as

$$\text{ABCDE} = bc + b^2 g(\theta) = bc \left[1 + \frac{b}{c} g(\theta) \right] \qquad (Q.8)$$

Light Reflection and Transmission

Light in a medium is characterized by propagation and attenuation, which are classically characterized by a complex refractive index:

$$\tilde{n} = n + i\kappa \tag{R.1}$$

The different contributions of n (propagation) and κ (attenuation) can be seen in the expression of the electric field of a plane electromagnetic wave traveling in the x-direction. A complex wave number can be defined, with λ_0 the light wavelength in a vacuum:

$$\tilde{k} = \frac{2\pi\tilde{n}}{\lambda_0} \tag{R.2}$$

The electric field can be written, with ω the light angular frequency:

$$E(x,t) = E_0 e^{i\left(\tilde{k}x - \omega t\right)} = E_0 e^{-\frac{2\pi}{\lambda_0}\kappa x} e^{i\left(\frac{2\pi n}{\lambda_0}x - \omega t\right)} \tag{R.3}$$

It corresponds to the light intensity $P \propto |E|^2$ such as, with $P_0 \propto |E_0|^2$

$$P = P_0 e^{-\frac{4\pi}{\lambda_0}\kappa x} \tag{R.4}$$

The light transmittance τ over length x for wavelength λ_0 can thus be defined as

$$\tau(x) = e^{-\frac{4\pi}{\lambda_0}\kappa x} \tag{R.5}$$

Equation R.5 can also be written as

$$\tau(x) = e^{-\alpha x} \tag{R.6}$$

It defines the attenuation or absorptivity coefficient

$$\alpha = \frac{4\pi}{\lambda_0}\kappa \tag{R.7}$$

© Springer Nature Switzerland AG 2022
D. Beysens, *The Physics of Dew, Breath Figures and Dropwise Condensation*,
Lecture Notes in Physics 994, https://doi.org/10.1007/978-3-030-90442-5

and the corresponding penetration depth

$$l = \alpha^{-1} = \frac{\lambda_0}{4\pi\kappa} \tag{R.8}$$

The reflection coefficient r_m at the interface vacuum/medium (m) under normal incidence is classically written as, with P_r the reflected intensity (see, e.g., Born and Wolf 1999)

$$r_m = \left|\frac{n + i\kappa - 1}{n + i\kappa + 1}\right|^2 = \frac{(n-1)^2 + \kappa^2}{(n+1)^2 + \kappa^2} \tag{R.9}$$

Since both n and κ depend on wavelength λ_0, one can also consider Eq. R.1 the wavelength dependence. One now deals with a spectral refractive index:

$$\tilde{n}_\lambda = n_\lambda + i\kappa_\lambda \tag{R.10}$$

It follows from the spectral absorptivity coefficient, with $\alpha_\lambda = \lambda_0/4\pi\kappa_\lambda$ the spectral transmittancy

$$\tau_\lambda(x) = e^{-\frac{4\pi}{\lambda_0}\kappa_\lambda x} = e^{-\alpha_\lambda x} \tag{R.11}$$

A spectral reflection coefficient $r_{m\lambda}$ follows from Eqs. R.9 and R.10:

$$r_{m\lambda} = \frac{(n_\lambda - 1)^2 + \kappa_\lambda^2}{(n_\lambda + 1)^2 + \kappa_\lambda^2} \tag{R.12}$$

Bibliography

Aarts, D.G.A.L., Lekkerkerker,H.N.W., Guo,H., Wegdam, G.H., Bonn, D.: Hydrodynamics of droplet coalescence. Phys. Rev. Lett. **95**, 164503 (2005)

Abdelsalam, M.E., Bartlett, P.N., Kelf, T., Baumberg, J.: Wetting of regularly structured gold surfaces. Langmuir **21**, 1753–1757 (2005)

Abu-Orabi, M.: Modeling of heat transfer in dropwise condensation. Int. J. Heat Mass Trans. **41**, 81–87 (1998)

Aitken, J.: Breath figures. Proc. R. Soc. Edinb. **20**, 94–97 (1893)

Alam, T., Li, W., Yang, F.Q., Chang, W., Li, J., Wang, Z., Khan, J., Li, C.: Force analysis and bubble dynamics during flow boiling in silicon nanowire microchannels. Int. J. Heat Mass Trans. **101**, 915–926 (2016)

Alduchov, O.A., Eskridge, R.E.: Improved Magnus form approximation of saturation vapor pressure. J. Appl. Meteor. **35**, 601–609 (1996)

Alizadeh-Birjandi, E., Alshehri, A., Kavehpour, H.P.: Condensation on surfaces with biphilic topography: experiment and modeling. Front. Mech. Eng. **5**, 38 (2019)

Allain C., Cloitre M.: Capillary aggregation at an interface. In: Jullien, R., Peliti, L., Rammal, R., Boccara, N. (eds.) Universalities in Condensed Matter, vol. 32. Springer Proceedings in Physics. Springer, Berlin, Heidelberg (1988a)

Allain, C., Cloitre, M.: Horizontal cylinders at an interface: equilibrium, shape of the meniscus and capillary interaction. Ann. Phys. **13**, 141 (1988b)

Al-Sharafi, A., Yilbas, B.S., Sahin, A.Z., Ali, H., Al-Qahtani, H.: Heat transfer characteristics and internal fluidity of a sessile droplet on hydrophilic and hydrophobic surfaces. Appl. Therm. Eng. **108**, 628–640 (2016)

Anand, S., Son, S.Y.: Sub-micrometer dropwise condensation under superheated and rarefied vapor condition. Langmuir **26**, 17100–17110 (2010)

Anand, S., Paxson, A.T., Dhiman, R., Kripa, J.D., Varanasi, K.: Enhanced condensation on lubricant-impregnated nanotextured surfaces. ACS Nano **6**, 10122–10129 (2012)

Anand, S., Varanasi, K.K.: Articles and methods for levitating liquids on surfaces, and devices incorporating the same. WO 2013188702 A1 (2013)

Anand, S., Rykaczewski, K., Subramanyam, S.B., Beysens, D., Varanasi, K.K.: How droplets nucleate and grow on liquids and liquid impregnated surfaces. Soft Matter **11**, 69–80 (2015)

Anderson, D.M., Gupta, M.K., Voevodin, A.A., Hunter, C.N., Putnam, S.A., Tsukruk, V.V., Fedorov, A.G.: Using amphiphilic nanostructures to enable long-range ensemble coalescence and surface rejuvenation in dropwise condensation. ACS Nano **6**, 3262–3268 (2012)

Andrieu, C., Beysens, D.A., Nikolayev, V.S., Pomeau, Y.J.: Coalescence of sessile drops. J. Fluid. Mech. **453**, 427 (2002)

Angstrom, A.: Über die Gegenstrahlung der Atmosphare. Meteor. Z. **33**, 529–538 (1916)

Anonymous.: The Kelvin equation (2020). https://fr.wikipedia.org/wiki/%C3%89 (In French)

Antoine, C.: Tensions des vapeurs; nouvelle relation entre les tensions et les températures (Vapor Pressure: a new relationship between pressure and temperature). Comptes Rendus des Séances de l'Académie des Sciences (in French) **107**, 681–684, 778–780, 836–837

© Springer Nature Switzerland AG 2022

D. Beysens, *The Physics of Dew, Breath Figures and Dropwise Condensation*,
Lecture Notes in Physics 994, https://doi.org/10.1007/978-3-030-90442-5

Antonini, C., Bernagozzi, I., Jung, S., Poulikakos, D., Marengo, M.: Water drops dancing on ice: How sublimation leads to drop rebound. Phys. Rev. Lett. **111**, 014501 (2013)

Arias, G.I., Reza, J., Trejo, A.: Temperature and sodium chloride effects on the solubility of anthracene in water. J. Chem. Thermodyn. **42**, 1386–1392 (2010)

Artus, G.R.J., Jung, S., Zimmermann, J., Gautschi, H.P., Marquardt, K., Seegar, S.: Silicone nanofilaments and their application as superhydrophobic coatings. Adv. Mater. **18**, 2758–2762 (2006)

ASTM G173-03.: Standard tables for reference solar spectral irradiances: direct normal and hemispherical on 37_Tilted Surface. Active Standard ASTM G173 j Developed by Subcommittee: G03.09. Book of Standards, vol. 14, no. 04 (2012)

Azimi, G., Dhiman, R., Kwon, H.M., Paxson, A.T., Varanasi, K.K.: Hydrophobicity of rare-earth oxide ceramics. Nat. Mater. **12**, 315–320 (2013)

Balaur, E., Macak, J.M., Tsuchiya, H., Schmuki, P.: Wetting behaviour of layers of TiO_2 nanotubes with different diameters. J. Mater. Chem. **15**, 4488–4491 (2005)

Baldacchini, T., Carey, J.E., Zhou, M., Mazur, E.: Superhydrophobic surfaces prepared by microstructuring of silicon using a femtosecond laser. Langmuir **22**, 4917–4919 (2006)

Barati, S.B., Pinoli, J.C., Valette, S., Gavet, Y.: Differential and average approaches to Rose and Mei dropwise condensation models. Int. J. Math. Model. Methods Appl. Sci. **11**, 40–46 (2017)

Baratian, D., Dey, R., Hoek, H., van den Ende, D., Mugele, F.: Breath figures under electrowetting: electrically controlled evolution of drop condensation patterns. Phys. Rev. Lett. **120**, 214502 (2018)

Barthes, M., Reynard, C., Santini, R., Tadrist, L.: Non-condensable gas influence on the Marangoni convection during a single vapour bubble growth in a subcooled liquid. EPL **77**, 14001 (2007)

Batchelor, G.K.: An Introduction to Fluid Dynamics. Cambridge University Press, Cambridge NJ (2000)

Ben, A.M., Cummings, L.J., Pomeau, Y.: Transition of a moving contact line from smooth to angular. Phys. Fluids **15**, 2949–2960 (2003)

Berdahl, P., Fromberg, R.: The thermal radiance of clear skies. Sol. Energy **29**, 299–314 (1982)

Berdahl, P., Martin, M.: Emissivity of clear skies. Sol. Energy **32**, 663–664 (1984)

Berger, X., Buriot, D., Gamier, F.: About the equivalent radiative temperature for clear skies. Sol. Energy **32**, 725–733 (1984)

Berger, X., Bathiebo, J., Kieno, F., Awanou, C.N.: Clear sky radiation as a function of altitude. Renew. Energy **2**, 139–157 (1992)

Bergeron, V.V., Langevin, D.: Monolayer Spreading of Polydimethylsiloxane Oil on Surfactant Solutions (1996)

Berthier, J., Brakke, K.A., Gosselin, D., Huet, M., Berthier, E.: Metastable capillary filaments in rectangular cross-section open microchannels. AIMS Biophys. **1**, 31–48 (2014a)

Berthier, J., Brakke, K.A., Berthier, E.: A general condition for spontaneous capillary flow in uniform cross-section microchannels. Microfluid. Nanfluid. **16**, 779–785 (2014b)

Betz, A.R., Jenkins, J., Kim, C.J., Attinger, D.: Boiling heat transfer on superhydrophilic, superhydrophobic, and superbiphilic surfaces. Int. J. Heat Mass Transf. **57**(2013), 733–741 (2013)

Beysens, D.: Dew nucleation and growth. C R Phys. **7**, 1082 (2006). and references therein

Beysens, D.: Estimating dew yield worldwide from a few meteo data. Atmos. Res. **167**, 146–155 (2016)

Beysens, D.: Dew Water. Rivers Publisher, Gistrup (2018)

Beysens, D., Knobler, C.M.: Growth of breath figures. Phys. Rev. Lett. **57**, 1433–1436 (1986)

Beysens, D., Narhe, R.: Contact line dynamics in the late-stage coalescence of diethylene glycol drops. J. Phys. Chem. B **110**, 22133–22135 (2006)

Beysens, D., Knobler, C.M., Schaffar, H.: Scaling in the growth of aggregates on a surface. Phys. Rev. B **41**, 9814–9818 (1990)

Beysens, D., Steyer, A., Guenoun, P., Fritter, D., Knobler, C.M.: How does dew form? Phase Trans. **31**, 219–246 (1991)

Beysens, D., Broggini, F., Milimouk-Melnytchouk, I., Ouazzani, J., Tixier, N.: New architectural forms to enhance dew collection. Chem. Eng. Trans. **34**, 79–84 (2013)

Beysens, D.: Dew songs. CEA Internal Report (1991)

Biance, A.L., Clanet, C., Quéré, D.: Leidenfrost drops. Phys. Fluids **15**, 1632–1637 (2003)

Bico, J., Marzolin, C., Quere, D.: Pearl drops. Europhys. Lett. **47**(2), 220–226 (1999)

Billingham, J., King, A.C.: Surface-tension-driven flow outside a slender wedge with an application to the inviscid coalescence of drops. J. Fluid Mech. **533**, 193–221 (2005)

Bintein, P.-B., Lhuissier, H., Mongruel, A., Royon, L., Beysens, D.: Grooves accelerate dew shedding. Phys. Rev. Lett. **122**, 098005 (2019)

Blake, T.D., Ruschak, J.: A maximum speed of wetting. Nat. Lond. **282**, 489–491 (1979)

Blaschke, J., Lapp, T., Hof, B., Vollmer, J.: Breath figures: nucleation, growth, coalescence, and the size distribution of droplets. Phys. Rev. Lett. **109**, 068701 (2012)

Blasius, H.: Grenzschichten in Flüssigkeiten mit kleiner Reibung (Boundary layers in fluids with little friction). Z. Math. Phys. **56**, 1–37 (1908). (In German)

Bliss, R.A.: Atmospheric radiation near the surface of the ground. Sol. Energy **5**, 103–120 (1961)

Bonn, D., Eggers, J., Indekeu, J., Meunier, J., Rolley, E.: Wetting and spreading. Rev. Mod. Phys. **81**, 739–805 (2009)

Boreyko, J.-B., Chen, C.H.: Self-propelled dropwise condensate on superhydrophobic surfaces. Phys. Rev. Lett. **103**, 184501-1–184501-4 (2009)

Boreyko, B., Chen, C.H.: Vapor chambers with jumping-drop liquid return from superhydrophobic condensers. Int. J. Heat Mass Transf. **61**, 409–418 (2013a)

Boreyko, J.B., Collier, C.P.: Delayed frost growth on jumping-drop superhydrophobic surfaces. ACS Nano **7**, 1618–1627 (2013b)

Bormashenko, E., Stein, T., Whyman, G., Bormashenko, Y., Pogreb, R.: Wetting properties of the multiscaled nanostructured polymer and metallic superhydrophobic surfaces. Langmuir **22**, 9982–9985 (2006)

Bormashenko, E., Frenkel, M., Vilk, A., Fedorets, A.A., Aktaev, N.E., Dombrovsky, L.A., Nosonovsky, M.: Characterization of self-assembled 2D patterns with Voronoi entropy. Entropy **20**, 956 (2018)

Born, M., Wolf, E.: Principles of Optics, 7th edn. Cambridge University Press, Cambridge, UK (1999)

Bouchaud, J.-P., Georges, A.: Anomalous diffusion in disordered media: statistical mechanisms, models and physical applications. Phys. Rep. **195**, 127–293 (1990)

Bouverot-Dupuis O., Herbaut, R., Mongruel, A., Beysens, D.: Influence of pinning on drop coalescence. PMMH Report (2021)

Bravo, J., Zhai, L., Wu, Z., Cohen, R.E., Rubner, M.F.: Transparent superhydrophobic films based on silica nanoparticles. Langmuir **23**(13), 7293–7298 (2007)

Bray, A.J., Derrida, B., Godreche, C.: Non-trivial algebraic decay in a soluble model of coarsening. Europhys. Lett. **27**, 175–180 (1994)

Brin, A., Mérigoux, R.: Adsorption d'une couche d'eau à la surface de l'acide oléique (Adsorption of a water layer at the surface of oleic acid). C.R. Acad. Sci. (Paris) **238**, 1808–1809 (1954) (In French)

Briscoe, B.J., Galvin, K.P.: The evolution of a 2-D constrained growth system of droplets-breath figures. J. Phys. D **23**, 422–428 (1990)

Briscoe, B.J., Galvin, K.P.: Growth with coalescence during condensation. Phys. Rev. A **43**, 1906–1917 (1991)

Brochard, F.: Motions of droplets on solid surfaces induced by chemical or thermal gradients. Langmuir **5**, 432–438 (1989)

Brunt, D.: Notes on radiation in the atmosphere. Quart. J. R. Meteorol. Soc. **58**, 389–420 (1932)

Brunt, D.: Radiation in the atmosphere. Quart. J. R. Meteorol. Soc. **66**, 34–40 (1940)

Brutsaert, W.: On a derivable formula for long-wave radiation from clear skies. Water Resour. Res. **11**, 742–744 (1975)

Buff, F.R., Lovett, R.A., Stillinger, F.H., Jr.: Interfacial density profile for fluids in the critical region. Phys. Rev. Lett. **15**, 621–623 (1965)

Buller, A.: Researches on Fungi. Longmans, Green, and Co., London (1909)

Burnside, B.M., Hadi, H.A.: Digital computer simulation of dropwise condensation from equilibrium droplet to detectable size. Int. J. Heat Mass Transf. **42**, 3137–3146 (1999)

Carlson, A., Kim, P., Amberg, G., Stone, H.A.: Short and long time drop dynamics on lubricated substrates. EPL **104**, 34008 (2013)

Carslaw, H., Jaeger, J.: Conduction of Heat in Solids, 2nd edn, p. 77. Oxford University Press, Oxford, UK (1959)

Cha, H., Wu, A., Kim, M.-K., Saigusa, K., Liu, A., Miljkovic, N.: Nanoscale-agglomerate-mediated heterogeneous nucleation. Nano Lett. **17**, 7544–7551 (2017)

Chan, D.Y.C., Henry, J.D., White, L.R.: The interaction of colloidal particles collected at fluid interfaces. J. Colloid Interface Sci. **79**, 410–418 (1981)

Chatterjee, R., Beysens, D., Anand, S.: Delaying ice and frost formation using phase-switching liquids. Adv. Mater. 1807812 (2019)

Chaudhury, M.K., Whitesides, G.M.: How to make water run uphill. Science **256**, 1539–1541 (1992)

Chaudhury, M.K., Chakrabarti, A., Tibrewal, T.: Coalescence of drops near a hydrophilic boundary leads to long range directed motion. Extreme Mech. Lett. **1**, 104–113 (2014)

Chavan, S., Cha, H., Orejon, D., Nawaz, K., Singla, N., Yeung, Y.F., Park, D., Kang, D.H., Chang, Y., Takata, Y., Miljkovic, N.: Heat transfer through a condensate droplet on hydrophobic and nanostructured superhydrophobic surfaces. Langmuir **32**, 7774–7787 (2016)

Che, Q., Lu, Y., Wang, F., Zhao, X.: Effect of substrate wettability and flexibility on the initial stage of water vapor condensation. Soft Matter **15**, 10055 (2019)

Chen, W., Fadeev, A.T., Hsieh, M.C., Oner, D., Youngblood, J., McCarthy, J.: Ultrahydrophobic and ultralyophobic surfaces: some comments and examples. Langmuir **15**, 3395–3399 (1999)

Chen, A.C., Peng, X.S., Koczkur, K., Miller, B.: Super-hydrophobic tin oxide nanoflowers. Chem. Commun. **17**, 1964–1965 (2004)

Chen, R., Lu, M.C., Srinivasan, V., Wang, Z., Cho, H.H., Majumdar, A.: Nanowires for enhanced boiling heat transfer. Nano Lett. **9**, 548–553 (2009)

Chen, X.M., Wu, J., Ma, R.Y., Hua, M., Koratkar, N., Yao, S.H., Wang, Z.K.: Nanograssed micropyramidal architectures for continuous dropwise condensation. Adv. Funct. Mater. **21**, 4617–4623 (2011)

Chen, X., Patel, R.S., Weibel, J.A., Garimella, S.V.: Coalescence-induced jumping of multiple condensate droplets on hierarchical superhydrophobic surfaces. Sci. Rep. **6**, 18649 (2016)

Chen, B., Kasker, J., Maloney, J., Gigis, G. A., Clark, D.: Determination of the clear sky emissivity for use in cool storage roof and roof pond applications. In: Proceedings of the ASES Proceedings, Denver, CO (1991)

Chen, C.-H., Cai, Q., Tsai, C., Chen, C.L., Xiong, G., Yu, Y., Ren, Z.: Dropwise condensation on superhydrophobic surfaces with two-tier roughness. Appl. Phys. Lett. **90**, 173108-1–173108-3 (2007)

Cheng, Y.T., Rodak, D.E., Angelopoulos, A., Gacek, T.: Microscopic observations of condensation of water on lotus leaves. Appl. Phys. Lett. **87**, 194112–194113 (2005)

Christian, J.W.: The Theory of Transformations in Metals and Alloys. Newnes (1975)

Chou, S.Y., Krauss, P.R., Renstrom, P.J.: Imprint lithography with 25-nanometer resolution. Science **272**(5258), 85–87 (1996)

Clark, G., Allen, C.P.: The estimation of atmospheric radiation for clear and cloudy skies. In: Proceedings of the Second National Passive Solar Conference, Philadelphia, Part 2, p. 676 (1978)

Cox, R.G.: The dynamics of the spreading of liquids on a solid surface. Part 1. Viscous Flow. J. Fluid Mech. **168**, 169–194 (1986)

Dai, X., Sun, N., Nielsen, S.O., Stogin, B.B., Wang, J., Yang, S., Wong, T.-S.: Hydrophilic directional slippery rough surfaces for water harvesting. Sci. Adv. **4**, eaaq0919 (2018)

Daniel, S., Chaudhury, M.K.: Rectified motion of liquid drops on gradient surfaces induced by vibration. Langmuir **18**, 3404–3407 (2002)

Daoud, S., Mongruel, A., Royon, L., Beysens, D.: PMMH internal report (2019)

Davis, S.H.: Moving contact lines and rivulet instabilities. Part 1. The Static Rivulet. J. Fluid Mech **98**, 225–242 (1980)

Delavoipière, J., Heurtefeu, B., Teisseire, J., Chateauminois, A., Tran, Y., Fermigier, M., Verneuil, E.: Swelling dynamics of surface-attached hydrogel thin films in vapor flows. Langmuir **34**, 15238–15244 (2018)

Derby, M.M., Chatterjee, A., Peles, Y., Jensen, M.K.: Flow condensation heat transfer enhancement in a mini-channel with hydrophobic and hydrophilic patterns. Int. J. Heat Mass Transf. **68**, 151–160 (2014)

Derrida, B., Godreche, C., Yekutieli, I.: Stable distributions of growing and coalescing droplets. Europhys. Lett. **12**, 385–389 (1990)

Derrida, B., Godreche, C., Yekutieli, I.: Phys. Rev. A **44**, 6241–6251 (1991)

Derrida, B., Godreche, C., Yekutieli, I.: Non-trivial exponents in the zero temperature dynamics of the ID Ising and Potts models. J. Phys.: Math. Gen. **27**, L357–L361 (1994)

Dietrich, O.: Diffusion Coefficients of Water (2020). http://www.dtrx.de/od/diff/

Diez, J.A., González, A.G., Kondic, L.: On the breakup of fluid rivulets. Phys. Fluids **21**, 082105 (2009)

Doppelt, E., Beysens, D.: PMMH internal report (2013)

Dorrer, C., Rühe, J.: Advancing and receding motion of droplets on ultrahydrophobic post. Surf. Langmuir **22**, 7652–7657 (2006)

Dorrer, C., Rühe, J.: Condensation and wetting transitions on microstructured ultrahydrophobic surfaces. Langmuir **2007**(23), 3820–3824 (2007)

Dorrer, C., Rühe, J.: Wetting of silicon nanograss: from superhydrophilic to superhydrophobic surfaces. Advanced **29**, 159–163 (2008)

Dorrer, C., Rühe, J.: Some thoughts on superhydrophobic wetting. Soft Matter **5**, 51–61 (2009)

Downing, H., Williams, D.: Optical constant of water in the infrared. J. Geophys. Res. **80**, 1656–1661 (1975)

Du Roure, O., Saez, A., Buguin, A., Austin, R.H., Chavrier, P., Silbrzan, P., Ladoux, B.: Force mapping in epithelial cell migration. PNAS **15**, 2390–2395 (2005)

Duineveld, P.C.: The stability of ink-jet printed lines of liquid with zero receding contact angle on a homogeneous substrate. J. Fluid Mech. **477**, 175–200 (2003)

Dupuis, A., Yeomans, J.M.: Modeling droplets on superhydrophobic surfaces: equilibrium states and transitions. Langmuir **21**, 2624–2629 (2005)

Eddi, A., Winkels, K.G., Snoeijer, J.H.: Influence of droplet geometry on the coalescence of low viscosity drops. Phys. Rev. Lett. **111**, 144502 (2013)

Eggers, J.: Nonlinear dynamics and breakup of free-surface flows Jens Eggers. Rev. Mod. Phys. **69**, 865–930 (1997)

Eggers, J., Lister, J.R., Stone, H.A.: Coalescence of liquid drops. J. Fluid Mech. **401**, 293–310 (1999)

Elsasser, W.M.: Heat transfer by infrared radiation in the atmosphere. In: Harvard Meteorological Studies, no. 6. Harvard University, Blue Hill Meteorological Observatory, Milton, MA, USA (1942)

ElSherbini, A.I., Jacobi, A.M.: Retention forces and contact angles for critical liquid drops on non-horizontal surfaces. J. Colloid Interface Sci. **299**, 841–849 (2006)

Enright, R., Miljkovic, N., Sprittles, J., Nolan, K., Mitchell, R., Wang, E.N.: How coalescing droplets jump. ACS Nano **8**, 10352–10362 (2014)

Erbil, H.Y., Cansoy, C.E.: Range of applicability of the Wenzel and Cassie-Baxter equations for superhydrophobic surfaces. Langmuir **25**, 14135–14145 (2009)

Erbil, H.Y., Demirel, A.L., Avci, Y., Mert, O.: Transformation of a simple plastic into a superhydrophobic surface. Science **299**, 1377–1380 (2003)

Eslami, F., Elliott, J.A.W.: Thermodynamic Investigation of the Barrier for heterogeneous nucleation on a fluid surface in comparison with a rigid surface. J. Phys. Chem. B **2011**(115), 10646–10653 (2011)

Extrand, C.W., Gent, A.N.: Retention of liquid drops by solid surface. J. Colloid Interface Sci. **138**, 431–442 (1990)

Extrand, C.W., Kumagai, Y.: Liquid drops on an inclined plane: the relation between contact angles, drop shape, and retentive force. J. Colloid Interface Sci. **170**, 515–521 (1995)

Family, F., Meakin, P.: Scaling of the droplet-size distribution in vapor-deposited thin films. Phys. Rev. Lett. **61**, 428–431 (1988)

Family, F., Meakin, P.: Kinetics of droplet growth processes: simulations, theory, and experiments. Phys Rev. A **40**, 3836–3854 (1989)

Fang, W.J., Mayama, H., Tsujii, K.: Spontaneous formation of fractal structures on triglyceride surfaces with reference to their super water-repellent properties. J. Phys. Chem. B **111**, 564–571 (2007)

Fedosov, A.I.: Thermocapillary motion. Zh. Fiz. Khim. **30**, 366–373 (1956)

Feng, L., Song, Y., Zhai, J., Liu, B., Xu, J., Jiang, L., Zhu, D.: Creation of a superhydrophobic surface from an amphiphilic polymer. Angew. Chem. Int. Ed. **42**, 800–802 (2003)

Feng, X.J., Feng, L., Jin, M.H., Zhai, J., Jiang, L., Zhu, D.B.: Reversible super-hydrophobicity to super-hydrophilicity transition of aligned ZnO nanorod films. J. Am. Chem. Soc. **126**, 62–63 (2004)

Flura, D., Saint-Jean, S., Huber, L., Chelle, M., Beysens, D., Jacquemoud, S.: Liquid water on leaves surfaces: development of a method of nonintrusive measurement of condensation for applications to biophysics or chemical surface processes. PROJET INNOVANT 2007–2008, Département Environnement et Agronomie, Rapport de restitution. (In French) (2009)

Fonash, S.J.: An overview of dry etching damage and contamination effects. J. Electrochem. Soc. **137**, 3885–3892 (1990)

Foster, E.L., De Leon, A.C.C., Mangadlao, J., Advincula, R.: Electropolymerized and polymer grafted superhydrophobic, superoleophilic, and hemi-wicking coatings. J. Mater. Chem. **22**, 11025–11031 (2012)

Fritter, D., Knobler, C.M., Roux, D., Beysens, D.: Computer simulations of the growth of breath figures. J. Stat. Phys. **52**, 1447–1459 (1988)

Fritter, D., Knobler, C.M., Beysens, D.: Experiments and simulation of the growth of droplets on a surface (breath figures). Phys. Rev. A **43**, 2558–2869 (1991)

Frymyer, B., Vahedi, N., Neti, S., Oztekin, A.: Thermal modeling with diffusion for dropwise condensation of humid air. Int. J. Heat Mass Transf. **151**, 119413 (2020)

Fürstner, R., Barthlott, W., Neinhuis, C., Walzel, P.: Wetting and self-cleaning properties of artificial superhydrophobic surfaces. Langmuir **2005**(21), 956 (2005)

Gao, N., Geyer, F., Pilat, D.W., Wooh, S., Vollmer, D., Butt, H.J., Berger, R.: How drops start sliding over solid surfaces. Nat. Phys. **14**, 191–196 (2018)

Gao, S., Chu, F., Zhang, X., Wu, X.: Behavior of condensed droplets growth and jumping on superhydrophobic surface. In: ICCHMT 2019-E3S Web of Conferences, vol. 128, p. 07003 (2019)

Garimella, M., Koppu, S., Kadlaskar, S.S., Pillutla, V., Abhijeet, Choi, W.: Difference in growth and coalescing patterns of droplets on bi-philic surfaces with varying spatial distribution. J. Colloid Interface Sci. **505**, 1065–1073 (2017)

de Gennes, P.-G., Brochart-Wyart, F., Quéré, D.: Capillarity and Wetting Phenomena: Drops, Bubbles, Pearls, Waves. Springer (2003)

George, S.M.: Atomic layer deposition: an overview. Chem. Rev. **110**, 111–131 (2010)

Gerasopoulos, K., Luedeman, W.L., Ölçeroglu, E., McCarthy, M., Benkoski, J.B.: Effects of engineered wettability on the efficiency of dew collection. ACS Appl. Mater. Interfaces **2018**(10), 4066–4076 (2018)

Gersten, K., Herwig, H.: Strömungsmechanik, Grundlagen der Impuls-, Wärme- und Stoff-Ubertragung aus Asymptotischer Sicht. (Fluid mechanics, basics of momentum, heat and mass transfer from an asymptotic point of view), (Vieweg+Teubner Verlag, Braunschweig-Wiesbaden). (In German) (1992)

Glicksman, L.R., Hunt, Jr., A.W., 1972. Numerical simulation of dropwise condensation. Int. J. Heat Mass

Gnanappa, A.K., O'Murchu, C., Slattery, O., Peters, F., Aszalo´ s-Kiss, B., Tofail, S.A.M.: Effect of annealing on hydrophobic stability of plasma deposited fluoropolymer coatings. Polym. Degrad. Stability **93**, 2119–2126 (2008)

Gose, E.E., Mucciardi, A.N., Baer, E.: Model for dropwise condensation on randomly distributed sites. Int. J. Heat Mass Transfer **10**, 15–22 (1967)

Graham, C.: The limiting heat transfer mechanisms of dropwise condensation. Ph.D. thesis, MIT (1969)

Guadarrama-Cetina J., Mongruel A., González-Viñas W., Beysens D.: Percolation-induced frost formation. EuroPhys. Lett. **101**, 16009 (2013a)

Guadarrama-Cetina J., Mongruel A., González-Viñas W., Beysens D.: Frost formation with salt. EuroPhys. Lett. **110**, 56002 (2013b)

Guadarrama-Cetina, J., Narhe, R.D., Beysens, D.A., Gonzalez- Viñas, W.: Droplet pattern and condensation gradient around a humidity sink. Phys. Rev. E 89, 012402 (2014); Erratum: Droplet pattern and condensation gradient around a humidity sink [Phys. Rev. E **89**, 012402 (2014). Phys. Rev. E **101**, 039901(E) (2020)

Guo, Z., Zhou, F., Hao, J., Liu, W.: Effects of system parameters on making aluminum alloy lotus. J. Colloid Interface Sci. **303**, 298–305 (2006)

Guyon, E., Hulin, J.-P., Petit, L.: Hydrodynamique physique. In: EDP Sciences, 3rd edn, Paris. (In French) (2012)

Hall, R.S., Board, S.J., Clare, A.J., Duffey, R.B., Playle, T.S., Poole, D.H.: Inverse leidenfrost phenomenon. Nature **1969**(224), 266–267 (1969)

Halperin, B.I., Nelson, D.R.: Theory of two-dimensional melting. Phys. Rev. Lett. **41**, 121–124 (1978); Erratum: Phys. Rev. Lett. **41**, 519

Haltiner, G.J., Martin, F.L.: Dynamical and Physical Meteorology, p. 52. McGraw-Hill, New York, NY, USA (1957)

Han, J.T., Xu, X.R., Cho, K.W.: Diverse access to artificial superhydrophobic surfaces using block copolymers. Langmuir **21**, 6662–6665 (2005)

Han, J.T., Jang, Y., Lee, D.Y., Park, J.H., Song, S.H., Ban, D.Y.: Fabrication of a bionic super-hydrophobic metal surface by sulfur-induced morphological development. J. Mater. Chem. **2005**(15), 3089–3092 (2005)

Hasimoto, H.: On the periodic fundamental solutions of the Stokes equations and their application to viscous flow past a cubic array of spheres. J. Fluid Mech. **5**, 317–328 (1959)

Hassel, S., Milenkovic, U.S., Greve, H., Zaporojtchenko, V., Adelung, R., Faupel, F.: Model systems with extreme aspect ratio, tunable geometry, and surface functionality for a quantitative investigation of the lotus effect. Langmuir **23**, 2091–2094 (2007)

He, B., Patankar, N.A., Lee, J.: Multiple equilibrium droplet shapes and design criterion for rough hydrophobic surfaces. Langmuir **19**, 4999–5003 (2003)

Heist, R.H., Reiss, H.: Investigation of the homogeneous nucleation of water vapor using a diffusion cloud chamber. J. Chem. Phys. **59**, 665–671 (1973)

Henkel, C., Snoeijer, J.H., Thiele, U.: Gradient-dynamics model for liquid drops on elastic substrates (2021). arXiv:2107.08397 [physics.flu-dyn]

Herminghaus, S., Brinkmann, M., Seemann, R.: Wetting and dewetting of complex surface geometries Annu. Rev. Mater. Res. **38**, 101 (2008)

Hernández-Sánchez, J.F., Lubbers, L.A., Eddi, A., Snoeijer, J.H.: Symmetric and asymmetric coalescence of drops on a substrate. Phys. Rev. Lett. **109**, 184502 (2012)

Hinds, W.C.: Aerosol Technology: Properties, Behavior, and Measurement of Airborne Particles, 2nd edn. Wiley, New York (1999)

Hinrichsen, E.L., Feder, J., Jossang, T.: Geometry of random sequential adsorption. J. Stat. Phys. **44**, 793–827 (1986). and references therein

Hiratsuka, M., Emoto, M., Konno, A., Ito, S.: Molecular dynamics simulation of the influence of nanoscale structure on water wetting and condensation. Micromachines **10**, 587 (2019)

Hosemann, R., Bagchi, S.N.: Direct Analysis of Diffraction by Matter. North-Holland, Amsterdam (1962)

Hosono, E., Fujihara, S., Honma, I., Zhou, H.: Superhydrophobic perpendicular nanopin film by the bottom-up process. J. Am. Chem. Soc. **127**, 13458–13459 (2005)

Hou, Y.M., Yu, M., Chen, X.M., Wang, Z.K., Yao, S.H.: Recurrent filmwise and dropwise condensation on a beetle mimetic surface. ACS Nano **9**, 71–81 (2015)

Hou, K., Li, X., Li, Q., Chen, X.: Tunable wetting patterns on superhydrophilic/superhydrophobic hybrid surfaces for enhanced dew-harvesting efficacy. Adv. Mater. Interfaces **2020**(7), 1901683 (2020)

Hsu, C.C., Long-Sheng Kuo, P.-S. Chen, P.H.: Classification of surface wettability and fabrication methods in surface modification: a review. Int. J. Microsc. Nanosc. Therm. Fluid Transp. Phenom. **1**, 301-327 (2011)

Idso, S.B., Jackson, R.D.: Thermal radiation from the atmosphere. J. Geophys. Res. **74**, 5397–5403 (1969)

Incropera, F.P., DeWitt, D.P.: Fundamental of Heat and Mass Transfer, 5th edn. Wiley, Hoboken, NJ (2002)

Ishino, C., Okumura, K., Quéré, D.: Wetting transitions on rough surfaces. Europhys. Lett. **68**, 419–425 (2004)

Ishizaki, T., Saito, N., Inoue, Y., Bekke, M., Takai, O.: Fabrication and characterization of ultra-water-repellent alumina silica composite films. J. Phys. D: Appl. Phys. **40**(1), 192–197 (2007)

Izumi, M., Kumagai, S., Shimada, R., Yamakawa, N.: Heat transfer enhancement of dropwise condensation on a vertical surface with round shaped grooves. Exp. Therm. Fluid Sci. **28**, 243–248 (2004)

Izumi, M., Shinmura, T., Isobe, Y., Yamakawa, N., Ohtani, S., Westwater, J.W.: Drop and filmwise condensation on a horizontally scratched rough surface. Kagaku Kogaku Ronbunshu **12**, 647–653 (1986) (In Japanese; translation by Yamakawa, N., Dept. of Resource Chern., Faculty of Engineering, Iwate University, Ueda 4–3–5, Morioka, Iwate)

Jaber, A., Schlenoff, J.B.: Recent developments in the properties and applications of polyelectrolyte multilayers. Curr. Opin. Colloid Interface Sci. **11**, 324–329 (2006)

Jain, S., Yamano, T.: Double power law in the Japanese financial market. Int. J. Product. Manag. Assess. Technol. **7**, 28–35 (2019)

Jeffreys, H., Jeffreys, B.S.: Methods of Mathematical Physics, 2nd edn. Cambridge University Press, New York (1950)

Jeong, Y.-M., Lee, J.-K., Jun, H.-W., Kim, G.-R., Choe, Y.: Preparation of super-hydrophilic amorphous titanium dioxide thin film via PECVD process and its application to dehumidifying heat exchangers. J. Ind. Eng. Chem. **15**, 202–206 (2009)

Jiang, X., Zhao, B., Chen, L.: Sessile microdrop coalescence on partial wetting surfaces: effects of surface wettability and stiffness. Langmuir **35**, 12955–12961 (2019)

Jin, M.H., Feng, L., Feng, X.J., Sun, T.L., Zhai, J., Li, T.J., Jiang, L.: Superhydrophobic aligned polystyrene nanotube films with high adhesive force. Adv. Mater. **17**, 1977–1981 (2005)

Ju, J., Bai, H., Zheng, Y., Zhao, T., Fang, R., Jiang, L.: A multi-structural and multi-functional integrated fog collection system in cactus. Nat. Commun. **3**, 1247 (2012)

Jung, Y.C., Bhushan, B.: Wetting behaviour during evaporation and condensation of water microdroplets on superhydrophobic patterned surfaces. J. Microsc. **229**, 127–140 (2008)

Kajiya, T., Schellenberger, F., Papadopoulos, P., Vollmer, D., Butt, H.J.: 3D imaging of waterdrop condensation on hydrophobic and hydrophilic lubricant-impregnated surfaces. Sci. Rep. **6**, 23687 (2016)

Karpitschkaa,S., Pandeya, A., Lubbersb, L.A., Weijsc, J.H., Bottod, L., Dase, S., Andreotti, B., Snoeijer, J.H.: Liquid drops attract or repel by the inverted Cheerios effect. PNAS 7403–7407 (2016)

Kashchiev, D.: Nucleation. In: Basis Theory with Application. Butterworth-Heinemann Ltd, Oxford, Boston (2000)

Kawai, A., Nagata, H.: Wetting behavior of liquid on geometrical rough surface formed by photolithography. J. Appl. Phys. **33L**, 1283–1285 (1994)

Keller, J.B., Milewski, P.A., Vanden-Broeck, J.M.: Merging and wetting driven by surface tension. Eur. J. Mech. B Fluids **19**, 491–502 (2000)

Kertesz, J., Kiss, L.B.: The noise spectrum in the model of self-organised criticality. J. Phys.: Math. Gen. **23**(L433), L440 (1990)

Khodabocus, I., Sellier, M., Nock, V.: Scaling laws of droplet coalescence: theory and numerical simulation. Adv. Math. Phys. **12**, 1–16 (2018)

Kim, B.S., Shin, S., Lee, D.H., Choi, G., Lee, H.J., Kim, K.M., Cho, H.H.: Stable and uniform heat dissipation by nucleate-catalytic nanowires for boiling heat transfer. Int. J. Heat Mass Tran. **70**, 23–32 (2014)

Kim, E., Lee, C., Kim, H., Kim, J.: Simple approach to superhydrophobic nanostructured Al for practical antifrosting application based on enhanced self-propelled jumping droplets. ACS Appl. Mater. Interfaces **2015**(7), 7206–7213 (2015)

Kim, S., Kim, K.J.: Dropwise condensation modeling suitable for superhydrophobic surfaces. J. Heat Transf. **133**, 081502 (2011)

Kim, T.-Y., Ingmar, B., Bewilogua, K., Hwan, Oh, K., Lee, K.-R.: Wetting behaviours of a-C:H:Si:O film coated nano-scale dual rough surface. Chem. Phys. Lett. **436**, 199–203 (2007)

Kim, M.-K., Kim, E.C., Ahn, J., Kim, Y.S., Cha, H., Miljkovic, N.: Condensation limits on biphilic surfaces. In: Proceedings of the 6th Micro and Nano Flows Conference Atlanta, USA (2018)

Klopp, C., Eremin, A.: On droplet coalescence in quasi-two-dimensional fluids. Langmuir **36**, 10615–10621 (2020)

Knobler, C.M., Beysens, D.: Growth of breath figures on fluid surfaces. Europhys. Lett. **6**, 7–712 (1988)

Kondratyev, K.Y.: Radiation in the Atmosphere. Academic Press, New York, NY, USA (1969)

Kou, L., Gao, C.: Making silica nanoparticle-covered graphene oxide nanohybrids as general building blocks for large-area superhydrophilic coatings. Nanoscale **3**, 519–528 (2011)

Koynov, S., Brandt, M.S., Stutzmann, M.: Black nonreflecting silicon surfaces for solar cells. Appl. Phys. Lett. **88**, 203107 (2006)

Kralchevsky, P.A., Nagayama, K.: Capillary interactions between particles bound to interfaces, liquid films and biomembranes. Adv. Coll. Interface. Sci. **85**, 145–192 (2000)

Kulinich, S.A., Farzaneh, M.: Hydrophobic properties of surfaces coated with fluoroalkylsiloxane and alkylsiloxane monolayers. Surf. Sci. **573**, 379–390 (2004)

Kumagai, S., Tanaka, S., Katsuda, H., Shimada, R.: On the enhancement of filmwise condensation heat transfer by means of the coexistence with dropwise condensation sections. Exp. Heat Transf. **4**, 71–82 (1991)

Kumagai, S., Yamauchi, A., Fukushima, H., Takeyama, T.: ASME-JSME Thermal Energy Joint Conference, 2n edn, vol. 4, pp. 409–415 (1987)

Kumar, G.U., Suresh, S., Thansekhar, M.R., Babu, P.D.: Effect of diameter of metal nanowires on pool boiling heat transfer with FC-72. Appl. Surf. Sci. **423**, 509–520 (2017)

Lafaurie, B., Nardone, C., Scardovelli, R., Zaleski, S., Zanetti, G.: Modelling merging and fragmentation in multiphase flows with SURFER. J. Comput. Phys. **113**, 134 (1994)

Lafuma, A., Quéré, D.: Superhydrophobic states. Nat. Mater. **2**, 457–460 (2003)

Landau, L.D., Lifshitz, E.M.: Statistical Physics. Pergamon Press, London-Paris (1958)

Lau, K.K.S., Bico, J., Teo, K.B.K., Chhowalla, M., Amaratunga, G.A.J., Milne, W.I., McKinley, G.H., Gleason, K.K.: Superhydrophobic carbon nanotube forests. Nano Lett. **3**, 1701–1705 (2003)

Lawrence, M.G.: The relationship between relative humidity and the dewpoint temperature in moist air. A simple conversion and applications. Bull. Am. Phys. Soc. **86**, 225–233 (2005)

Leach, R.N., Stevens, F., Langford, S.C., Dickinson, J.T.: Dropwise condensation: experiments and simulations of nucleation and growth of water drops in a cooling system. Langmuir **22**, 8864–8872 (2006)

Lee, S., Kwon, T.H.: Effects of intrinsic hydrophobicity on wettability of polymer replicas of a superhydrophobic lotus leaf. J. Micromech. Microeng. (17), 687–692 (2007)

Lee, W., Jin, M.K., Yoo, W.C., Lee, J.K.: Nanostructuring of a polymeric substrate with well-defined nanometer-scale topography and tailored surface wettability. Langmuir **20**, 7665–7669 (2004)

Lee, S.-M., Lee, H.S., Kim, D.S., Kwon, T.H.: Fabrication of hydrophobic films replicated from plant leaves in nature. Surf. Coat. Technol. **201**, 553–559 (2006)

Lee, A., Moon, M.-W., Lim, H., Kim, W.-D., Kim, H.-Y.: Water harvest via dewing. Langmuir **28**, 10183–10191 (2012)

Lee, M.W., Kang, D.K., Yoon, S.S., Yarin, A.L.: Coalescence of two drops on partially wettable substrates. Langmuir **28**, 3791–3798 (2012)

Lee, J., Shao, B., Won, Y.: Droplet jumping on superhydrophobic copper oxide nanostructured surfaces. IEEE Trans. Compon. Packag. Manuf. Technol. **9**, 1075–1081 (2019)

Lee, J., Lee, S., Lee, J.: Improved humid air condensation heat transfer through promoting condensate drainage on vertically stripe patterned bi-philic surfaces. Int. J. Heat Mass Transf. **160**, 120206 (2020)

Lefevre, E.J., Rose, J.W.: A theory of heat transfer by dropwise condensation. In: Proceedings of the Third International Heat Transfer Conference, American Institute of Chemical Engineers, vol. 2, pp. 362–375, New York (1966)

Leroy, C.: Mémoire sur l'Elévation et la Suspension de l'Eau dans l'Air, et sur la Rosée (Dissertation on the Elevation and the Suspension of Water in the Air, and on Dew). Mémoires de l'Académie Royale des Sci. 481–518 (In French) (1751)

Lhuissier, H., Bintein, P.-B., Mongruel, A., Royon, L., Beysens, D.: Report on dew collection by post and grooved substrates. PMMH Internal Report (2015)

Li, Y., Duan, C.: Bubble-regulated silicon nanowire synthesis on micro-structured surfaces by metal-assisted chemical etching. Langmuir **31**, 12291–12299 (2015)

Li, Y., Cai, W., Cao, B., Duan, G., Sun, F., Li, C., Jia, L.: Two-dimensional hierarchical porous silica film and its tunable superhydrophobicity. Nanotechnology **17**, 238–243 (2006a)

Li, J., Fu, J., Cong, Y., Wu, Y., Xue, L.J., Han, Y.C.: Macroporous fluoropolymeric films templated by silica colloidal assembly: a possible route to super-hydrophobic surfaces. Appl. Surf. Sci. **252**, 2229–2234 (2006b)

Li, X.H., Chen, G., Ma, Y., Feng, L., Zhao, H.: Preparation of a super-hydrophobic poly(vinyl chloride) surface via solvent-nonsolvent coating. Polymer **47**, 506–509 (2006c)

Li, Y., Huang, X.J., Heo, S.H., Li, C.C., Choi, Y.K., Cai, W.P., Cho, S.O.: Superhydrophobic bionic surfaces with hierarchical microsphere/SWCNT composite array. Langmuir **23**, 2169–2174 (2007)

Li, C., Wang, Z., Wang, P.I., Peles, Y., Koratkar, N., Peterson, G.P.: Nanostructured copper interfaces for enhanced boiling. Small **4**, 1084–1088 (2008)

Li, D., Wu, G.S., Wang, W., Wang, Y.D., Liu, D., Zhang, D.C., Chen, Y.F., Peterson, G.P., Yang, R.: Enhancing flow boiling heat transfer in microchannels for thermal management with monolithically-integrated silicon nanowires. Nano Lett. **12**, 3385–3390 (2012)

Lide D.R. (ed.): Handbook of Chemistry and Physics, 79th edn. CRC Press (1998)

Limat, L., Podgorski, T., Flesselles, J.-M., Fermigier, M., Moal, S., Stone, H.A., Wilson, S.K., Andreotti, B.: Shape of drops sliding down an inclined surface. In: Proceedings 2001 European Coating Symposium Advances in Coating Processes, Bruxelles, Belgique (2001)

Lipschutz, S., Spiegel, M., Liu, J.: Schaum's Outlines: mathematical Handbook of Formulas and Tables. McGraw-Hill, New York (2008)

Liu, X., Trosseille, J., Mongruel, A., Marty, F., Basset, P, Laurent, J., Royon, L., Beysens, D., Cui, T., Bourouina, T.: PMMH Internal Report (2020)

Liu, X., Trosseille, J., Mongruel, A., Marty, F., Basset, P, Laurent, J., Royon, L., Beysens, D., Cui, T., Bourouina, T.: Tailoring silicon for dew water harvesting panels. ISCIENCE (2021). https://doi.org/10.1016/j.isci.2021.102814

Liu, X., He, J.: Hierarchically structured superhydrophilic coatings fabricated by self-assembling raspberry-like silica nanospheres. J. Colloid Interface Sci. **314**, 341–345 (2007)

Lo, C.W., Wang, C.C., Lu, M.C.: Spatial control of heterogeneous nucleation on the superhydrophobic nanowire array. Adv. Funct. Mater. **24**, 1211–1217 (2014a)

Lo, C.W., Wang, C.C., Lu, M.C.: Scale effect on dropwise condensation on superhydrophobic surfaces. ACS Appl. Mater. Interfaces **6**, 14353–14359 (2014b)

Love, C., Gates, B.D., Wolfe, D.B., Paul, K.E., Whitesides, G.M.: Fabrication and wetting properties of metallic half-shells with submicron diameters. Nano Lett. **2**, 891–894 (2002)

Lu, X.Y., Zhang, C.C., Han, Y.C.: Macromol. Low-density polyethylene superhydrophobic surface by control of its crystallization behavior. Macromol. Rapid Commun. **25**, 1606–1610 (2004)

Lu, M.C., Chen, R., Srinivasan, V., Carey, V.P., Majumdar, A.: Critical heat flux of pool boiling on Si nanowire array-coated surfaces. Int. J. Heat Mass Tran. **54**, 5359–5367 (2011)

Lu, M.C., Lin, C.C., Lo, C.W., Huang, C.W., Wang, C.C.: Superhydrophobic Si nanowires for enhanced condensation heat transfer. Int. J. Heat Mass Tran. **111**, 614–623 (2017)

Lyons, C.G.: The angles of floating lenses. J. Chem. Soc. 623–634 (1930)

Majee, S.B. (ed.): Emerging concepts in analysis and applications of hydrogels (Intech, Rijeka) (2016)

Mahapatra, P.S., Ghosh, A., Ganguly, R., Meg, C.M.: Key design and operating parameters for enhancing dropwise condensation through wettability patterning. Int. J. Heat Mass Transf. **92**, 877–883 (2016)

Maheshwari, N., Kottantharayil, A., Kumar, M., Mukherji, S.: Long term hydrophilic coating on poly(dimethylsiloxane) substrates for microfluidic applications. Appl. Surf. Sci. **257**, 451–457 (2010)

Malik, F.T., Clement, R.M., Gethin, D.T., Beysens, D., Cohen, R.E., Krawszik, W., Parker, A.R.: Dew harvesting efficiency of four species of cacti. Bioinspiration and Biomimetics **10**, 036005 (2015)

Marcos-Martin, M., Beysens, D., Bouchaud, J.-P., Godreche, C., Yekutieli, I.: Self-diffusion and 'visited' surface in the droplet condensation problem (breath figures). Phys. A **214**, 396–412 (1995)

Marcos-Martin, M.-A., Bardat, A., Schmitthaeusler, R., Beysens, D.: Sterilization by vapor condensation. Pharm. Technol. Eur. **8**, 24–32 (1996)

Marek, R., Straub, J.: Analysis of the evaporation coefficient and the condensation coefficient of water. Int. J. Heat Mass Transf. **44**, 39–53 (2001)

Martines, E., Seunarine, K., Morgan, H., Gadegaard, N., Wilkinson, D.W., Riehle, M.O.: Superhydrophobicity and superhydrophilicity of regular nanopatterns. Nano Lett. **5**, 2097 (2005)

McCarthy, M., Gerasopoulos, K., Maroo, S.C., Hart, A.J.: Materials, fabrication, and manufacturing of micro/nanostructured surfaces for phase-change heat transfer enhancement. Nanosc. Microsc. Thermophys. Eng. **18**(288–310), 43 (2014)

Meakin, P.: Dropwise condensation: the deposition growth and coalescence of fluid droplets. Phys. Scripta **44**, 31–41 (1992)

Meakin, P., Family, F.: Scaling in the kinetics of droplet growth and coalescence: heterogeneous nucleation. J. Phys. A: Math. Gen. **22**(1989), L225–L230 (1989)

Medici, M.-G., Mongruel, A., Royon, L., Beysens, D.: Edge effects on water droplet condensation. Phys. Rev. E **90**, 062403 (2014)

Mei, M., Yu, B., Cai, J., Luo, L.: A fractal analysis of dropwise condensation heat transfer. Int. J. Heat Mass Transf. **52**, 4823–4828 (2009)

Mei, M., Hu, F., Han, C., Cheng, Y.: Time averaged droplet size distribution in steady-state dropwise condensation. Int. J. Heat Mass Transf. **88**, 338–345 (2015)

Melchor Centeno, V.: New formulae for the equivalent night sky emissivity. Sol. Energy **28**, 489–498 (1982)

Mérigoux, R.: Recherches sur la contamination du verre par les corps gras (Research on the contamination of glass by fats). Rev. Opt. **9**, 281–296 (1937). (In French)

Mérigoux, R.: Various structures of dew deposited by breath on certain fats. Différentes structures de la buée déposée par le souffle sur certains corps gras. C.R. Acad. Sci. (paris) **207**, 47–48 (1938). (In French)

Merkel, T.C., Bondar, V.I., Nagai, K., Freeman, B.D., Pinnau, I.: Gas sorption, diffusion, and permeation in poly(dimethylsiloxane). J. Polym. Sci Part B: Polym. Phys. **38**, 415–434 (2000)

Miljkovic, N., Enright, R., Wang, E.N.: Effect of droplet morphology on growth dynamics and heat transfer during condensation on superhydrophobic nanostructured surfaces. ACS Nano **2**, 1776–1785 (2012)

Miljkovic, N., Preston, D.J., Enright, R., Wang, E.N.: Electric-field-enhanced condensation on superhydrophobic nanostructured surfaces. ACS Nano **7**, 11043–11054 (2013a)

Miljkovic, N., Enright, R., Nam, Y., Lopez, K., Dou, N., Sack, J., Wang, E.N.: Jumping-droplet-enhanced condensation on scalable superhydrophobic nanostructured surfaces. Nano Lett. **13**, 179–187 (2013b)

Milne, A.J.B., Amirfazli, A.: The cassie equation: how it is meant to be used. Adv. Coll. Interface Sci. **170**, 48–55 (2012)

Ming, W., Wu, D., van Benthem, R., de With, G.: Superhydrophobic films from raspberry-like particles. Nano Lett. **5**, 2298–2301 (2005)

MIT.: (2016). http://web.mit.edu/16.unified/www/FALL/thermodynamics/notes/node64.html

MODTRAN.: The MODTRAN 2/3 report and LOWTRAN 7 MODEL. In: Abreu, L.W., Anderson, G.P. (eds.) (1996). http://web.gps.caltech.edu/_vijay/pdf/modrept.pdf

Mohammadi, R., Wassink, J., Amirfazli, A.: Effect of surfactants on wetting of super-hydrophobic surfaces. Langmuir **20**, 9657–9662 (2004)

Mouterde, T., Lehoucq, G., Xavier, S., Checco, A., Black, C.T., Rahman, A., Midavaine, T., Clanet, C., Quéré, D.: Antifogging abilities of model nanotextures. Nat. Mater. **6**, 658–663 (2017)

Mukherjee, R., Berrier, A.S., Murphy, K.R., Vieitez, J.R., Boreyko, J.B.: How surface orientation affects jumping-droplet condensation. Joule **3**, 1360–1376 (2019)

Mulroe, M.D., Srijanto, B.R., Ahmadi, S.F., Collier, C.P., Boreyko, J.B.: Tuning superhydrophobic nanostructures to enhance jumping-droplet condensation. ACS Nano **11**, 8499–8510 (2017)

Nakajima, A., Abe, K., Hashimoto, K., Watanabe, T.: Thin Solid Films **376**, 140–143 (2000)

Narhe, R.D., Beysens, D.A.: Nucleation and growth on a superhydrophobic grooved surface. Phys. Rev. Lett. **93**, 076103-1–076103-4 (2004a)

Narhe, R., Beysens, D., Nikolayev, V.S.: Contact line dynamics in drop coalescence and spreading. Langmuir **20**, 1213–1221 (2004b)

Narhe, R.D., Beysens, D.A.: Water condensation on a super-hydrophobic spike surface. Europhys. Lett. **75**, 1–7 (2006)

Narhe, R.D., Beysens, D.A.: Growth dynamics of water drops on a square-pattern rough hydrophobic surface. Langmuir **23**, 6486–6489 (2007)

Narhe, R.D., Beysens, D.A., Pomeau, D.: Dynamic drying in the early-stage coalescence of droplets sitting on a plate. Europhys. Lett. **81**, 46002 (2008)

Narhe, R., Anand, S., Rykaczewski, K., Medici, M.G., González-Viñas, W., Varanasi, K.K., Beysens, D.: Inverted leidenfrost-like effect during condensation. Langmuir **31**, 5353–5363 (2015)

Narhe, R.D., Khandkar, M.D., Shelke, P.B., Limaye, A.V., Beysens, D.A.: Condensation-induced jumping water drops. Phys. Rev. E **80**, 031604 (2009)

Narhe, R.: Water condensation on ultrahydrophobic flexible micro pillar. EPL 36002 (2016)

Narita, M., Kasuga, T., Kiyotani, A.: Super water-repellent aluminum by electrolytic etching and chemical adsorption. Keikinzoku **50**(11), 594–597 (2000)

Nath, S., Boreyko, J.B.: On localized vapor pressure gradients governing condensation and frost phenomena. Langmuir **32**, 8350–8365 (2016)

Nath, S., Bisbano, C.E., Yue, P., Boreyko, J.B.: Duelling dry zones around hygroscopic droplets. J. Fluid Mech. **853**, 601–620 (2018)

Nelson, D.R., Rubinstein, M., Spaepen, F.: Order in two-dimensional binary random arrays. Philos. Mag. A **46**, 105–126 (1982)

Nguyen, K.N., Abi-Saab, D., Basset, P., Richalot, E., Malak, M., Pavy, N., Flourens, F., Marty, F., Angelescu, D., Leprince-Wang, Y., Bourouina, T.: Study of black silicon obtained by cryogenic plasma etching: approach to achieve the hot spot of a thermoelectric energy harvester. Microsyst. Technol. **18**, 1807–1814 (2012)

Nicolson, M.M.: The interaction between floating particles. Proc. Cambridge Philos. Soc. **45**, 288–295 (1949)

Nikolayev, V.S., Beysens, D.A.: Boiling crisis and non-equilibrium drying transition. Europhys. Lett. **47**, 345–351 (1999)

Nikolayev, V.S., Beysens, D.A.: Relaxation of nonspherical sessile drops towards equilibrium. Phys. Rev. E **65**, 046135 (2001)

Nikolayev, V.S., Beysens, D.A.: Relaxation of nonspherical sessile drops towards equilibrium. Phys. Rev. E **65**, 46135 (2002)

Nikolayev, V.S., Sibille, P., Beysens, D.: Coherent light transmission by a dew pattern. Optics Commun. **150**, 263–269 (1998)

Nikolskij, B.P., Grigorov, O.N., Posin, M.E. (eds.): Spravotchnik himika (Directory chemist). (Goskhimizdat, Moscow-Leningrad), vol. III (1964)

Niu, Y., Liu, Y., Liu, H., Hu, Y.: Time-dependent density functional study for nanodroplet coalescence. AIChE J. **66**, e16810 (2020)

Nurujjaman, M., Hossain, A., Ahmed, D.P.: A review of fractals properties: mathematical approach. Sci. J. Appl. Math. Stat. **5**, 98–105 (2017)

Oh, J., Zhang, R., Shetty, P.P., Krogstad, J.A., Braun, P.V., Miljkovic, N.: Thin film condensation on nanostructured surfaces. Adv. Funct. Mater. **28**, 1707000 (2018)

Okabe, A., Boots, B., Sugihara, K.: Spatial Tessellations Concepts and Applications of Voronoi Diagram. Wiley, New-York (1992)

Öner, D., McCarthy, T.J.: Ultrahydrophobic surfaces. Effects of topography length scales on wettability. Langmuir **16**, 7777–7782 (2000)

Onuki, A.: Phase Transition Dynamics. Cambridge University Press, Cambridge (2002)

OPUR.: (2021). http://www.opur.fr/

Orejon, D., Shardt, O., Waghmare, P.R., Gunda, N.S.K., Takataab, Y., Mitra, S.K.: Droplet migration during condensation on chemically patterned micropillars. RSC Adv. **2016**(6), 36698 (2016)

Park, K.-C., Fox, D., Hoang, M., McManus, B., Aizenberg, J.: Dropwise condensation on hydrophobic cylinders 00378 (2016a). arXiv:1603

Park, K.-C., Kim, P., Grinthal, A., He, N.,Fox, D., Weaver, J.C., Aizenberg, J.: Condensation on slippery asymmetric bumps. Nature **531**,78–82 (2016b)

Parker, A.R., Lawrence, C.R.: Water capture by a desert beetle. Nature **414**, 33–34 (2001)

Patankar, N.A.: On the modeling of hydrophobic contact angles on rough surfaces. Langmuir **19**, 1249–1253 (2003)

Patankar, N.A.: Mimicking the lotus effect: influence of double roughness structures and slender pillars. Langmuir **20**, 8209–8213 (2004)

Paulsen, J.D., Burton, J.C., Nagel, S.R., Appathurai, S., Harris, M.T., Basaran, O.A.: The inexorable resistance of inertia determines the initial regime of drop coalescence. PNAS **109**, 6857–6861 (2012)

Paulsen, J.D., Burton, J.C., Nagel, S.R.: Viscous to inertial crossover in liquid drop coalescence. Phys. Rev. Lett. **106**, 114501 (2011)

Pawar, N.D., Bahga, S.S., Kale, S.R., Kondaraju, S.: Symmetric and asymmetric coalescence of droplets on a solid surface in the inertia dominated regime. Phys. Fluids **31**, 092106 (2019)

Paxson, A.T., Yague, J.L., Gleason, K.K., Varanasi, K.K.: Stable dropwise condensation for enhancing heat transfer via the initiated chemical vapor deposition (iCVD) of grafted polymer films. Adv. Mater. **26**, 418–423 (2014)

Peng, B., Ma, X., Lan, Z., Xu, W., Wen, R.: Analysis of condensation heat transfer enhancement with dropwise-filmwise hybrid surface: droplet sizes effect. Int. J. Heat Mass Transf. **77**, 785–794 (2014)

Peng, B., Ma, X., Lan, Z., Xu, W., Wen, R.: Experimental investigation on steam condensation heat transfer enhancement with vertically patterned hydrophobic–hydrophilic hybrid surfaces. Int. J. Heat Mass Transf. **83**, 27–38 (2015)

Peng, B., Wang, S., Lan, Z., Xu, W., Wen, R., Maa, X.: Analysis of droplet jumping phenomenon with lattice Boltzmann simulation of droplet coalescence. Appl. Phys. Lett. **102**, 151601 (2013)

Peterson, P.F., Schrock, V.E., Kageyama, T.: Diffusion layer theory for turbulent vapor condensation with noncondensable gases. J. Heat Transf. **115**, 998–1003 (1993)

Petrov, G., Sedev, R.V.: On the existence of a maximum speed of wetting. Colloids Surf. **13**, 313–322 (1985)

Phadnis, A., Rykaczewski, K.: Dropwise condensation on soft hydrophobic coatings. Langmuir **33**, 12095–12101 (2017)

Phan, C.M.: Stability of a floating water droplet on an oil surface. Langmuir **30**, 768–773 (2014)

Phan, C.M., Allen, B., Peters, L.B., Le, T.N., Tade, M.O.: Can water float on oil? Langmuir **28**, 4609–4613 (2012)

Picknett, R.G., Bexon, R.: The evaporation of sessile or pendant drops in still air. J. Colloid Interface Sci. **61**, 336 (1977)

Planck, M.: The Theory of Heat Radiation. Masius, M. (transl.), 2nd edn. P. Blakiston's Son & Co (1914)

Plateau, J.A.F.: Statique expérimentale et théorique des liquides soumis aux seules forces moléculaires, vol. 2. Gauthier Villars, Paris (1873).(In French)

Pomeau, Y.: Recent progress in the moving contact line problem: a review. C. R. Mec. **330**(2002), 207–222 (2002)

Pomeau, Y.: Representation of the mobile contact line in the equations of fluid mechanics. C.R. Acad. Sci. (Paris) Serie II, **328**, 411–416 (In French) (2000)

Porcheron, F., Monson, P.A.: Mean-field theory of liquid droplets on roughened solid surfaces: application to superhydrophobicity. Langmuir **22**, 1595–1601 (2006)

Pothier, J.C., Lewis, L.J.: Molecular-dynamics study of the viscous to inertial crossover in nanodroplet coalescence. Phys. Rev. B **85**, 115447 (2012)

Pozzato, A., Zilio, S.D., Fois, G., Vendramin, D., Mistura, G., Belotti, M., Chen, Y., Natali, M.: Superhydrophobic surfaces fabricated by nanoimprint lithography. Microelectron. Eng. **83**, 884–888 (2006)

Prata, A.J.: A new long-wave formula for estimating downward clearsky radiation at the surface. Q. J. R. Meteorol. Soc. **122**, 1127–1151 (1996)

Preston, D.J., Mafra, D.L., Miljkovic, N., Kong, J., Wang, E.N.: Scalable graphene coatings for enhanced condensation heat transfer. Nano Lett. **15**, 2902–2909 (2015)

Pringle, A., Patek, S.N., Fischer, M., Stolze, J., Money, N.P.: The captured launch of a ballistospore. Mycologia **97**, 866–871 (2005)

Projectile Motion.: (2020). https://en.wikipedia.org/wiki/Projectile_motion

Qian, B., Shen, Z.: Fabrication of superhydrophobic surfaces by dislocation-selective chemical etching on aluminum, copper, and zinc substrates. Langmuir **21**, 9007–9009 (2005)

Qu, M., Zhang, B., Song, S., Chen, L., Zhang, J., Cao, X.: Fabrication of superhydrophobic surfaces on engineering materials by a solution-immersion process. Adv. Funct. Mater. **17**, 593–596 (2007)

Quéré, D.: Non-sticking drops. Rep. Prog. Phys. **68**, 2495–2532 (2005)

Quéré, D.: Leidenfrost dynamics. Annu. Rev. Fluid Mech **45**, 197–215 (2013)

Rao, A.V., Kulkarni, M.M.: Hydrophobic properties of TMOS/TMES-based silica aerogels. Mater. Res. Bull. **37**, 1667–1677 (2002)

Rayleigh, L.: On the instability of jets. Proc. Roy. Soc. London **10**, 4–13 (1879)

Rayleigh, L.: Breath figures. Nature **86**, 416–418 (1911)

Rio, E., Daerr, A., Andreotti, B., Limat, L.: Boundary conditions in the vicinity of a dynamic contact line: experimental investigation of viscous drops sliding down an inclined plane. Phys. Rev. Lett. **94**, 024503 (2005)

Ristenpart W.D., McCalla P.M., Roy R.V., Stone H.A.: Coalescence of spreading droplets on a wettable substrate. Phys. Rev. Lett. **97**, 064501 (2006)

Roach, P., Shirtcliffe, N.J., Newton, M.I.: Progess in superhydrophobic surface development. Soft Matter **2008**(4), 224–240 (2008)

Rodríguez-Hernández, J., Bormashenko, E.: Breath Figures. Mechanisms of Multi-scale Patterning and Strategies for Fabrication and Applications of Microstructured Functional Porous Surfaces. Springer International Publishing (2020)

Rogers, T.M., Elder, K.R., Desai, R.C.: Droplet growth and coarsening during heterogeneous vapor condensation. Phys. Rev. A **38**, 534–5309 (1988)

Rohsenow, W.M.: Heat transfer and temperature distribution in laminar film condensation. Trans. ASME **78**, 1645–1648 (1956)

Rohsenow, W.M., Webber, J.H., Ling, A.T.: Effect of velocity on laminar and turbulent-film condensation. Trans. ASME **78**, 1637–1643 (1956)

Rohsenow, W.M., Hartnett, J.R., Cho, Y.I.: Handbook of Heat Transfer, 3rd (edn). Mc Graw-Hill (1998)

Rose, J.W., Glicksman, L.R.: Dropwise condensation-the distribution of drop sizes. Int. J. Heat Mass Transf. **16**, 411–425 (1973)

Royon, L., Bintein, P.-B., Lhuissier, H., Mongruel, A., Beysens, D.: Micro grooved surface improve dew collection. In: Proceedings of the 12th International Conference on Heat Transfer, Fluid Mechanics and Thermodynamics. Spain (2016)

Rykaczewski, K.: Microdroplet growth mechanism during water condensation on superhydrophobic surfaces. Langmuir **28**, 7720–7729 (2012a)

Rykaczewski, K., Paxson, A.T., Anand, S., Chen, X., Wang, Z., Varanasi, K.K.: Multimode multidrop serial coalescence effects during condensation on hierarchical superhydrophobic surfaces. Langmuir **2013**(29), 881–891 (2012b)

Rykaczewski, K., Landin, T., Walker, M.L., Scott, J.H.J., Varanasi, K.K.: Direct imaging of complex nano- to microscale interfaces involving solid, liquid, and gas phases. ACS Nano **6**, 9326–9334 (2012c)

Rykaczewski, K., Osborn, W.A., Chinn, J., Walker, M.B., Scott, J.H.J., Jones, W., Hao, C., Yaod, S., Wang, Z.: How nanorough is rough enough to make a surface superhydrophobic during water condensation? Soft Matter **8**, 8786–8794 (2012d)

Saab, D.A., Basset, P., Pierotti, M.J., Trawick, M.L., Angelescu, D.E.: Static and dynamic aspects of black silicon formation. Phys. Rev. Lett. **113**, 265502 (2014)

Sawamura, S., Egoshi, N., Setoguchi, Y., Matsuo, H.: Solubility of sodium chloride in water under high pressure. Fluid Phase Equilib. **254**, 158–162 (2007)

Schäfle, C., Leiderer, P., Bechinger, C.: Subpattern formation during condensation processes on structured substrates. Europhys. Lett. **63**, 394–400 (2003)

Schlichting, H.: Boundary Layer Theory, 9th edn. Springer, Berlin, Heidelberg (2017)

Seemann, R., Brinkmann, M., Kramer, E., Lange, F., Lipowski, R.: Wetting morphologies at microstructured surfaces. Proc. Natl. Acad. Sci. U.S.A. **102**, 1848–1852 (2005)

Seiwert, J., Clanet, C., Quéré, D.: Coating of a textured solid. J. Fluid Mech. **669**, 55–63 (2011a)

Seiwert, J., Maleki, M., Clanet, C., Quéré, D.: Drainage on a rough surface. Europhys. Lett. **94**, 16002-1 –16002-5 (2011b)

Self-Sustained Cascading Coalescence in Surface Condensation

Seo, D., Shim, J., Lee, C., Nam, Y.: Brushed lubricant-impregnated surfaces (BLIS) for long-lasting high condensation heat transfer. Sci. Rep. **10**, 2959 (2020)

Seppecher, P.: Moving contact lines in the Cahn-Hilliard theory. Int. J. Eng. Sci. **34**, 977–992 (1996)

Sharma, C.S., Lam, C.W.E., Milionis, A., Eghlidi, H., Poulikakos, D.: Self-sustained cascading coalescence in surface condensation. ACS Appl. Mater. Interfaces **11**, 27435−27442 (2019a)

Sharma, C.S., Wing, C., Lam, E., Milionis, A., Eghlidi, H., Poulikakos, D.: Self-sustained cascading coalescence in surface condensation. ACS Appl. Mater. Interfaces **11**, 27435−27442 (2019b)

Shi, F., Song, Y.Y., Niu, H., Xia, X.H., Wang, Z.Q., Zhang, X.: Facile method to fabricate a large-scale superhydrophobic surface by galvanic cell reaction. Chem. Mater. **18**(5), 1365–1368 (2006)

Shi, B., Wang, Y., Chen, K.: Pool boiling heat transfer enhancement with copper nanowire arrays. Appl. Therm. Eng. **75**, 115–121 (2015)

Shibuchi, S., Onda, T., Satoh, N., Tsujii, K.: Super water-repellent surfaces resulting from fractal structure. J. Phys. Chem. **100**, 19512–19517 (1996)

Shibuichi, S., Yamamoto, T., Onda, T., Tsujii, K.: Super water- and oil-repellent surfaces resulting from fractal structure. J. Colloid Interface Sci. **208**, 287–294 (1998)

Shim, D.I., Choi, G., Lee, N., Kim, T., Kim, B.S., H.H.: Enhancement of pool boiling heat transfer using aligned silicon nanowire arrays. ACS Appl. Mater. Interfaces **9**, 17595–17602 (2017)

Shirtcliffe, N.J., McHale, G., Newton, M.I., Perry, C.C.: Intrinsically superhydrophobic organosilica sol−gel foams. Langmuir **2003**(19), 5626–5631 (2003)

Shirtcliffe, N.J., McHale, G., Newton, M.I., Perry, C.C.: Wetting and wetting transitions on copper-based super-hydrophobic surfaces. Langmuir **21**, 937–943 (2005)

Shiu, J.Y., Kuo, C.W., Chen, P.L., Mou, C.Y.: Fabrication of tunable superhydrophobic surfaces by nanosphere lithography. Chem. Mater. **16**, 561–564 (2004)

Sikarwar, B.S., Khandekar, S., Agrawal, S., Kumar, S., Muralidhar, K.: Dropwise condensation studies on multiple scales. Transf. Eng. **33**, 301–341 (2012)

Sikarwar, B.S., Battoo, N.K., Khandekar, S., Muralidhar, K.: Dropwise condensation underneath chemically textured surfaces: simulation and experiments. J. Heat **133**, 021501-1–021501-15 (2011)

Simões-Moreira, J.R.: A thermodynamic formulation of the psychrometer constant. Meas. Sci. Technol. **10**, 302–311 (1999)

Singh, R.A., Yoon, E.S., Kim, H.J., Kim, J., Jeong, H.E., Suh, K.Y.: Mater. Sci. Eng. C **27**, 875–879 (2007)

Singh, M., Pawar, N.D., Kondaraju, S., Bahga, S.S.: Modeling and simulation of dropwise condensation: a review. J. Indian Inst. Sci. **99**, 157–171 (2019)

Smith, J.D., Dhiman, R., Anand, S., Reza-Garduno, E., Cohen, R.E., McKinleya, G.H., Varanasi, K.K.: Droplet mobility on lubricant-impregnated surfaces. Soft Matter **9**, 1772–1780 (2013)

Snoeijer, J.H., Le Grand-Piteira, N., Limat, L., Stone, H.A., Eggers, J.: Cornered drops and rivulets shape of drops sliding down an inclined surface. Phys. Fluids **19**, 042104 (2007)

Snoeijer, J.H., Peters, I., Limat, L., Daerr, A.: Simple views on cornered contact lines near instability. In: Proceedings of the 23rd Canadian Congress of Applied Mechanics (CANCAM 2011), pp. 172–174 (2011)

Sokuler, M., Auernhammer, G.K., Roth, M., Liu, C., Bonaccurrso, E., Butt, H.G.: The softer the better: fast condensation on soft surfaces. Langmuir **26**, 1544–1547 (2010a)

Sokuler, M., Auernhammer, Liu, C.J., Bonaccurso, E., Butt, H.-J.: Dynamics of condensation and evaporation: effect of inter-drop spacing. EuroPhys. Lett. **89**, 36004 (2010b)

Solomon, H.: Random packing density. In: Proceedings Fifth Berkeley Symposium on Mathematcs, Statistics and Probability, vol. 3, pp. 119–134. University of California Press (1967)

Somwanshi, P.M., Muralidhar, K., Khandekar, S.: Coalescence dynamics of sessile and pendant liquid drops placed on a hydrophobic surface. Phys. Fluids **30**, 092103 (2018)

Song, Y.S., Adler, D., Xu, F., Kayaalp, E., Nureddin, A., Anchan, R.M., Maas, R.L., Demirci, U.: Vitrification and levitation of a liquid droplet on liquid nitrogen. Proc. Natl. Acad. Sci. u.s.a. **107**, 4596–4600 (2010)

Staley, D.O., Jurica, G.M.: Effective atmospheric emissivity under clear skies. J. Appl. Meteorol. **11**, 349–356 (1972)

Standard normal distribution.: (2019). https://en.wikipedia.org/wiki/Normal_distribution

Steyer, A., Guenoun, P., Beysens, D., Fritter, D., Knobler, C.M.: Growth of droplets on a one-dimensional surface: experiments and simulation. Europhys. Lett. **12**, 211–215 (1990)

Steyer, A., Guenoun, P., Beysens, D., Knobler, C.M.: Growth of droplets on a substrate by diffusion and coalescence. Phys. Rev. A **44**, 8271–8277 (1991)

Steyer, A., Guenoun, P., Beysens, D.: Spontaneous jumps of a droplet. Phys. Rev. Lett. **68**, 1869–1871 (1992a)

Steyer, A., Guenoun, P., Beysens, D.: Coalescence-induced $1/f^2$ noise. Phys. Rev. Lett. **68**, 64–66 (1992b)

Steyer, A., Guenoun, P., Beysens, D.: Hexatic and fat-fractal structures for water droplets condensing on oil. Phys. Rev. **E48**, 428–431 (1993)

Straub, J.: Boiling heat transfer and bubble dynamics in microgravity. Adv. Heat Transfer **35**, 57–172 (2001)

Su, B., Li, M., Shi, Z., Lu, Q.: From superhydrophilic to superhydrophobic: controlling wettability of hydroxide zinc carbonate film on zinc plates. Langmuir **25**, 3640–3645 (2009)

Su, J., Charmchi, M., Sun, H.A.: Study of drop-microstructured surface interactions during dropwise condensation with quartz crystal microbalance. Sci Rep **6**, 35132 (2016)

Sun, M.H., Luo, C.X., Xu, L.P., Ji, H., Qi, O.Y., Yu, D.P., Chen, Y.: Artificial lotus leaf by nanocasting. Langmuir **21**, 8978–8981 (2005)

Sun, C., Ge, L.Q., Gu, Z.Z.: Fabrication of super-hydrophobic film with dual-size roughness by silica sphere assembly.thin Solid Films **515**, 4686–4690 (2007)

Sun, J., Weisensee, P.B.: Microdroplet self-propulsion during dropwise condensation on lubricant-infused surfaces. Soft Matter. **15**, 4808–4817 (2019)

Tanaka, H.: A theoretical study of dropwise condensation. J. Heat Transf. **97**, 72–78 (1975)

Tanasawa, I.: Advances in condensation heat transfer. Adv. Heat Transf. **21**, 55–139 (1991)

Tanasawa, I., Ochiai, J.: Experimental study on heat transfer during dropwise condensation. Bull. Jap. Soc. Mech. Eng. **16**, 1184–1197 (1973)

Tanasawa, I., Tachibana, F.: A synthesis of the total process of dropwise condensationusing the method of computer simulation. In: Proceedings of the 4th International Heat Transfer Conference 6, Paper Cs 1.3 (1970)

Tang, R., Etzion, Y., Meir, I.A.: Estimates of clear night sky emissivity in the Negev Highlands, Israel. Energy Convers. Manag. **45**, 1831–1843 (2004)

Tanner, L.H.: The spreading of silicone oil drops on horizontal surfaces. Phys. D: Appl. Phys. **12**, 1473–1484 (1979)

Tavana, H., Amirfazli, A., Neumann, A.W.: Fabrication of superhydrophobic surfaces of *n*-hexatriacontane. Langmuir **22**, 5556–5559 (2006)

Teshima, K., Sugimura, H., Inoue, Y., Takai, O., Takano, A.: Ultra-water-repellent poly (ethylene terephthalate) substrates. Langmuir **19**, 10624–10627 (2003)

Teshima, K.A., Takano, M., Akada, Y., Inoue, H., Sugimura, O., Takai, O.: Hyomen Gijutsu **56**(9), 524–527 (2005)

Texier, B.D., Laurent, P., Stoukatch, S., Dorbolo, S.: Wicking through a confined micropillar array. Microfluid. Nanofluid. **20**(53), 1–9 (2016)

Thoroddsen, S.T., Takehara, K., Etoh, T.G.: The coalescence speed of a pendent and a sessile drop. J. Fluid Mech. **527**, 85–114 (2005)

Tian, J, Zhu, J., Guo, H.Y., Li, J., Feng, X.Q., Gao, X.: Efficient self-propelling of small-scale condensed microdrops by closely packed ZnO nanoneedles. J. Phys. Chem. Lett. **5,** 2084–2088 (2014)

Trosseille, J., Mongruel, A., Royon, L., Medici, M.-G., Beysens, D.: Roughness-enhanced collection of condensed droplets. Eur. Phys. J. E **42**, 1–9 (2019)

Trosseille, J., Mongruel, A., Royon, L., Beysens, D.: Radiative cooling for dew condensation. Int. J. Heat Mass Transf. **172**, 121160 (2021a)

Trosseille, J., Mongruel, A., Royon, L., Beysens, D.: Effective surface emissivity during dew water condensation. Int. J. Heat Mass Transf. **183**, 122078 (2021b)

Trosseille, J.: Refroidissement radiative et texturation de surfaces pour condenseurs de rosée à haut rendement. Thesis (Sorbonne Université, France) (2019)

Tsai, P.S., Yang, Y.M., Lee, Y.L.: Fabrication of hydrophobic surfaces by coupling of langmuir–blodgett deposition and a self-assembled monolayer. Langmuir **22**, 5660–5665 (2006)

Tsuge, Y., Kim, J., Sone, Y., Kuwaki, O., Shiratori, S.: Fabrication of transparent TiO_2 film with high adhesion by using selfassembly methods: application to superhydrophilic film. Thin Solid Films **516**, 2463–2468 (2008)

Tsujii, K., Yamamoto, T., Onda, T., Shibuichi, S.: Super oil-repellent surfaces. Angew. Chem Int. Ed. Engl. **36**, 1011–1012 (1997)

Tung, P.H., Kuo, S.W., Jeong, K.U., Cheng, S.Z.D., Huang, C.F.: Formation of honeycomb structures and superhydrophobic surfaces by casting a block copolymer from selective solvent mixtures macromol. Rapid Commun. **28**, 271–275 (2007)

Turner, J., Webster, J.: Mass and momentum transfer on the small cale: how do mushrooms shed their spores? Chem. Eng. Sci. **46**, 1145–1149 (1991)

Twomey, S.: Experimental test of the volmer theory of heterogeneous nucleation. J. Chem. Phys. **30**, 941–943 (1959)

Ulrich, S., Stoll, S., Pefferkorn, E.: Computer simulations of homogeneous deposition of liquid droplets. Langmuir **20**, 1763–1771 (2004)

Umberger, D.K., Farmer, J.D.: Fat fractals on the energy surface. Phys. Rev. Lett. **55**, 661–664 (1985)

Umur, A., Griffith, P.: Mechanism of dropwise condensation. J Heat Transf **87**, 275–282 (1965)

Urbina, R., Lefavrais, S., Royon, L., Mongruel, A., W.González-Viñas, W., Beysens, D.: Gel-induced dew condensation. J. Hydrol. **599**, 126263 (2021)

Vahabi, H., Wang, W., Davies, S., Mabry, J.M., Kota, A.K.: Coalescence-induced self-propulsion of droplets on superomniphobic surfaces. ACS Appl. Mater. Interfaces **9**, 29328–29336 (2017)

Vakarelski, I.U., Patankar, N.A., Marston, J.O., Chan, D.Y.C., Thoroddsen, S.T.: Stabilization of leidenfrost vapour layer by textured superhydrophobic surfaces. Nature **489**, 274–277 (2012)

Varanasi, K.K., Hsu, M., Bhate, N., Yang, W., Deng, T.: Spatial control in the heterogeneous nucleation of water. Appl. Phys. Lett. **95**, 094101 (2009)

Veeramasuneni, S., Drelich, J., Miller, J.D., Yamauchi, G.: Hydrophobicity of ion-plated PTFE coatings. Prog. Org. Coat. **1997**, 265–270 (1997)

Verbrugghe, N.: Condensed water collection on rough vertical substrates. PMMH report (2016)

Vicsek, T., Family, F.: Dynamic scaling for aggregation of clusters. Phys. Rev. Lett. **52**, 1669–1672 (1984)

Vincenti, W.G., Kruger, C.H.: Introduction to Physical Gas Dynamics. Krieger Publishing Company (1965)

Viovy, J.L., Beysens, D., Knobler, C.M.: Scaling description for the growth of condensation patterns on surface. Phys. Rev. **A37**, 4965–4970 (1988)

Voinov, V.: Hydrodynamics of wetting (English translation). Fluid Dyn. **11**, 714–721 (1976)

Volmer, M.: Kinetic der phasebildung (Kinetics of phase transition), Th. Steinkopff, Dresden and Leipzig (In German) (1938)

von Bahr, M., Tiberg, F., Zhmud, B.: Oscillations of sessile drops of surfactant solutions on solid substrates with differing hydrophobicity. Langmuir **19**, 10109 (2003)

Vourdas, N., Tserepi, A.E., Gogolides, E.: Nanotextured super-hydrophobic transparent poly (methyl methacrylate) surfaces using high-density plasma processing. Nanotechnology **18**, 125304 (2007)

Wagner, P., Fürstner, R., Barthlott, W., Neinhuis, C.: Quantitative assessment to the structural basis of water repellency in natural and technical surfaces. J. Exp. Bot. **54**, 1295–1303 (2003)

Wallace, J.M., Hobbs, P.W.: Atmospheric Science, 2nd edn. Elsevier (2006)

Wang, X., Wen, Y.Q., Feng, X.J., Song, Y.L., Jiang, L.: Control over the wettability of colloidal crystal films by assembly temperature. Macromol. Rapid Commun. **27**, 188–192 (2006)

Wang, M., Raghunathan, N., Ziaie, B.: A nonlithographic top-down electrochemical approach for creating hierarchical (micro−nano) superhydrophobic silicon surfaces. Langmuir **23**, 2300–2303 (2007)

Wang, W., Li, D., Tian, M., Lee, Y., Yang, R.: Wafer-scale fabrication of silicon nanowire arrays with controllable dimensions. Appl. Surf. Sci. **258**, 8649–8655 (2012)

Wang, F.-C., Yang, F., Zhao, Y.P.: Size effect on the coalescence-induced self-propelled droplet. Appl. Phys. Lett. **98**, 053112 (2011)

Wen, R.F., Lan, Z., Peng, B.L., Xu, W., Ma, X.H.: Droplet dynamics and heat transfer for dropwise condensation at lower and ultra-lower pressure. Appl. Therm. Eng. **88**, 265–273 (2015)

Wen, R.F., Li, Q., Wang, W., Latour, B., Li, C.H., Li, C., Lee, Y.-C., Yang, R.G.: Enhanced bubble nucleation and liquid rewetting for highly efficient boiling heat transfer on two-level hierarchical surfaces with patterned copper nanowire arrays. Nano Energy **38**, 59–65 (2017a)

Wen, R.F., Xu, S.S., Zhao, D.L., Lee, Y.C., Ma, X.H., Yang, R.G.: Hierarchical superhydrophobic surfaces with micropatterned nanowire arrays for highefficiency jumping droplet condensation. ACS Appl. Mater. Interfaces **9**, 44911–44921 (2017b)

Wen, R.F., Li, Q., Wu, J.F., Wu, G.S., Wang, W., Chen, Y.F., Ma, X.H., Zhao, D.L., Yang, R.G.: Hydrophobic copper nanowires for enhancing condensation heat transfer. Nano Energy **33**, 177–183 (2017c)

Wen, R.F., Xu, S.S., Ma, X.H., Lee, Y.C., Yang, R.G.: Three-dimensional superhydrophobic nanowire networks for enhancing condensation heat transfer. Joule **2**, 269–279 (2018a)

Wen, R., Ma, X., Lee, Y.-C., Yang, R.: Liquid-vapor phase-change heat transfer on functionalized nanowired surfaces and beyond. Joule **2**, 1–41 (2018b)

Wen, R., Zhou, X., Peng, B., Lan, Z., Yang, R., Ma, X.: Falling-droplet-enhanced filmwise condensation in the presence of non-condensable gas. Int. J. Heat Mass Transf. **140**, 173–186 (2019)

Wen, R.F., Lan, Z., Peng, B.L., Xu, W., Ma, X.H., Cheng, Y.Q.: Droplet departure characteristic and dropwise condensation heat transfer at low steam pressure. J. Heat Tran. **138**, 071501 (2016)

Whitehouse, D.: Surfaces and their Measurement. Butterworth-Heinemann, Boston (2004)

Wichterle, O., Lím, D.: Hydrophilic gels for biological use. Nature **185**, 117–118 (1960)

Wier, K.A., McCarthy, T.J.: Condensation on ultrahydrophobic surfaces and its effect on droplet mobility: ultrahydrophobic surfaces are not always water repellant. Langmuir **22**, 2433–2436 (2006)

Winkels, K.G.: Fast contact line motion: fundamentals and applications. Thesis. University of Twente, Nederland (2013)

Wong, T.S., Kang, S.H., Tang, S.K.Y., Smythe, E.J., Hatton, B.D., Grinthal, A., Aizenberg, J.: Bioinspired self-repairing slippery surfaces with pressure-stable omniphobicity. Nature **477**, 43–447 (2011)

Wu, X.D., Zheng, L.J., Wu, D.: Fabrication of superhydrophobic surfaces from microstructured ZnO-based surfaces via a wet-chemical route. Langmuir **21**, 2665–2667 (2005)

Wu, J., Zhang, L., Wang, L., Wang, P.: Efficient and anisotropic fog harvesting on a hybrid and directional surface. Adv. Mater. Interfaces **4**, 1600801 (2017)

Xiao, R., Chu, K., Wang, E.N.: Multilayer liquid spreading on superhydrophilic nanostructured surfaces. Appl. Phys. Lett. **94**, 19310 (2009)

Xiao, R., Miljkovic, N., Enright, R., Wang, E.N.: Immersion condensation on oil-infused heterogeneous surfaces for enhanced heat transfer. Sci. Rep. **3**, 1988 (2013)

Xie, J., She, Q., Xu, J., Liang, C., Li, W.: Mixed dropwise-filmwise condensation heat transfer on biphilic surface. Int. J. Heat Mass Transf. **150**, 119273 (2020)

Xu, C., Yu, H., Peng, S., Lu, Z., Lei, L., Lohsecdand, D., Zhang, X.: Collective interactions in the nucleation and growth of surface drop. Soft Matter. **13**, 937–944 (2017)

Xu, Z., Zhang, L., Wilke, K., Wang, E.N.: Multiscale dynamic growth and energy transport of droplets during condensation. Langmuir **34**(30), 9085–9095 (2018)

Yadav, S., Kumar, P.: Pollutant scavenging in dew water collected from an urban environment and related implications. Air Qual. Atmos. Health **7**, 559–566 (2014a)

Yadav, A.R., Sriramb, R., Cartere, J.A., Millerb, B.L.: Comparative study of solution phase and vapor phase deposition of aminosilanes on silicon dioxide surfaces. Mater. Sci. Eng. C Mater. Biol. Appl. **35**, 283–290 (2014b)

Yamanaka, M., Sada, K., Miyata, M., Hanabusa, K., Nakano, K.: Construction of superhydrophobic surfaces by fibrous aggregation of perfluoroalkyl chain-containing organogelators. Chem. Commun. **2006**(21), 2248–2250 (2006)

Yamauchi, A., Kumagai, S., Takeyama, T.: Condensation heat transfer on various dropwise-filmwise coexisting surfaces. Heat Transf. Jpn. Res. **15**, 50–64 (1986)

Yan, X., Zhang, L., Sett, S., Feng, L., Zhao, C., Huang, Z., Vahabi, H., Kota, A.K., Chen, F., Miljkovic, N.: Droplet jumping: effects of droplet size, surface structure, pinning, and liquid properties. ACS Nano **13**, 1309–1323 (2019)

Yang, J., Zhang, Z., Xu, X., Zhu, X., Men, X., Zhou, X.: Superhydrophilicsuperoleophobic coatings. J. Mater. Chem. **22**, 2834–2837 (2012)

Yang, K.-S., Lin, K.-H., Tu, C.-W., He, Y.-Z., Wang, C.-C.: Experimental investigation of moist air condensation on hydrophilic, hydrophobic, superhydrophilic, and hybrid hydrophobic-hydrophilic surfaces. Int. J. Heat Mass Transf. **115**, 1032–1041 (2017)

Yao, Z., Lu, Y.W., Kandlikar, S.G.: Effects of nanowire height on pool boiling performance of water on silicon chips. Int. J. Therm. Sci. **50**, 2084–2090 (2011a)

Yao, Z., Lu, Y., Kandlikar, S.G.: Direct growth of copper nanowires on a substrate for boiling applications. Micro Nano Lett. **6**, 563–566 (2011b)

Yersak, A.S., Lewis, R.J., Liew, L.A., Wen, R., Yang, R., Lee, Y.C.: Atomic layer deposited coatings on nanowires for high temperature water corrosion protection. ACS Appl. Mater. Interfaces **8**, 32616–32623 (2016)

Yoshimitsu, Z., Nakajima, A., Watanabe, T., Hashimoto, K.: Effects of surface structure on the hydrophobicity and sliding behavior of water droplets. Langmuir **18**, 5818–5822 (2002)

Yu, X., Dorao, C.A., Fernandino, M.: Droplet evaporation during dropwise condensation due to deposited volatile organic compounds. AIP Adv. **11**, 085202 (2021)

Yu, Y.-S., Zhao, Y.-P.: Elastic deformation of soft membrane with finite thickness induced by a sessile liquid droplet. J. Colloid Interface Sci. **339**, 489–494 (2009)

Yuan, Z., Chen, H., Tang, J., Chen, X., Zhao, D., Wang, Z.: Facile method to fabricate stable superhydrophobic polystyrene surface by adding ethanol. Surf. Coat. Technol. **201**, 7138–7142 (2007)

Zappoli, B., Beysens, D., Garrabos, Y.: Heat Transfer and Related Phenomena in Supercritical Fluids. Springer, Berlin, Heidelberg (2015)

Zhai, L., Cebeci, F.C., Cohen, R.E., Rubner, M.F.: Stable superhydrophobic coatings from poly-electrolyte multilayers. Nano Lett. **4**, 1349 (2004)

Zhang, J.L., Li, J.A., Han, Y.C.: Superhydrophobic PTFE Surfaces by Extension. Macromol. Rapid Commun. **25**, 1105–1108 (2004)

Zhang, G., Wang, D.Y., Gu, Z.Z., Mohwald, H.: Fabrication of superhydrophobic surfaces from binary colloidal assembly. Langmuir **21**, 9143–9148 (2005)

Zhang, X., Jin, M., Liu, Z., Nishimoto, S., Saito, H., Murakami, T., Fujishima, A.: Preparation and photocatalytic wettability conversion of TiO$_2$-based superhydrophobic surfaces. Langmuir **22**, 9477–9479 (2006a)

Zhang, L., Zhou, Z., Cheng, B., DeSimone, J.M., Samulski, E.T.: Superhydrophobic behavior of a perfluoropolyether lotus-leaf-like topography. Langmuir **22**, 8576–8580 (2006b)

Zhang, L., Chen, H., Sun, J., Shen, J.: Layer-by-layer deposition of Poly(diallyldimethylammonium chloride) and sodium silicate multilayers on silica-sphere-coated substrate—Facile method to prepare a superhydrophobic surface. Chem. Mater. **19**, 948–953 (2007)

Zhang, B., Chen, X., Dobnikar, J., Wang, Z., Zhang, X.: SpontaneousWenzel to cassie dewetting transition on structured surfaces. Phys. Rev. Fluids **1**, 073904 (2016)

Zhao, H., Beysens, D.: From droplet growth to film growth on a heterogeneous surface: condensation associated with a wettability gradient. Langmuir **11**, 627–634 (1995)

Zhao, N., Xu, J., Xie, Q.D., Weng, L.H., Guo, X.L., Zhang, X.L., Shi, L.H.: Fabrication of biomimetic superhydrophobic coating with a micro-nano-binary structure. Macromol. Rapid Commun. **26**, 1075–1080 (2005)

Zhao, Y., Preston, D.J., Lu, Z., Zhang, L., Queeney, J., Wang, E.N.: Effects of millimetric geometric features on dropwise condensation under different vapor conditions. Int. J. Heat Mass Transf. **119**, 931–938 (2018)

Zheng, Q., Guo, H., Gunton, J.D.: Growth of breath figures and a possible relationship with ultradynamics. Phys. Rev. A **39**, 3181–3184 (1989)

Zheng, J., Shi, H., Chen, G., Huang, Y., Wei, H.: Relaxation of liquid bridge after droplets coalescence. AIP Adv. **6**, 115115 (2016)

Zhong, Y., Jacobi, A.M., Georgiadis, J.G.: Effects of surface chemistry and groove geometry on wetting characteristics and droplet motion of water condensate on surfaces with rectangular microgrooves. Int. J. Heat Mass Transf. **57**, 629–641 (2013)

Zhu, L., Feng, Y.Y., Ye, X.Y., Zhou, Z.Y.: Tuning wettability and getting superhydrophobic surface by controlling surface roughness with well-designed microstructures. Sens. Actuat. A **130**, 595–600 (2006a)

Zhu, W., Feng, X., Feng, L., Jiang, L.: UV-Manipulated wettability between superhydrophobicity and superhydrophilicity on a transparent and conductive SnO$_2$ nanorod film. Chem. Commun. 2753–2755 (2006b)

Zhu, F., Fang, W.Z., Zhang, H., Zhu, Z., New, T.H., Zhao, Y., Yang, C.: Water condensate morphologies on a cantilevered microfiber. J. Appl. Phys. **127**, 244902 (2020)

Index

Printed in the United States
by Baker & Taylor Publisher Services